Mercury in the Environment

Mercury in the Environment

An Epidemiological and Toxicological Appraisal

Editors:

Lars Friberg, M.D.
Department of Environmental Hygiene
The Karolinska Institute and the
National Environment Protection Board
Stockholm, Sweden

Jaroslav Vostal, M.D., Ph.D.
Department of Pharmacology and Toxicology
School of Medicine and Dentistry
University of Rochester
Rochester, New York

published by:

18901 Cranwood Parkway · Cleveland, Ohio 44128

ST. PHILIPS COLLEGE LIBRARY

This book represents information obtained from authentic and highly regarded sources. Reprinted material is quoted with permission, and sources are indicated. A wide variety of references is listed. Every reasonable effort has been made to give reliable data and information, but the author and the publisher cannot assume responsibility for the validity of all materials or for the consequences of their use.

Copyright © 1972 The Chemical Rubber Co.
Second printing November 1974
CRC Press, Inc.

International Standard Book Number 087819-004-X
Library of Congress Catalog Card Number 74-21911

THE EDITORS

Lars T. Friberg, M.D., is Professor and Chairman of the Department of Environmental Hygiene of the Karolinska Institute and the National Environment Protection Board (up to 1972, the National Institute of Public Health), Stockholm, Sweden.

Dr. Friberg's degrees are from the Karolinska Institute of Stockholm (M. D., 1945, and Dr. of Medical Sciences, 1950).

Prior to his present appointment, Dr. Friberg held different positions in internal medicine and industrial toxicology. He serves in various consulting capacities to Swedish Governmental agencies, e.g., the National Board of Health and Welfare and the National Environment Protection Board. He is a member of the World Health Organization advisory panel on Occupational Health and Chairman of the Sub-committee on the Toxicology of Metals under the Permanent Commission and International Association of Occupational Health. During 1967 he was Visiting Professor at the Department of Environmental Health of the University of Cincinnati in Ohio, U.S.

Dr. Friberg has published over 100 papers on epidemiology and toxicology. Most papers on toxicology concern metals, particularly mercury and cadmium. He was chairman of a Swedish expert committee for the evaluation of risks from methyl mercury in fish. He is one of the authors of the recent review, *Cadmium in the Environment*, published by the Chemical Rubber Company, Cleveland, Ohio, U. S.

Jaroslav J. Vostal, M.D., Ph.D., is Associate Professor in Pharmacology and Toxicology and in Preventive Medicine and Community Health at the University of Rochester School of Medicine and Dentistry, Rochester, N.Y., U.S.

Dr. Vostal obtained his M.D. degree (1951) at Charles University in Prague and his Ph.D. degree (1961) in Medical Sciences at the Czechoslovak Academy of Sciences in Prague, Czechoslovakia.

Prior to his appointment at the University of Rochester, Dr. Vostal was a member of the Institute of Occupational Diseases and Industrial Hygiene in Prague, Czechoslovakia, and in 1967, Visiting Research Fellow at the International Institute of Public Health, Stockholm, Sweden. He served repeatedly as consultant to agencies such as Federal Water Quality Administration, National Air Pollution Control Administration, Food and Drug Administration, U.S. Department of Justice, New York State Department of Health, and others.

Dr. Vostal is a member of the International Sub-committee on Toxicology of Metals at the Permanent Committee and International Association on Occupational Health and of the Committee on Biological Effects of Air Pollutants at the National Research Council, National Academy of Sciences, Washington, D.C., U.S. Dr. Vostal was Chairman and one of the authors of the recently published NRC-NAS Report on Fluoride.

Dr. Vostal's research is oriented towards problems in heavy metals exposures and in mechanisms of their toxic action. His list of publications involves about 50 papers, mainly in industrial toxicology.

CONTRIBUTORS

Lars Friberg, M.D.
Department of Environmental Hygiene
The Karolinska Institute and the
National Environment Protection Board
Stockholm, Sweden

Gösta Lindstedt, Ph.D.
Associate Professor
Chemistry Department
National Institute of Occupational Health
Stockholm, Sweden

Gunnar F. Nordberg, M.D.
Research Assistant
Department of Environmental Hygiene
The Karolinska Institute and the
National Environment Protection Board
Stockholm, Sweden

Claes Ramel, Ph.D.
Member of Environmental Toxicology Group
University of Stockholm
Stockholm, Sweden

Staffan Skerfving, M.D.
Department of Nutrition and Food Hygiene
National Food Administration
Stockholm, Sweden

Jaroslav Vostal, M.D., Ph.D.
Department of Pharmacology and Toxicology
School of Medicine and Dentistry
University of Rochester
Rochester, New York

TABLE OF CONTENTS

Chapter 1
Introduction   1
Lars Friberg

Chapter 2
Methods of Analysis   3
Gösta Lindstedt and Staffan Skerfving
    2.1   Mercury in Air   3
       2.1.1   Air Sampling Methods   3
       2.1.2   Direct-reading Methods   3
    2.2   Mercury in Biological Material   4
       2.2.1   Total Mercury   4
           2.2.1.1   Methods of Analysis   4
                2.2.1.1.1   Colorimetric Methods   4
                     2.2.1.1.1.1   Wet Digestion and Extraction with Dithizone and Related Compounds   4
                     2.2.1.1.1.2   Colorimetric Precipitation Methods   5
                2.2.1.1.2   Atomic Absorption Analysis   5
                     2.2.1.1.2.1   Combustion Methods   5
                     2.2.1.1.2.2   Stannous Reduction Methods   6
                2.2.1.1.3   Neutron Activation Analysis   7
                2.2.1.1.3.1   Nondestructive Analysis   7
                2.2.1.1.3.2   Destructive Analysis   8
                2.2.1.1.4   Micrometric Method   8
           2.2.1.2   Interlaboratory Comparisons   8
           2.2.1.3   Discussion   10
       2.2.2   Specific Methods for Inorganic or Organic Mercury   11
           2.2.2.1   Specific Methods for Inorganic Mercury in the Presence of Organic Mercury   11
           2.2.2.2   Specific Methods for Organic Mercury   11
                2.2.2.2.1   Methods of Analysis   11
                2.2.2.2.2   Interlaboratory Comparisons   12
                2.2.2.2.3   Discussion   13

Chapter 3
Transport and Transformation of Mercury in Nature and Possible Routes of Exposure   15
Jaroslav Vostal
    3.1   Natural Sources and Transport of Mercury in the Environment   15
       3.1.1   Geographical Occurrence of Mercury   15
       3.1.2   Modes of Entry of Mercury into Various Media of the Natural Geocycle   15
           3.1.2.1   Environmental Transport of Mercury into the Atmosphere   15
                3.1.2.1.1   Vaporization Processes   15
                3.1.2.1.2   Volatilization Processes   16
           3.1.2.2   Environmental Transport of Mercury into the Hydrosphere — Dissolution Processes   17
           3.1.2.3   Environmental Transport of Mercury into the Pedosphere — Weathering, Precipitation, Sedimentation, and Biodegradation   18
    3.2   Man-made Sources and Transport of Mercury in the Environment   18
       3.2.1   Industrial Sources   20
       3.2.2   Agricultural Sources   20
       3.2.3   Other Sources   21
    3.3   Possible Routes of Environmental Exposure and Levels of Mercury in the Environment   21
       3.3.1   Possible Routes of Environmental Exposure through Atmosphere   21

3.3.2	Possible Routes of Environmental Exposure through Hydrosphere	22
3.3.3	Possible Routes of Environmental Exposure through Food Chains	23
3.3.3.1	Aquatic Food Chains	23
3.3.3.2	Terrestrial Food Chains	26
3.3.3.3	Foodstuffs Other Than Fish	27

Chapter 4
Metabolism　29
Gunnar F. Nordberg and Staffan Skerfving

4.1　Absorption	29
4.1.1　Inorganic Mercury	29
4.1.1.1　Elemental Mercury	29
4.1.1.1.1　Respiratory Uptake	29
4.1.1.1.1.1　In Animals	29
4.1.1.1.1.2　In Human Beings	29
4.1.1.1.2　Gastrointestinal Uptake	30
4.1.1.1.3　Skin Absorption	30
4.1.1.1.4　Placental Transfer	31
4.1.1.2　Inorganic Mercury Compounds	31
4.1.1.2.1　Respiratory Uptake	31
4.1.1.2.2　Gastrointestinal Absorption	31
4.1.1.2.3　Skin Absorption	32
4.1.1.2.4　Placental Transfer	33
4.1.2　Organic Mercury Compounds	33
4.1.2.1　Alkyl Mercury Compounds	33
4.1.2.1.1　Respiratory Uptake	33
4.1.2.1.1.1　In Animals	33
4.1.2.1.1.2　In Human Beings	33
4.1.2.1.2　Gastrointestinal Absorption	33
4.1.2.1.2.1　In Animals	33
4.1.2.1.2.2　In Human Beings	34
4.1.2.1.3　Skin Absorption	34
4.1.2.1.3.1　In Animals	34
4.1.2.1.3.2　In Human Beings	34
4.1.2.1.4　Placental Transfer	34
4.1.2.1.4.1　In Animals	34
4.1.2.1.4.2　In Human Beings	35
4.1.2.2　Aryl Mercury Compounds	35
4.1.2.2.1　Respiratory Uptake	35
4.1.2.2.2　Gastrointestinal Absorption	35
4.1.2.2.3　Skin Absorption	35
4.1.2.2.3.1　In Animals	35
4.1.2.2.3.2　In Human Beings	35
4.1.2.2.4　Placental Transfer	35
4.1.2.3　Alkoxyalkyl Mercury Compounds	36
4.1.2.3.1　Respiratory Uptake	36
4.1.2.3.2　Gastrointestinal and Skin Absorption	36
4.1.2.3.3　Placental Transfer	36
4.1.2.4　Other Organic Mercury Compounds	36
4.1.3　Summary	36
4.2　Biotransformation and Transport	37
4.2.1　Inorganic Mercury	37

- 4.2.1.1 Oxidation Forms of Mercury and Their Interconversions — 37
- 4.2.1.2 Transport of Elemental Mercury in Blood and into Tissues — 38
- 4.2.1.3 Transport of Mercuric Mercury in Blood — 38
- 4.2.2 Organic Mercury Compounds — 39
  - 4.2.2.1 Alkyl Mercury Compounds — 39
    - 4.2.2.1.1 In Animals — 39
      - 4.2.2.1.1.1 Methyl Mercury Compounds — 39
        - 4.2.2.1.1.1.1 Transport — 39
        - 4.2.2.1.1.1.2 Biotransformation — 39
      - 4.2.2.1.1.2 Ethyl and Higher Alkyl Mercury Compounds — 41
        - 4.2.2.1.1.2.1 Transport — 41
        - 4.2.2.1.1.2.2 Biotransformation — 41
    - 4.2.2.1.2 In Human Beings — 41
      - 4.2.2.1.2.1 Methyl Mercury — 41
      - 4.2.2.1.2.2 Ethyl Mercury — 42
  - 4.2.2.2 Aryl Mercury Compounds — 42
    - 4.2.2.2.1 Transport — 42
    - 4.2.2.2.2 Biotransformation — 43
  - 4.2.2.3 Alkoxyalkyl Mercury Compounds — 44
    - 4.2.2.3.1 Transport — 44
    - 4.2.2.3.2 Biotransformation — 44
  - 4.2.2.4 Other Organic Mercury Compounds — 44
- 4.2.3 Summary — 45
- 4.3 Distribution — 46
  - 4.3.1 Inorganic Mercury — 46
    - 4.3.1.1 In Animals — 46
      - 4.3.1.1.1 Mercuric Mercury — 46
      - 4.3.1.1.2 Mercurous Mercury — 51
      - 4.3.1.1.3 Elemental Mercury — 51
    - 4.3.1.2 In Human Beings — 51
  - 4.3.2 Organic Mercury Compounds — 54
    - 4.3.2.1 Alkyl Mercury Compounds — 54
      - 4.3.2.1.1 In Animals — 54
        - 4.3.2.1.1.1 Methyl Mercury Compounds — 54
        - 4.3.2.1.1.2 Ethyl and Higher Alkyl Mercury Compounds — 60
      - 4.3.2.1.2 In Human Beings — 62
        - 4.3.2.1.2.1 Methyl Mercury Compounds — 62
        - 4.3.2.1.2.2 Ethyl Mercury Compounds — 62
    - 4.3.2.2 Aryl Mercury Compounds — 62
    - 4.3.2.3 Alkoxyalkyl Mercury Compounds — 67
    - 4.3.2.4 Other Organic Mercury Compounds — 67
  - 4.3.3 Summary — 69
- 4.4 Retention and Excretion — 70
  - 4.4.1 Inorganic Mercury — 70
    - 4.4.1.1 Mercuric Mercury — 70
      - 4.4.1.1.1 In Animals — 70
        - 4.4.1.1.1.1 Retention and Risk of Accumulation at Repeated Exposure — 70
        - 4.4.1.1.1.2 Excretion — 71
          - 4.4.1.1.1.2.1 Urinary and Fecal Excretion — 71
          - 4.4.1.1.1.2.2 Mechanism for Fecal and Urinary Excretion — 73
          - 4.4.1.1.1.2.3 Other Routes of Elimination — 74
      - 4.4.1.1.2 In Human Beings — 75

- 4.4.1.2 Mercurous Mercury — 76
- 4.4.1.3 Elemental Mercury — 76
  - 4.4.1.3.1 In Animals — 76
    - 4.4.1.3.1.1 Retention and Risk of Accumulation at Repeated Exposure — 76
    - 4.4.1.3.1.2 Excretion — 77
  - 4.4.1.3.2 In Human Beings — 77
- 4.4.2 Organic Mercury Compounds — 78
  - 4.4.2.1 Alkyl Mercury Compounds — 78
    - 4.4.2.1.1 Methyl Mercury Compounds — 78
      - 4.4.2.1.1.1 In Animals — 78
        - 4.4.2.1.1.1.1 Retention — 78
        - 4.4.2.1.1.1.2 Excretion — 79
      - 4.4.2.1.1.2 In Human Beings — 80
        - 4.4.2.1.1.2.1 Retention — 80
        - 4.4.2.1.1.2.2 Excretion — 81
    - 4.4.2.1.2 Ethyl and Higher Alkyl Mercury Compounds — 82
      - 4.4.2.1.2.1 In Animals — 82
        - 4.4.2.1.2.1.1 Retention — 82
        - 4.4.2.1.2.1.2 Excretion — 82
      - 4.4.2.1.2.2 In Human Beings — 82
  - 4.4.2.2 Aryl Mercury Compounds — 83
    - 4.4.2.2.1 In Animals — 83
      - 4.4.2.2.1.1 Retention — 83
      - 4.4.2.2.1.2 Excretion — 83
    - 4.4.2.2.2 In Human Beings — 84
      - 4.4.2.2.2.1 Retention — 84
      - 4.4.2.2.2.2 Excretion — 84
  - 4.4.2.3 Alkoxyalkyl Mercury Compounds — 84
    - 4.4.2.3.1 In Animals — 84
      - 4.4.2.3.1.1 Retention — 84
      - 4.4.2.3.1.2 Excretion — 85
    - 4.4.2.3.2 In Human Beings — 85
  - 4.4.2.4 Other Organic Mercury Compounds — 85
    - 4.4.2.4.1 In Animals — 85
      - 4.4.2.4.1.1 Retention — 85
      - 4.4.2.4.1.2 Excretion — 85
    - 4.4.2.4.2 In Human Beings — 86
      - 4.4.2.4.2.1 Retention — 86
      - 4.4.2.4.2.2 Excretion — 86
- 4.4.3 Summary — 86
- 4.5 Indices of Exposure and Retention — 88
  - 4.5.1 Inorganic Mercury — 88
  - 4.5.2 Organic Mercury Compounds — 89
    - 4.5.2.1 Alkyl Mercury Compounds — 89
    - 4.5.2.2 Aryl Mercury Compounds — 90
    - 4.5.2.3 Alkoxyalkyl Mercury Compounds — 90
  - 4.5.3 Summary — 90

Chapter 5
Symptoms and Signs of Intoxication — 93
Staffan Skerfving and Jaroslav Vostal
- 5.1 Inorganic Mercury — 93

    5.1.1   Prenatal Intoxication —   93
    5.1.2   Postnatal Intoxication   93
        5.1.2.1   Acute Poisoning   93
            5.1.2.1.1   Elemental Mercury Vapor   93
                5.1.2.1.1.1   In Human Beings   93
                5.1.2.1.1.2   In Animals   94
            5.1.2.1.2   Inorganic Mercury Salts   94
                5.1.2.1.2.1   In Human Beings   94
                5.1.2.1.2.2   In Animals   94
        5.1.2.2   Chronic Poisoning   95
            5.1.2.2.1   Nonspecific Signs and Symptoms   95
            5.1.2.2.2   Oropharyngeal Symptoms   96
            5.1.2.2.3   Symptoms Related to Central Nervous System   96
                5.1.2.2.3.1   Asthenic-vegetative Syndrome   96
                5.1.2.2.3.2   Mercurial Tremor   96
                5.1.2.2.3.3   Mercurial Erethism   97
            5.1.2.2.4   Renal Effects   97
            5.1.2.2.5   Ocular Symptomatology (Mercurialentis)   98
        5.1.2.3   Hypersensitivity or Idiosyncracy   99
    5.1.3   Summary   100
5.2   Organic Mercury Compounds   101
    5.2.1   Alkyl Mercury Compounds   101
        5.2.1.1   Prenatal Intoxication   101
            5.2.1.1.1   In Human Beings   101
            5.2.1.1.2   In Animals   102
        5.2.1.2   Postnatal Intoxication   102
            5.2.1.2.1   In Human Beings   102
                5.2.1.2.1.1   Local Effects   102
                5.2.1.2.1.2   Systemic Effects   102
            5.2.1.2.2   In Animals   104
    5.2.2   Aryl Mercury Compounds   105
        5.2.2.1   In Human Beings   105
            5.2.2.1.1   Local Effects   105
            5.2.2.1.2   Systemic Effects   105
            5.2.2.1.3   Hypersensitivity or Idiosyncracy   106
        5.2.2.2   In Animals   106
    5.2.3   Alkoxyalkyl Mercury Compounds   106
        5.2.3.1   In Human Beings   106
        5.2.3.2   In Animals   107
    5.2.4   Other Organic Mercury Compounds   107
    5.2.5   Summary   107

Chapter 6
"Normal" Concentrations of Mercury in Human Tissue and Urine   109
Staffan Skerfving

6.1   Introduction   109
6.2   Blood   109
    6.2.1   Data on Fish Consumption not Available   109
    6.2.2   Data on Fish Consumption Available   109
6.3   Hair   110
6.4   Brain, Liver, and Kidneys   110

  6.5 Urine   112
  6.6 Summary   112

Chapter 7
Inorganic Mercury — Relation between Exposure and Effects   113
Lars Friberg and Gunnar F. Nordberg
 7.1 In Human Beings   113
  7.1.1 Acute Effects   113
  7.1.2 Chronic Effects   113
   7.1.2.1 Relation between Mercury in Air and Effects   113
    7.1.2.1.1 Studies in General   113
    7.1.2.1.2 Russian Studies — Including Studies on Micromercurialism   115
   7.1.2.2 Relation between Mercury in Urine and Effects or Exposure   118
    7.1.2.2.1 Mercury in Urine and Effects   118
    7.1.2.2.2 Mercury in Urine and Exposure   122
   7.1.2.3 Relation between Mercury in Blood and Effects or Exposure   124
    7.1.2.3.1 Mercury in Blood and Effects   124
    7.1.2.3.2 Mercury in Blood and Exposure   125
   7.1.2.4 Relation between Mercury in Organs and Effects or Exposure   125
  7.1.3 Conclusions   126
 7.2 In Animals   126
  7.2.1 Acute Effects   126
   7.2.1.1 Injection   126
   7.2.1.2 Oral and Percutaneous Exposure   127
   7.2.1.3 Inhalation   127
  7.2.2 Chronic Effects   128
   7.2.2.1 Injection   128
   7.2.2.2 Oral and Percutaneous Exposure   128
   7.2.2.3 Inhalation   128
    7.2.2.3.1 Studies in General   128
    7.2.2.3.2 Russian Studies — Including Studies on Micromercurialism   130
  7.2.3 Conclusions   138

Chapter 8
Organic Mercury Compounds — Relation between Exposure and Effects   141
Staffan Skerfving
 8.1 Alkyl Mercury Compounds   141
  8.1.1 Prenatal Exposure   141
   8.1.1.1 In Human Beings   141
    8.1.1.1.1 Methyl Mercury   141
    8.1.1.1.2 Ethyl Mercury   142
   8.1.1.2 In Animals   142
    8.1.1.2.1 Methyl Mercury   142
    8.1.1.2.2 Ethyl Mercury   143
   8.1.1.3 Conclusions   143
  8.1.2 Postnatal Exposure   144
   8.1.2.1 In Human Beings   144
    8.1.2.1.1 Relation between Organ Levels and Effects   144
     8.1.2.1.1.1 Blood   144
      8.1.2.1.1.1.1 Methyl Mercury Exposure   144
      8.1.2.1.1.1.2 Ethyl Mercury Exposure   146
     8.1.2.1.1.2 Hair   147

	8.1.2.1.1.2.1	Methyl Mercury Exposure	147
	8.1.2.1.1.2.2	Ethyl Mercury Exposure	149
8.1.2.1.1.3	Brain, Liver, and Kidney		149
	8.1.2.1.1.3.1	Methyl Mercury Exposure	149
	8.1.2.1.1.3.2	Ethyl Mercury Exposure	149
8.1.2.1.1.4	Conclusions		151

8.1.2.1.2 Relation between Exposure and Effects — 152
    8.1.2.1.2.1 Methyl Mercury Exposure — 152
    8.1.2.1.2.2 Ethyl Mercury Exposure — 152
    8.1.2.1.2.3 Conclusions — 153
8.1.2.1.3 Relation between Exposure and Organ Levels — 153
8.1.2.2 In Animals — 155
  8.1.2.2.1 Single Administration — 155
  8.1.2.2.2 Repeated Administration — 155
    8.1.2.2.2.1 Methyl Mercury Exposure — 155
    8.1.2.2.2.2 Ethyl Mercury Exposure — 155
    8.1.2.2.2.3 Other Alkyl Mercury Exposure — 155
8.2 Aryl Mercury Compounds — 155
  8.2.1 Prenatal Exposure — 155
  8.2.2 Postnatal Exposure — 157
    8.2.2.1 In Human Beings — 157
    8.2.2.2 In Animals — 164
8.3 Alkoxyalkyl Mercury Compounds — 164
  8.3.1 In Human Beings — 164
  8.3.2 In Animals — 164
8.4 Other Organic Mercury Compounds — 168

Chapter 9
Genetic Effects — 169
Claes Ramel

9.1 Introduction — 169
9.2 Effects on Cell Division — 169
  9.2.1 Mitotic Activity — 169
  9.2.2 C-mitosis — 170
  9.2.3 Dose-response Relationships of C-Mitosis — 170
  9.2.4 Mechanisms of C-Mitotic Action — 173
9.3 Radiomimetic Effects — 173
9.4 Effects on Meiosis — 174
  9.4.1 Cytological Observations — 174
  9.4.2 Nondisjunction in *Drosophila* — 174
    9.4.2.1 Standard X Chromosomes — 175
    9.4.2.2 Inversion Heterozygotes — 177
  9.4.3 Effects on Crossing Over and Chromosome Repair — 178
  9.4.4 Point Mutations — 179
9.5 Concluding Remarks — 181

Chapter 10
General Discussion and Conclusions — Need For Further Research — 183
Lars Friberg and Jaroslav Vostal

References — 187

Chapter 1

INTRODUCTION

Lars Friberg

This review of the toxicity of mercury has been performed under a contract between the U.S. Environmental Protection Agency and the Department of Environmental Hygiene of the Karolinska Institute, Sweden. The Project Officer has been Robert J. M. Horton, M.D., of the Air Pollution Control Office of the U.S. Environmental Protection Agency. The review has focused on information considered of special importance for understanding the toxic action of mercury and on quantitative information in regard to the relation between dose (exposure to mercury) and effects on human beings and animals. The intention has not been to give a complete review of all available data on mercury toxicity.

The report was originally intended to serve as a background for a future air quality criteria document on mercury. Particular attention has been given to information relevant for the evaluation of risks due to long-term exposure to low concentrations of mercury. Acute effects from short-term exposure to high concentrations are dealt with briefly.

The report is not limited to effects due to exposure via inhalation. A considerable amount of information, particularly from recent years, is available on mercury toxicity from exposure via the oral route. Such information should certainly be treated in a review to be used for future air quality criteria documents. Examining exposure via the oral route can give valuable evidence about the mode of action, distribution, and retention of mercury compounds in the body and about the relation between dose, measured, e.g., as blood levels, and the effects found. Furthermore, mercury in the air can contaminate other vehicles such as water and food.

Mercury is found in the environment in different chemical or physical forms. The most toxic of the mercury compounds is methyl mercury, which during the last decade has given rise to a great number of severe poisonings, several of them fatal, due to consumption of contaminated fish from waters with a very low mercury content. Of importance are the findings that nature can convert elemental mercury and mercury compounds into methyl mercury.

The data presented are based on a literature survey as well as on our own experience. Information has also been made available by correspondence or personal visits with scientists in several countries, including Japan, the U.S., and the U.S.S.R. Of special value has been the report by a Swedish expert group, *Methyl Mercury in Fish — A Toxicologic — Epidemiologic Evaluation of Risks,* which is referred to repeatedly when methyl mercury is discussed in this review.

In the report, the term "inorganic" refers to mercury in the form of elemental vapor, mercurous and mercuric salts, and those complexes in which mercuric ions can form reversible bonds to such tissue ligands as thiol groups on proteins. Those compounds in which mercury is linked directly to a carbon atom by a covalent bond are classified as organomercurial compounds and mercury in this state of combination will be described as "organic mercury."

The final responsibility toward the U.S. Environmental Protection Agency for this report is held by Dr. Lars Friberg. Dr. Jaroslav Vostal was invited to act as coeditor and in all other respects the two editors share the responsibility. Although the different chapters have their own authors, all the work has been done in close collaboration with the editors, who are in accord with all conclusions drawn.

We express our thanks to Miss Pamela Boston for assistance in editing the English of the report.

Chapter 2

METHODS OF ANALYSIS

Gösta Lindstedt and Staffan Skerfving

An appraisal of experimental and epidemiological data concerning mercury cannot be made without an evaluation of the reliability of the analytical methods used. This chapter is not a complete treatise but a brief description of some important analytical methods used in the toxicological work referred to in subsequent chapters. Only the two matters of pertinence for the entire review, the analysis of mercury in air and in biological material, have been considered.

Much of the data has been taken from a recently published review on methyl mercury toxicity (Berglund et al., 1971). When available, data on the reliability of the methods have been given special consideration. Detection limit, or sensitivity, is then defined as the smallest total amount or concentration that the method is able to determine. The precision (reproducibility) of a method is the standard deviation (or coefficient of variation) of a number of analyses made of the same sample. Accuracy denotes the systematic deviation from the true value.

2.1 Mercury in Air

The determination of mercury vapor in air is of great importance for evaluating the health hazards of industrial atmospheres, e.g., in chloralkali plants. In certain cases analyses of mercury particles or of organic mercurials in the air may also be of interest. Two different types of analytical methods can be named: air sampling methods and direct-reading methods.

2.1.1 Air Sampling Methods

These methods require the collection of mercury from the air before analysis. Impinger flasks containing potassium permanganate-sulfuric acid solutions are generally preferred (IUPAC, 1969) but iodine-potassium iodide solution is also recommended (AIHA, 1969). These 2 sampling methods are excellent for the collection of elemental mercury vapor, but not all organic mercury compounds are absorbed quantitatively. Iodine monochloride solution is a more effective absorbent for methyl and ethyl mercury compounds (Linch, Stalzer, and Lefferts, 1968). Only permanganate can be used in connection with final mercury determination by atomic absorption, since iodine interferes with this analysis.

Isopropanol has been used to collect di-butyl mercury from air (Quino, 1962). Sodium carbonate phosphate solution has been employed as a specific absorbent for mono-methyl and mono-ethyl mercury in the presence of metallic mercury (Kimura and Miller, 1960).

Solid adsorbents, such as impregnated charcoal, can also be used to collect mercury from air (Sergeant, Dixon, and Lidzey, 1957, and Moffitt and Kupel, 1970). The mercury is liberated again when the charcoal is heated. Adsorption tubes containing a small amount of charcoal are much easier to transport to the laboratory than are impingers or other sampling devices containing liquids.

In the laboratory, the mercury collected by the methods described is analyzed either by chemical methods (dithizone, etc.) or by atomic absorption. These methods will be discussed in Sections 2.2.1.1.1 and 2.2.1.1.2.

Detection limit — The sensitivity of the sampling methods can be adjusted at will by collecting an air volume of sufficient size. If 1 μg of mercury can be determined by the dithizone method used, a 20 l. air sample will be required to cover 50 μg of elemental mercury vapor/m^3 of air. The atomic absorption determination of mercury is far more sensitive, and air samples of less than 1 l. can be used. Generally, however, this method is applied to direct mercury analysis in air without any sampling, as will be described in Section 2.1.2.

2.1.2 Direct-Reading Methods

Some methods have been developed for immediate semiquantitative estimation of elemental mercury vapor in air. Indicator papers have been described, but gas-detecting tubes, manufactured by some firms (Draeger, MSA, etc.) are more commonly used. Their sensitivity is not very high, but they are quick and simple to use in pilot investigations.

Elemental mercury vapor is monoatomic and

absorbs light at certain resonance wavelengths, as do other free atoms. Long before atomic absorption analysis was heard of, the strong ultraviolet light absorption at 253.7 nm was utilized to measure the elemental mercury vapor in air (Woodson, 1939). Several instrument makers have introduced portable "mercury detectors" (Kruger-Beckman, General Electric, Engelhard-Hanovia, Incentive, Perkin-Elmer, Coleman, and many others). All these instruments measure no forms of mercury other than elemental mercury vapor. They contain a mercury arc lamp, a gas absorption cell, a photomultiplier or a phototube, and a direct-reading instrument, which is calibrated to show the mercury level of the air pumped through the gas cell. Thus the mercury content is monitored immediately on the spot, which makes this type of analysis practical and inexpensive.

Detection limit – For the most modern types of "mercury detectors," it is about 2 µg of mercury/m^3.

Any volatile substance present in the air and absorbing light at 253.7 nm interferes with the analysis. On the other hand, for such substances as sulfur dioxide, nitrous oxides, or aromatic hydrocarbon vapors, about 100-fold molar excesses are needed to get a similar reading as for mercury. To correct for this "nonatomic absorption," double-beam detectors have been constructed which split the air stream into 2 branches. In 1 of these branches a filter is inserted which absorbs mercury vapor specifically (gold, silver, etc.). The difference in light absorption between the 2 air streams is measured by the apparatus, being proportional to the mercury content of the air (James and Webb, 1964). A more ingenious way of correcting for "nonatomic absorption" by purely optical means (Lorentz broadening of mercury emission lines) has been devised by Barringer, 1966, and Ling, 1967.

Another type of interference is caused by the strong magnetic fields prevailing in some industrial buildings, such as chloralkali plants. These magnetic fields interfere with the electronics of the "mercury detectors" to such an extent that the analysis may be impossible (Smith et al., 1970). In such cases, sampling techniques must be applied.

To sum up, "mercury detectors" are very convenient to work with, but attention must be paid to other vapors and gases present in the atmosphere under analysis, as well as to other possible interferences. If such sources of error can be eliminated, fractions of the MAC (TLV) for elemental mercury vapor are easily detected.

The atomic absorption principle for mercury analysis has been used extensively for the analysis of biological samples (Section 2.2.1.1.2).

2.2 Mercury in Biological Material
2.2.1 Total Mercury

The rapid developments in the methods of analysis for total mercury during recent years have enabled a higher degree of sensitivity and precision. Now, hundredths of an ng/g can be determined routinely.

2.2.1.1 Methods of Analysis
2.2.1.1.1 Colorimetric Methods
2.2.1.1.1.1 Wet Digestion and Extraction with Dithizone and Related Compounds

For about 25 years the dithizone method was the predominating analytical method for determination of mercury in biological material. Several hundred papers describing modifications of this method have been published since 1940.

Dithizone, C_6H_5-NH-NH-CS-N=N-C_6H_5, is a green compound soluble in chloroform and in other organic solvents. It creates strongly colored chelates with most heavy metals. By variation of pH and addition of complexing agents (cyanide, citrate, etc.) many metals can be extracted separately from aqueous solutions as chelates and determined colorimetrically. A related compound, di-β-naphtyl-thiocarbazone, can be used as well (Cholak and Hubbard, 1946).

Mercury is extracted by dithizone in chloroform from a strongly acid aqueous solution. Copper, silver, gold, palladium, and platinum are also extracted but can be eliminated in different ways. The mercury dithizonate is orange and has an absorption maximum at about 490 nm in chloroform.

Biological samples must be wet digested before mercury analysis can be carried out. Generally, strong acid mixtures or potassium permanganate-sulfuric acid are used. The volatility of mercury and its compounds makes the digestion a somewhat hazardous operation. To avoid losses, flasks with reflux condensers are generally recommended (Analytical Methods Committee, 1965).

A few standard works on dithizone analysis of mercury in biological material may be referred to. In 1965, the Analytical Methods Committee described a standard method for determining mercury in organic matter, especially food. Analysis of mercury in urine has been treated by Nobel, 1961. A titration method, based upon di-β-naphtylthiocarbazone, has been proposed (Truhaut and Boudène, 1959). A similar method, working with ordinary dithizone, has been accepted for urine (IUPAC, 1969). Dithizone analysis can be carried out with rather inexpensive equipment, available in all analytical laboratories. Its main disadvantage is the large amount of manual work required for each analysis and the relatively low sensitivity compared to modern physical methods.

Detection limit — About 0.5 µg Hg in 10 g samples (Analytical Methods Committee, 1965).

Precision — Four to five percent (Smart and Hill, 1969).

2.2.1.1.1.2 Colorimetric Precipitation Methods

Methods based on the formation of colored compounds of mercury with copper and iodine are widely used in the U.S.S.R. for analysis of mercury in air and urine. Since much data obtained by these methods are presented in Chapter 7, and since no accounts of the methods are available in English, the procedures will be described in some detail. The following section is based upon personal communications to Gunnar F. Nordberg from Drs. Kournossov, Moscow, and Korshun, in Kiev. These investigators have had considerable experience with the methods to be described.

A procedure for analysis of mercury in air, presently used and generally accepted in the U.S.S.R., was described by Poleshajev in 1956. Air is passed through a glass apparatus in which it is mixed with iodine vapor. The mercury-iodine mixture is absorbed in a solution of iodine and potassium iodide in water. A solution of Na_2SO_3 and $CuCl_2$ is added. Pink-orange $Cu_2(HgI)_4$ precipitates together with white Cu_2I_2. The mercury content is estimated by subjective comparison, using the naked eye, of the color of the precipitate with a standard scale of precipitates. A modification of the procedure was described by Barnes in 1946.

Detection limit — The U.S.S.R. MAC value for air in the general environment, 0.3 µg/m^3, is checked by this method. In the standard procedure, 70 l. of air are sampled (one l./min for 70 min). At 0.3 µg/m^3, the total content of mercury in the sample is 20 ng. The limit of detection is considered to be 20 ng.

Precision and accuracy — Since no data are available it is not possible to evaluate these aspects of the method. As the readings are made by visual comparison, it seems likely that the precision and the accuracy will be influenced very much by the person making the readings.

By a modification, according to Ginzberg in 1948, mercury in urine can be determined. Ovalbumin is added to the urine. The protein (containing the mercury) is precipitated by adding trichloracetic acid and heating. The precipitation is filtered off and dissolved into a solution of iodine in potassium iodide. The mercury content is evaluated by precipitation, as described above.

Detection limit — According to Trachtenberg and Korshun (personal communication), 1.9 µg in 0.5 l. of urine (3.7 µg/l.).

Precision and accuracy — The interval between the steps in the standard scale corresponds to 3.7 µg/l. (Trachtenberg and Korshun, personal communication). The error must then be at least 1.8 µg/l. No further data are available.

2.2.1.1.2 Atomic Absorption Analysis

The mercury of the sample is converted into vapor, after which the mercury is determined by atomic absorption (see Section 2.1.2). A long series of variants of this principle have been used in the analysis of biological material. The essential difference among the various methods is the way in which the mercury is converted into an elemental vapor phase.

2.2.1.1.2.1 Combustion Methods

Methods based upon release of mercury vapor from urine by direct injection of urine into a flame or a furnace and atomic absorption analysis of the combustion gases have been proposed by Lindström, 1959, and Hayes, Muir, and Whitby, 1970.

Detection limit — About 50 µg/l. of urine (Hayes, Muir, and Whitby, 1970).

Precision and accuracy — Precision, 24.5 µg/l. in the range 50 to 500 µg/l. No significant difference from the dithizone method in the range 50 to 500 µg/l. (Hayes, Muir, and Whitby, 1970).

Jacobs et al., 1960, described a procedure with a wet digestion of the sample (a few g) and subsequent extraction of Hg^{2+} with dithizone in chloroform. Mercury dithizonate is pyrolyzed through heating, and the mercury vapor formed is measured by atomic absorption. This procedure has been used widely in the U.S. and Japan. By means of a slight modification of the method, Jacobs, Goldwater, and Gilbert, 1961, reduced the amount of the sample (blood) to 0.1 ml.

Detection limit — About 10 ng/g (Jacobs et al., 1960).

Precision and accuracy — No data available.

Lidums and Ulfvarson, 1968a, have carried out a direct combustion procedure. Combustion takes place with oxygen, which is passed through a combustion tube. The mercury is collected on a gold filter, driven off in a rapid operation, and passed through the atomic absorption photometer. When tested also with methyl mercury as standard, the method gave complete yield (Ulfvarson, personal communication). The amount of the sample must be small, about 20 to 200 mg. Another direct combustion method has been used by Schütz, 1969. The combustion gases from samples up to about 3 g are passed through a tube furnace at 950° C, in which complete combustion of the distillation products occurs. The absorption of mercury takes place in a potassium permanganate solution. The permanganate is reduced with hydroxylamine, after which elemental mercury vapor is liberated with tin (II) chloride (see below).

Detection limit — Down to a few tenths of an ng for samples of about 0.2 g (Lidums and Ulfvarson, 1968b).

Precision and accuracy — With regard to fish, see Section 2.2.2.1. Lidums and Ulfvarson, 1968b, compared the results of 2 to 4 analyses of the same sample (0.2 to 0.4 g) of 6 whole blood samples in the concentration range 3 to 98 ng/g and 6 plasma samples in the concentration range 2 to 260 ng/g with activation analysis (single analysis according to Sjöstrand, 1964). Deviation from the common mean value for all 25 single analyses may be estimated at $\leq 10\%$. Schütz, 1969, has reported a comparison of the results of a single analysis of 10 blood cell samples (about 1 g), in the concentration range 5 to 25 ng/g, with activation analysis (Sjöstrand, 1964). The deviations from the common mean values were in 9 cases $\leq \pm 10\%$ and always $\leq \pm 20\%$. From the reported results of the analyses, the precision of the methods for samples of about 1 g may be estimated at 1 to 5% in the concentration range 5 to 100 ng/g. The accuracy has been checked in various organs from animals treated with labeled mercury and has been found to be within $\pm 10\%$ (Nordberg and Schütz, personal communication).

2.2.1.1.2.2 Stannous Reduction Methods

Another way of liberating mercury from a digested sample is the reduction of Hg^{2+} to elemental mercury with Sn^{2+} ions, followed by volatilization of the mercury by aid of a gas stream. No elevated temperature is needed, and the evaporation of mercury is completed within a few minutes. The final determination is made by atomic absorption. Pioneer work on this method was done by Poluektov, Vitkun, and Zelyukova, 1964.

Methods for analysis of mercury in urine by this principle have been published by Rathje, 1969, and Lindstedt, 1970. The former author uses nitric acid for the digestion, the latter, permanganate-sulfuric acid, both at room temperature. Magos and Cernik, 1969, reduced mercury in urine with Sn^{2+} in alkaline solution, without digestion. The latter method works even in the presence of iodide, which interferes with the acid Sn^{2+} reduction. Noble metals, which are more easily reduced than mercury, interfere with the analysis, but they are met with rather seldomly in biological samples. A very similar method, applicable to food and biological fluids, has been worked out by Thorpe, 1970.

Lindstedt and Skare, 1971, have constructed an automatic apparatus which analyzes 60 digested samples in 2 hours without supervision. In addition to urine, other biological samples such as blood, fish, meat, or organs can be digested by special methods and analyzed in this apparatus (Skare, in press). Malaiyandi and Barrette, 1970, utilize an autoanalyzer in combination with an atomic absorption spectrophotometer.

Detection limit — Two ng/ml for urine with permanganate digestion (Lindstedt, 1970); 3 ng/g for blood (0.2 ml samples), and 5 ng/g for fish meat (Skare, in press).

Precision — Two percent for a urinary level of 0.17 μg/ml, and 7% for a level of 0.04 μg/ml (Lindstedt, 1970); 15% for blood of the 20 ng/g level (Skare, in press).

Accuracy — Lindstedt, 1970, found good agreement with dithizone analysis of urine (r = 0.98; n = 110) and Skare (in press) likewise with activation analysis for blood (r < 0.99; n = 9) and for fish (r < 0.99; n = 19), using the automatic equipment.

The literature about stannous chloride atomic absorption methods is rapidly growing and it seems to be the most popular method at present.

It is by far the quickest and cheapest analytical method for microdetermination of mercury in biological material.

Atomic absorption analysis was used for the determination of mercury in biological material obtained in connection with the epidemics of methyl mercury poisoning in Niigata (Kawasaka et al., 1967). The sensitivity was stated to be "more than 100 times greater than that of the dithizone method." The description of the method is not sufficient for estimating its precision and accuracy.

2.2.1.1.3 Neutron Activation Analysis

The sample is sealed in quartz or polyethylene vials and irradiated with neutrons. The gamma radiation emitted by the ^{197}Hg formed is measured by spectrometry in relation to a known standard. A number of variations have been published, but there are 2 main principles. On the one hand, there are instrumental techniques in which the intact irradiated sample is measured (nondestructive analysis), and on the other hand, there are techniques involving different kinds of chemical procedures by which the constituents of the sample are separated before measurement (destructive analysis). Generally lower detection limits and higher degrees of specificity can be achieved by the latter methods.

2.2.1.1.3.1 Nondestructive Analysis

Instrumental procedures have been described by a number of authors (e.g., Westermark and Sjöstrand, 1960, Filby et al., 1970, and Nadkarni and Ehmann, 1971).

Detection limit — Westermark and Sjöstrand, 1960, reported 100 to 500 ng/g in a 0.3 g sample. Filby et al., 1970, reported 3.5 ng/g in a 5 g blood sample.

Precision — 0.4 µg in the range 3 to 30 µg (Westermark and Sjöstrand, 1960). Filby et al., 1970, found 6 to 11% in the range 0.06 to 0.2 mg/kg. Nadkarni and Ehmann, 1971, reported 6 to 19% in the range 0.06 to 3.9 mg/kg.

An interlaboratory comparison was organized by IAEC (Tugsavul, Merten, and Suschny, 1970). Three laboratories used nondestructive neutron activation analysis in analyzing the standards, 2 samples of flour, 1 with and 1 without mercury added. The results are presented in Figure 2:1. Repeated analysis was made by only 2 of the laboratories and only on the treated sample. The precision of these laboratories can be calculated from the figures given by Tugsavul, Merten, and Suschny, 1970, at 2 and 22%, respectively, of the means of all analyses, 5.1 mg/kg and 80 ng/g, respectively.

Accuracy — In the interlaboratory comparison

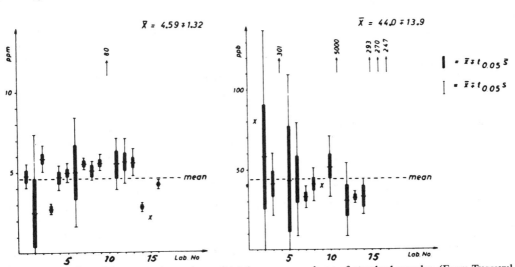

FIGURE 2:1. Inter-laboratory comparison of total mercury analyses of standard samples. (From Tugsavul, Merten and Suschny, 1970.) Laboratories no. 9, 11, and 16 used nondestructive activation analysis. The rest used activation analysis including a chemical separation step with the exception of laboratory no. 1, which used a chemical method for the treated sample. Each laboratory made 1 to 6 analyses of each sample. The overall average of all laboratories is shown by the dotted line, the individual laboratory averages by horizontal lines. The 95% confidence limits of single determinations and of means within laboratories are shown by thin and thick vertical lines, respectively. In the calculation of the overall mean the extreme values (arrows) were excluded.

reported by Tugsavul, Merten, and Suschny, 1970 (Figure 2:1), only 1 mean of 1 laboratory using nondestructive activation analysis was used in the calculation of the overall average for all laboratories. That laboratory had a mean of 5.1 mg/kg for the treated sample as compared to the overall mean 4.6 mg/kg. The rest of the results deviated heavily.

2.2.1.1.3.2 Destructive Analysis

In the destructive analysis, different principles have been employed for the separation of mercury. Sjöstrand, 1964, performed a wet digestion, added Hg^{2+} carrier, distilled the mercury as $HgCl_2$, and precipitated by electrolysis on a gold foil. This method has been used widely in Sweden in the epidemiological work in connection with the presence of methyl mercury in fish. Kim and Silverman, 1965, used an isotope exchange method in which ^{197}Hg was accumulated in a mercury droplet. A similar technique has been used by other authors (Brune, 1966, Brune and Jirlow, 1967, and Brune, 1969). Rottschafer, Jones, and Mark, 1971, separated the mercury on an ion exchange resin. Other procedures have included extraction, displacement, sulfide precipitation, and reduction (Tugsavul, Merten, and Suschny, 1970).

Detection limit — Ljunggren et al., 1969, reported for Sjöstrand's 1964 method 0.1 to 0.3 ng absolute in biological material, which means 0.1 to 0.3 ng/g in a 1 g sample. Rottschafer, Jones, and Mark, 1971, reported 3 ng/g in a 1 g sample.

Precision — Sjöstrand's 1964 method had a coefficient of variation of less than 2 and 6% in analysis of samples of 0.16 (kale) and 10 (fish) mg/kg, respectively (Ljunggren et al., 1971). For analyses of whole blood, blood cells, and plasma with the same method, a precision of 1.1 ng/g has been obtained in the concentration range 5 to 50 ng/g, corresponding to 22 and 2.2% at the terminal points of the interval, and 2.2 ng/g in the range 25 to 250 ng/g, corresponding to 8.7 and 0.9% (Birke et al., to be published). Kim and Silverman, 1965, reported 7 and 14% in analyses of tobacco containing 0.07 and 0.47 µg/g, respectively. Brune, 1966, found 6% in blood samples containing 3 to 24 ng/g. Rottschafer, Jones, and Mark, 1971, reported 10% for analysis of fish ranging 0.05 to 10 mg/kg.

In the interlaboratory comparison reported by Tugsavul, Merten, and Suschny, 1970, and illustrated in Figure 2:1, 13 laboratories used methods including some kind of separation step. For the treated sample (overall mean 4.6 mg/kg) the precision varied between 2 and 75% for different laboratories. Ten laboratories were at or below 5% and 12 were below 25%. For the untreated sample (overall mean 56 ng/g) the precision ranged 1 to 53%. The lowest value was obtained from a laboratory with a mean of all analyses deviating 50 times from the overall mean! Four other laboratories were at or below 10%, and 10 were below 20%.

Accuracy — On a testing (Bowen, 1969, see Section 2.2.1.2) based on 31 determinations by activation analysis made at 7 laboratories, the mean value for the analyses according to Sjöstrand's (1964) method did not show any deviations from the best value based on the results of all 7 laboratories. This means that the accuracy approaches the precision, i.e., 2% (Ljunggren et al., 1971).

In the above mentioned interlaboratory comparison (Tugsavul, Merten, and Suschny, 1970, Figure 2:1), the overall mean of the treated sample was 4.6 mg/kg. Of the 13 laboratories using activation analysis, 4 had means within ± 10% of the overall mean, and 10 within ± 30%. For the untreated sample the overall mean was 44 ng/g. Of 14 laboratories, 4 had means within ± 10% and 9 within ± 30%. The mean of 1 laboratory deviated 50 times from the overall mean!

During the epidemic of methyl mercury intoxication in Niigata, activation analysis was used (nondestructive and destructive) in biological material. Sensitivity, precision, and accuracy were not reported.

2.2.1.1.4 Micrometric Method

In the method used by Stock and Zimmermann, 1928a and b, and Stock, 1938, mercury in the sample was reduced to elemental mercury, which, in the form of a drop, was measured under a microscope. This method was applied, among other things, for the analysis of biological material; however, it does not seem to have come into general use. Nonetheless, the results reported show good agreement with the levels found subsequently in samples of different types.

2.2.1.2 Interlaboratory Comparisons

Comparisons between the analyses made with dithizone and those made with an activation method by the Department of Pharmacology and the Institute of Hygienic Chemistry and Legal Chemistry at the University of Tokyo, with regard to 2 materials consisting of hair of the head, can be made on the basis of data in the Niigata Report (Kawasaka et al., 1967). Duplicate analyses in the range of 0.5 to 500 µg/g show on a statistical analysis rank correlation coefficients of 0.91 and 0.79, respectively. As a rule, the results of the activation analyses are 20% and 8% higher, respectively, than those of the dithizone method. In several cases the methods show a difference of 100% or more calculated with regard to the lowest value.

An attempt to evaluate different methods of analysis used in Sweden was made in 1968. Samples were taken from 3 different fish. Two laboratories used activation analysis (Sjöstrand, 1964, and Brune and Jirlow, 1967) and 1 used atomic absorption (Lidums and Ulfvarson, 1968b). The precision as estimated for the entire material (levels 100 to 1,000 ng/g), was 41 to 86 ng/g for the different laboratories. The differences for both the mean values and the precision errors among the laboratories were statistically significant ($p < 0.01$). It should be emphasized, however, that precision is greatly dependent upon the level in the sample. The material was too small for the complete elucidation of this question. Table 2:1 shows data on deviations of individual analyses from the mean value for all of the analyses. It is evident that 50% of the analyses were within ± 10%, whereas over 90% were within ± 20%, and all of them were within ± 40%.

Bowen, 1969, organized a test in which a kale powder was analyzed by neutron activation analysis in 7 different laboratories and by colorimetric method in 1. The number of analyses performed at each laboratory was 2 to 9. The "best mean value" was 0.16 mg/kg. The mean of the activation analyses from different laboratories ranged 0.14 to 0.18 mg/kg while the colorimetric method gave only 0.012 mg/kg.

Analyses were compared (data quoted by Berglund et al., 1971) in an investigation of Japanese and Swedish fish between laboratories in Sweden and Japan for total mercury and alkyl mercury (Section 2.2.2.2.2). The total mercury analyses were made in Sweden by a laboratory using activation analysis according to Sjöstrand, 1964, and in Japan by a laboratory using an atomic absorption method. In 5 samples of Japanese fish the Japanese analyses (0.4 to 4.6 mg/kg) in every case were lower than the Swedish (64 to 82% of the levels found in Sweden). In the 3 samples of pike from Sweden, the Japanese laboratory found higher total mercury levels (0.1 to 1.2 mg/kg) than the Swedish laboratory, 109 to 123% of the Swedish values.

An interlaboratory comparison of laboratories using neutron activation analysis of flour (Tugsavul, Merten, and Suschny, 1970) has been discussed in Section 2.2.1.1.3.

In an interlaboratory comparison by Uthe,

TABLE 2:1

Comparison Among Analyses of Total Mercury Made by 3 Swedish Laboratories (Table from Berglund et al., 1971, based on data from Working Team for Coordination of Investigations of Mercury in Fish, 1968.)

Fish no.	Mean level mg Hg/kg	Number of analyses	Distribution of individual analytical values in intervals from mean level			
			±10%	±20%	±30%	±40%
1	0.13	23	15	22	23	23
2	0.85	23[x]	12	20	22	23
3	0.62	24	8	24	24	24
Total number of analyses		70	35	66	69	70
% of analyses		100	50	94	99	100

[x]One zero-value is excluded.

Armstrong, and Tam, 1971, 29 laboratories in Canada and the U.S. analyzed 3 homogenates of fish. Nineteen of the laboratories used different variants of wet digestion followed by atomic absorption (14 flameless and 5 flame), 2 used pyrolysis followed by flameless atomic absorption, 5 used neutron activation analysis, and 2, dithizone methods. The results of 3 laboratories (the 2 using pyrolysis followed by flameless atomic absorption and 1 using a dithizone method) were excluded from the statistical treatment because of obvious separation from the rest of the results (deviation greater than 50%). A summary of the combined results is given in Table 2:2. Neutron activation, flameless atomic absorption, and flame atomic absorption gave overall averages close together but the last mentioned method had much lower precision than the other 2. The coefficient of variation of the combined material of analyses of samples A and C (about 1.3 and 4.1 mg/kg) was 19 and 20%, respectively, while it was 83% for sample B (about 0.1 mg/kg). Graphic analysis of the results of samples A and C showed that most laboratories tended to obtain either high or low results with both samples and that several had more consistent results with the low fat sample A than with the high fat sample C. The coefficient of variation from the laboratory mean, for the laboratories reporting separate values, ranged 2 to 12% for samples A and C, and 12 to 36% for sample B, without any clearcut difference among methods.

2.2.1.3 Discussion

The data on limit of detection, precision, and accuracy given in Sections 2.2.1.1 and 2.2.1.2 generally refer to optimal conditions. At routine use the reliability might be lower. Also, data on the reliability of a method when used in one laboratory must be used only with greatest caution for evaluations of the reliability of analytical results obtained with the same method at other laboratories.

The different methods of analysis for mercury in air are of different sensitivity and reliability. The simplest and cheapest method is the semi-quantitative mercury determination by gas detector tubes. Levels of 0.1 mg of mercury per m^3 of air generally can be covered by them, but the precision is poor and they are used mainly for preliminary investigations.

Mercury detectors, based upon the light absorption of elemental mercury vapor, are rather expensive, but the cost of each analysis is low. Their sensitivity is high: 2 to 5 $\mu g/m^3$ generally are covered. The result is obtained immediately. When using them in industrial atmospheres, however, attention must be paid to other gases or vapors which may interfere with the mercury determination, as well as to other possible sources of error.

The most reliable method of analysis for mercury in air is the air sampling method. Either in combination with classical chemical (dithizone) or with atomic absorption methods for final mercury determination, its sensitivity can be in-

TABLE 2:2

Interlaboratory Comparisons of Analyses of 3 Fish Homogenate Samples (Data from Uthe, Armstrong, and Tam, 1971.)

Method of analysis	No. of laboratories	Sample A		Sample B		Sample C	
		Mean mg/kg	Coefficient of variation	Mean mg/kg	Coefficient of variation	Mean mg/kg	Coefficient of variation
Flameless atomic absorption	14	1.36	19	0.10	55	4.28	18
Flame atomic absorption	5	1.29	29			3.72	32
Neutron activation	6	1.37	19	0.11	55	4.06	16

creased to cover fractions of the MAC (TLV) value by increasing the sample volume. It is much less subject to chemical interferences than are the mercury detectors. On the other hand, the amount of work required is considerable, and the result generally is not obtained on the day of sampling. This type of analysis is by far the most expensive.

With regard to analyses of total mercury in biological material, the methods seem to have been hampered by a considerable degree of uncertainty until the middle of the 1960's. Thereafter, the reliability of the analyses has increased, especially within the higher concentration range.

From what has been stated above, it is evident that from a toxicological point of view most modern methods of analysis for total mercury in urine meet the demand for a reasonable degree of reliability. The same is true for total mercury in fish and other foods.

For the analysis of total mercury in blood, activation analysis and flameless atomic absorption spectrometry are the methods of choice. The sensitivity of these 2 methods is satisfactory, and the precision is acceptable. For the atomic absorption method, a precision within a few percent has been reported for concentrations in the range of 5 to 100 ng/g. For activation analysis, the error seems to be of about the same magnitude. Comparison between the 2 methods has shown acceptable agreement. No data are available on the reliability of hair analyses. The mercury level in hair is 2 orders of magnitude higher than that in blood.

For the evaluation of toxicity of short chain alkyl mercury compounds, the total mercury levels in blood reported in patients poisoned by methyl mercury-contaminated fish in Niigata in Japan are of great importance (Chapter 8). Most of the analyses were made by a dithizone method. It is not possible to assess the methods used because they are not reported in detail. The blood levels in the patients were relatively high, which probably implies a reasonably high degree of analytical certainty, but it is possible that systematic errors occurred. The repeated analyses reported for the same patients indicate, however, a relatively good analytical precision (Section 8.1.2.1.1.1.1). In addition to blood, hair also was analyzed. The difficulties in the evaluation of the reliability of the results are the same as for blood. As the levels in hair at methyl mercury exposure are about 300 times higher than those in blood (Section 4.5.2.1), it is reasonable to assume that the reliability of the hair analyses was higher than that of the blood analyses.

From the information available on the colorimetric precipitation methods widely used in the U.S.S.R. for analysis of air and urine, it must be assumed that the results are greatly dependent upon the skill of the laboratory personnel.

2.2.2 Specific Methods for Inorganic or Organic Mercury

2.2.2.1 Specific Methods for Inorganic Mercury in the Presence of Organic Mercury

Westöö, 1966a, 1967a, and 1968a, separated inorganic mercury and organomercurials in biological material by thin layer chromatography. Similar systems have been used by Takeda et al., 1968a, and Östlund, 1969b, for estimation of labeled inorganic mercury formed from alkyl mercury compounds in experimental animal studies.

Clarkson and Greenwood, 1968, described an isotope exchange method for measurements of nonradioactive inorganic mercury in tissues. Clarkson, 1969, Norseth, 1969b, and Norseth and Clarkson, 1970a, used the same principle for estimation of labeled inorganic mercury in the presence of organic mercury in biological material. The method is based on the fact that the exchange of inorganic mercury with elemental mercury vapor in a sample is much faster than that of covalently bound mercury. The radioactive elemental mercury vapor is collected in a metallic mercury drop and measured.

Clarkson and Greenwood, 1970, have utilized stannous chloride reduction to differentiate between inorganic and organic mercury in tissues after administration of compounds labeled with radioactive mercury. Without preceding digestion, only inorganic mercury is reduced by Sn^{2+} ions and can be carried away by air. Gage and Warren, 1970, based a similar method upon the reduction of organomercurials by stannous ions after treatment with cysteine. Without cysteine, alkyl and alkoxy-alkyl mercury salts are not reduced, and inorganic mercury can be determined separately.

2.2.2.2 Specific Methods for Organic Mercury
2.2.2.2.1 Methods of Analysis

A few workers have used methods for estimation of *organic mercury* in biological material. Miller, Lillis, and Csonka, 1958, determined phenyl mercury by oxidation with alkaline permanganate, extraction with dithizone in chloroform, and spectrophotometrical reading of the extract. The method, later modified for ethyl mercury (Miller et al., 1961), is not very sensitive. Gage, 1961b, analyzed phenyl and alkyl mercury with a more sensitive method involving acidification, extraction with benzene, re-extraction with aqueous sodium sulfide, oxidation with acid permanganate, and determination of mercury by a titration procedure.

Reviews of the available methods for quantitative analysis of *specific organic mercury compounds* in biological material have been published recently (Fishbein, 1970, and Berglund et al., 1971). Of special toxicological interest are methods for analysis of short chain alkyl mercury compounds, particularly methyl mercury.

The generally used procedures for alkyl mercury analysis have included addition of a halogenhydrogen acid to a homogenate of the sample, which causes the alkyl mercury originally bound to the biological material to form alkyl mercury halide. This is extracted with some organic solvent. After purifying, concentrating, and drying if necessary, the extract is analyzed quantitatively by gas-liquid chromatography. Several variations of this main route have been published (Westöö, 1966a, 1967a, and 1968a, Sumino, 1968a, Tatton and Wagstaffe, 1969, Ueda, Aoki, and Nishimura, 1971, Newsome, 1971, and Westöö and Rydälv, 1971). Methods which do not include purification after the extraction have been used by Kitamura et al., 1966, and Takizawa and Kosaka, 1966.

For fish meat a yield of methyl mercury over 90% has been reported (Westöö, 1966a, 1967a, and 1968a). Substantial losses may occur in other samples. Various modifications of the purification method, however, may increase the yield (Westöö, 1968a, 1969a and b). By some procedures considerable losses may occur even in analysis of fish samples.

Detection limit — According to Westöö's method (1968a), 1 to 5 ng Hg as methyl mercury/g for a sample of 10 g.

Precision — Three percent for levels over 0.05 mg Hg as methyl mercury/kg of fish for a 10 g sample (Westöö and Rydälv, 1969). See also Section 2.2.2.2.2.

Accuracy — Comparison of a great number of analyses of methyl mercury in fish by the methods of Westöö (1966a, 1967a, and 1968a) and total mercury determinations by neutron activation analysis according to Sjöstrand (1964) has shown a good agreement, the average methyl mercury level making up 94% of the total mercury (Westöö and Rydälv, 1971). This favors high accuracy. See also Section 2.2.2.2.2.

2.2.2.2.2 Interlaboratory Comparisons

An attempt to evaluate different variants of methyl mercury analysis was carried out in Sweden in 1968 by 4 different laboratories. The samples consisted of untreated white dorsal muscles from 3 pike (levels 0.1 to 1 mg Hg/kg). Statistical analysis showed that the difference among the mean values obtained by the various laboratories was significant ($p < 0.01$). In Table 2:3 data have been listed on the percentage deviation of individual analyses from the mean of all of them. It is evident that 80% of the analyses were within ± 10% of the mean value, and all were within 20%.

In 1968 there was an exchange of fish between Japan and Sweden in order to compare the results of analysis of the same fish samples (Westöö, 1968b and c, Westöö and Rydälv, 1969, and Kitamura et al., personal communication). Here, only the alkyl mercury analyses will be dealt with.

TABLE 2:3

Comparison Among Methyl Mercury Analyses Made in 4 Swedish Laboratories (Table from Berglund et al., 1971, based on data from Working Team for Coordination of Investigations of Mercury in Fish, 1968.)

Fish No.	Mean level mg Hg/kg[x]	Number of analyses	Distribution of individual analytical values in intervals from mean level	
			±10%	±20%
1	0.14	16	16	16
2	0.96	12[xx]	9	12
3	0.67	12[xx]	7	12
Total number of analyses		40	32	40
% of analyses		100	80	100

[x]Mean of all analytical values
[xx]One laboratory reported disturbances in the chromatograms. These values have been excluded.

For total mercury, see Section 2.2.1.2. The Japanese results were, for 5 samples of fish from Japan (total mercury levels of 0.5 to 7.2 mg/kg according to analyses by the method of Sjöstrand, 1964), lower than the Swedish results in 9 out of 10 determinations (methyl mercury 63 to 77%, ethyl mercury 50 to 100%, of the Swedish values). As judged from the Swedish analyses 81 to 94% of the mercury was present as alkyl mercury. In 3 Swedish fish (total mercury about 1 mg/kg) only methyl mercury was found. The Japanese values were 75 to 94% of the Swedish.

In 1971 a comparison among 6 laboratories in Scandinavia was reported (Nordic Committee on Food Analysis, to be published). Four samples of freeze-dried fish containing 0.1 to 4.2 mg mercury as methyl mercury/kg were analyzed 4 times at each laboratory by the method of Westöö, 1968a. The average of 1 laboratory deviated 20 to 80% from the common mean of the others in 3 of the samples. The means of the others were within ± 10%. The precision of the total number of analyses of these laboratories was 22% for the 0.1 mg/kg sample and 2 to 5% for the others. In 1 sample analyzed in 1 laboratory the coefficient of variation was 25%, and the rest were within ± 10%.

2.2.2.2.3 Discussion

Data available show that modern methods for analysis of methyl mercury have a reliability that satisfies the demands for use in toxicological evaluation of levels in fish and other foods, i.e., exceeding 0.02 mg/kg. Analyses of fish meat are somewhat simpler to carry out than the analyses of certain other types of biological material, e.g., liver and kidney, which give rise to more difficult extraction problems.

Of special toxicological interest is the proportion of mercury in fish present as methyl mercury (see also Section 3.3.3.1). Methyl mercury makes up almost all of the total mercury in flesh of Swedish fish (Westöö and Rydälv, 1969 and 1971) and of North American fish (Smith et al., 1971). Lower fractions have been reported in some samples of young fish analyzed by Bache, Gutenmann, and Lisk, 1971. In that study the whole fish, without evisceration, was chopped and ground before analysis. A large amount of Japanese material consisting of fish from different areas has shown that methyl mercury constitutes an average of about 25% (range 0 to 75%) of the total mercury (Kitamura, personal communication). Ueda, Aoki, and Nishimura, 1971, reported that 4 to 65% of the total mercury (dithizone method) in fish from mercury contaminated and noncontaminated rivers in Japan consisted of alkyl mercury.

Westöö, 1968b, has pointed out that there are reasons for assuming that the methyl mercury determinations were too low in methods used by some Japanese investigators (Kitamura et al., 1966, and Sumino, 1968a). A direct comparison between Swedish laboratories and a Japanese laboratory in 1968 showed, for the Japanese laboratory, a higher proportion of methyl mercury than that previously reported for Japanese fish.

Only a few methyl mercury levels in fish have been reported from the Japanese epidemics of intoxication in Minamata and Niigata. Although sufficiently detailed descriptions of the analytical procedures are not available, it is reasonable to assume that most of the mercury in fish in connection with these catastrophes was in the form of methyl mercury.

Chapter 3

TRANSPORT AND TRANSFORMATION OF MERCURY IN NATURE AND POSSIBLE ROUTES OF EXPOSURE

Jaroslav Vostal

The increasing threat of contamination of the environment by the widespread use of mercury and its compounds in industry and agriculture, and the potential hazard of high intake of toxic forms of mercury by large groups of the population, have focused a great deal of attention on the fate of mercury in the environment. Many environmental sources of mercury have been analyzed and evaluated in recent scientific meetings and reviews (Löfroth, 1969, Maximum Allowable Concentrations of Mercury Compounds — Report of an International Committee, 1969, Miller and Berg, 1969, Nordiskt Symposium, 1969, Stahl, 1969, Berglund et al., 1970, 1971, Environmental Mercury Contamination Conference, 1970, Keckes and Miettinen, 1970, U.S. Geological Survey, 1970, Jones, 1971, Mercury in Man's Environment Symposium, 1971, Mercury in the Western Environment Conference, 1971, Nelson et al., 1971, Wallace et al., 1971, Miller and Clarkson, in press, d'Itri, 1972, in press, and Mercury Contamination in Man and his Environment, to be published). A summary of the findings of these studies, cited many times under the names of the individual contributors, will be included in this chapter.

3.1 Natural Sources and Transport of Mercury in the Environment

3.1.1 Geographical Occurrence of Mercury

Mercury occurs in the natural state only in small amounts, estimated at 50 to 80 ppb of the earth's content. It exists mainly in the form of various sulfides, especially red sulfide (cinnabar). Primary deposits of this metal occur in practically all types of igneous, metamorphic, or sedimentary rocks in concentrations varying in general between 50 and 500 ppb (Jonasson, 1970, U.S. Geological Survey, 1970, and Shacklette, Boerngen, and Turner, 1971). Ninety-nine percent of the mercury mined in the world is concentrated in mercuriferous belts which correspond to the mobile zones of dislocation of the earth: the East Pacific Rise, involving the west coast of America and the eastern part of Asia, and the Mid-Atlantic Ridge (Jonasson and Boyle, 1971). All industrially-used deposits of mercury are located within these belts. The total world production from these sources amounts to 10,000 tons of mercury a year. The grades of ore differ considerably among the individual sources. The highest contents of mercury are reported from Spain, with an average of 60 lb of mercury/ton, and as high as 1,400 lb/ton in some places. Italian ores average 10 lb of mercury/ton. The U.S. and Canada report 4 to 5 lb of mercury content. The world reserves of mercury are estimated to be 200,000 tons, half of which are in Spain (*Minerals Yearbook, 1970*, in press).

3.1.2 Modes of Entry of Mercury into Various Media of the Natural Geocycle

Mercury can enter the geochemical cycle by simple transport in the form of metallic mercury vapors, or transformed into volatilized organic mercury compounds, and/or by chemical transformation into more soluble salts or mercury compounds.

3.1.2.1 Environmental Transport of Mercury into the Atmosphere

3.1.2.1.1 Vaporization Processes

Because of its ability to be vaporized at normal temperatures, the vaporing of metallic mercury constitutes the easiest way of transport into the atmosphere. The vapor pressure is high even at normal temperatures ($1.2 \cdot 10^{-3}$ mm Hg at 20°C) and rapidly increases with rising temperature. At 25°C the heat of vaporization is 14.67 cal/g atom. The saturation concentration of mercury in the air can be calculated from its vapor pressure. At room temperatures it amounts to 10 to 15 mg Hg/m^3.

Atmospheric data collected by McCarthy et al., 1970, revealed high levels of mercury in the air over the localities with ore deposits, whereas the atmosphere over nonmineralized areas showed low levels of mercury. In England, mercury concentrations in the air over regions with exceptionally high levels of mercury in the humus layers of topsoil (about 10 ppm) were reported to be in the range of 20 to 200 ng/m^3, compared with background levels of 5 ng/m^3 (Barber, Beauford, and Shieh, in press).

3.1.2.1.2 Volatilization Processes

Transition of ionized forms of mercury into the atmosphere by volatilization can occur theoretically by 3 processes: 1. chemical reduction into the elemental form, 2. reduction through the activity of microbes, plants, or other living organisms, and 3. biotransformation into volatile organomercury compounds, mainly short chain alkyl mecurials.

Although chemical reduction of ionized mercury into the elemental form by inorganic reducing agents is well defined in laboratory conditions, definite evidence has not been reported on its role in volatilization of mercury in nature. Volatilization of different mercury compounds by soil was studied by Kimura and Miller, 1964. Approximately 15% of added phenyl mercury acetate was converted to metallic mercury vapor in 28 days, ethyl mercury was decomposed only partly, and methyl mercury not at all. Ethyl and methyl mercury, however, evaporated in their original forms. Later, bacterial cultures (*Pseudomonas*) isolated from phenyl mercury contaminated soil were shown to convert solutions of methyl, ethyl, and phenyl mercury into metallic mercury vapors. Corresponding hydrocarbons were detected simultaneously by gas chromatography (Tonomura et al., 1968a and b, Tonomura, Maeda, and Futai, 1968, Tonomura and Kanzaki, 1969, Furukawa, Suzuki, and Tonomura, 1969, and Furukawa and Tonomura, 1971). Volatilization of inorganic mercury from humus-containing soil by bacterial activity has not been studied extensively. Barber, Beauford, and Shieh (in press) reported that the bacterial profile of the soil closely follows the profile of mercury. Moreover, laboratory experiments proved that bacteria isolated from this soil can induce volatilization of mercury. Barber (unpublished data) recently confirmed the volatilizing ability of bacteria by showing that suspensions of live *Pseudomonas* released 4 to 30 times as much mercury as did the dead control cells. Similar results were obtained with cultures of bacteria isolated directly from mercury-containing humus.

Microbial activity in volatilization of mercury from biological fluids had been proved earlier and mercury-volatilizing strains identified (Magos, Tuffery, and Clarkson, 1964). Volatilization of mercuric ion by plants also has been studied to a small extent. Low levels of uptake of inorganic mercury from the soil have been reported (Shacklette, 1965, and 1970, and Smart, 1968). Fukunaga and Tsukano, 1969, and Rissanen and Miettinen (to be published) reviewed the Japanese studies (Furutani and Osajima, 1965, 1967, Tomizawa, 1966, and Yamada, 1968) on the uptake by rice plants of labeled mercury from soils contaminated by organomercurial compounds. High accumulation rates of mercury from phenyl mercury-treated soil and phenyl mercury solutions were reported. Autoradiographic studies on spearmint (*Mentha spicata*) (Barber, Beauford, and Shieh, in press) after incubation of its roots in labeled mercuric chloride solution containing 0.4 ppm of mercury indicated that mercuric ion can enter the plant, accumulate in the vascular system of roots, and be transported into the leaves. The foliage concentrations, although clearly detectable, were 10 to 500 times lower than the root concentrations of mercury. No significant transpiration of mercury in the air by the plant was recorded.

As for animals, Clarkson and Rothstein, 1964, found radioactive mercury vapors in the air exhaled by rats injected with labeled mercuric ion, and proved that the animals are able to volatilize mercury. No similar evidence has been found in man.

Conversion of deposits of inorganic mercury into volatile organomercury compounds could be a more effective way of transporting mercury into the geocycle. Jensen and Jernelöv, 1969, reported that an unidentified microorganism in sludge from aquaria can methylate inorganic mercury. Formation of alkyl mercury compounds was a function of mercury concentrations of up to 100 ppm in the freshwater sediments. The authors stressed the fundamental importance of this process for the mobilization of mercury from the sediments into the general environment and proposed 2 pathways: either a direct formation of mono-methyl mercury, or primary synthesis of di-methyl mercury that is later converted into mono-methyl mercury.

Wood, Kennedy, and Rosen, 1968, showed that cell extracts of a strictly anaerobic methanogenic bacterium effectively convert inorganic mercury into methyl mercury using methyl cobalamin as substrate, and described the process as a combination of both pathways depending upon various pH conditions of the environment. The authors supported the concept that di-methyl mercury can be the predominant product of the reaction and under mild acid conditions is further converted to mono-methyl mercury. The rapidity of demethylation of the substrate in vitro suggested further that methyl transfer could occur in biological systems as well as a nonenzymatic, chemical reaction. Experiments performed without the presence of any bacteria showed that transfer of methyl groups occurs also by a nonenzymatic process. The authors suggested that this nonenzymatic process is enhanced in vivo by anaerobic conditions and by the presence of bacteria that synthesize alkyl cobalamins. The emphasis on the anaerobic character of both possible interconversion mechanisms led to the opinion that anaerobic conditions in sediments contaminated by mercury are required for the biotransformation of mercury. However, Fagerström and Jernelöv, 1971, found that hydrogen sulfide, ubiquitous in the natural environment under anaerobic conditions, binds deposited mercury into insoluble chemical form and decreases the availability of mercury for methylation. The authors pointed out that the methylation rate is generally more dependent on microbial activity than on anaerobic conditions. These conclusions conform with the observations that microorganisms producing hydrogen sulfide inhibit the volatilization of mercury from soil and biological materials (Booer, 1944, and Magos, Tuffery, and Clarkson, 1964) and with recently presented results indicating a complete lack of methylation activities in mud and soil under strict anaerobic conditions (Rissanen, Erkama, and Miettinen, 1970).

Landner, 1971, described another biochemical model for the methylation pathway in studies on the relationship between mercury resistance of *Neurospora crassa* and its ability to methylate inorganic mercury. He suggested a link between the ability to produce methyl mercury and methionine biosynthesis in these fungi. Preliminary experiments (Imura et al., 1971, and Jernelöv, in press) showed that a direct transmethylation involving methionine or S-adenosylmethionine is improbable. Therefore, mutants of *Neurospora* with high resistance to mercury were selected, and it was found that their methylation efficiency was much higher than that of other strains. The authors suggested that increased methylation rate is an induced detoxification process in resistant fungi, and that the methylation pathway is linked with the intracellular biosynthesis of methionine. Methylation of the mercuric ion thus might be regarded as an "incorrect synthesis" of methionine.

On the other hand, Bertilsson and Neujahr, 1971, reemphasized the nonenzymatic methyl transfer from methyl cobalamin to mercury. Methyl cobalamin was incubated with solutions of mercuric chloride and a rapid transfer of the methyl group occurred. The end products of the reaction were methyl mercuric ion and hydroxycobalamin. However, when mercuric ion was replaced in the incubation mixture by methyl mercury or other organomercurials, the reaction rate decreased. The results agreed with observations reported by Hill et al., 1971, and were interpreted by the authors as evidence against primary formation of di-methyl mercury as a predominant product of the reaction. In contrast, prevailing formation of di-methyl mercury as initial product of the reaction of mercuric ion with methyl cobalamin in vitro was reported by Imura et al., 1971. Different ratios of mono- and di-methylated products were formed under their experimental conditions, depending upon the molar ratios of reactants and reaction times; an immediate conversion of freshly formed di-methyl mercury into mono-methyl form by the action of mercuric ion was assumed.

Direct conversion of other organomercurials into methyl mercury by microbial activity does not appear to be a common process. Formation of other volatile products cannot be excluded, however. Although no methyl mercury was found among volatile mercury products after 10 days of incubation of phenyl mercury with sludge microorganisms, approximately 40% of all solvent-extractable metabolic products were detected in the form of volatile di-phenyl mercury (Matsumura, Gotoh, and Mallory Boush, 1971).

Although the biotransformation pathway has not been completely clarified, it might be concluded that the physicochemical properties of the newly formed compounds may contribute considerably to the release of mercury into the biosphere, since their high volatility and solubility make the transition of mercury into the environment easy.

3.1.2.2 Environmental Transport of Mercury into the Hydrosphere – Dissolution Processes

Solubility of metallic mercury in water is low

($2 \cdot 10^{-5}$ g/l., Hughes, 1957). Contact with oxygen-containing water increases the solubilization of mercury and the final solubility is limited in practice only by the saturation limits of the oxidation products (Stock, 1934). The solubilities of ionic mercury compounds depend upon the actual conditions of the solubilizing water, i.e., acidity, presence of complexing anions, other organometallic complexes, etc. As a result, mercury content of surface water depends upon the accessibility of mercury, time of contact, and conditions of the solubilizing media. Higher concentrations of mercury might occur in underground waters and geothermal springs, and mercury deposits in sediments of some thermal waters may reach very high levels (White, Hinkle, and Barnes, 1970). Mercury levels in ground waters have been reported to be in the range of 0.02 to 0.07 ppb (Stock and Cucuel, 1934a, Heide, Lerz and Böhm, 1957, Dall'Aglio, 1968, and Wiklander, 1968). Samples of ground water recently were analyzed in 73 areas of the U.S. Only 2 samples were higher than 5 ppb and 83% were lower than 1 ppb (Wershaw, 1970).

Seawater concentrations originally were reported at the level of 0.03 ppb (Stock and Cucuel, 1934a) but higher levels were reported later by other authors (Hamaguchi, Rokuro, and Hosohara, 1961, and Hosohara, 1961). It is supposed that mercury in the sea originated mainly from weathering of primary rocks and probably is in the form of chlorocomplexes (Sillén, 1961). There is no information on how large the contribution could be from the mercury transferred into the sea by air masses and precipitation. Rainwater may contain levels up to 200 ng/l. (Stock and Cucuel, 1934a).

3.1.2.3 Environmental Transport of Mercury into the Pedosphere — Weathering, Precipitation, Sedimentation, and Biodegradation

Both metallic mercury and mercuric sulfides, the most abundant forms of mercury in rocks and minerals, are resistant to solubilization through weathering and enter the geocycle often in the form of only mechanically degraded particulate matter. Consequently, the actual content of mercury in topsoil varies extensively, although normal rural areas do not usually exceed the concentration of 150 ng/g (Pierce, Botbol, and Learned, 1970, Cadigan, 1970, and Shacklette, Boerngen, and Turner, 1971). Several world locations (Eire in the U.K. and certain areas in the U.S.S.R.) might have levels up to 10,000 ng/g (Wallace et al., 1971, and Barber, Beauford, and Shieh, in press). No detailed studies have been reported on the mechanisms of the transfer of mercury or on the form of mercury in these soils. Mercury distribution in the soil has a characteristic profile. Low concentrations usually are found in subsoil; levels in topsoil are 5 to 10 times higher (Rissanen and Miettinen, to be published). Andersson, 1967, compared African and Swedish topsoil and found the average mercury content of the former to be 23 ng/g and of the latter, 60 ng/g. Similar levels were found by Warren and Delavault, 1969, in several British soils. Generally the natural mercury content in the soil is determined by many undefined factors: variations of pH, drainage, concentrations of humus, etc. Soil with higher humus content accumulates generally higher levels of mercury than more mineralized soils (Andersson, 1967). The upper limit for the natural release of mercury due to chemical weathering was estimated by comparison with sodium leaching into the surface water (Joensuu, 1971). The ratio of mercury to sodium in weathering rocks was assumed to be identical with the ratio of their terrestrial abundance, and 230 tons of mercury were estimated to be the upper limit of mercury released into the environment. The actual amount of leached mercury is expected to be less than this estimate, since more mercury than sodium is adsorbed on particulate matter and prevented from being dissolved in the surface water.

3.2 Man-made Sources and Transport of Mercury in the Environment

The role of human activities in the amount of mercury released into the environment can be deduced from the annual production rates of mercury. Although not all produced mercury is dissipated directly into the environment, only minor portions of the total production are stocked or recycled, and the rest of the mercury and its compounds is finally released in some way into the atmosphere, surface waters and soil, or ends in landfills, dumps, and refuse. Table 3:1 shows the relative participation of various types of industries and agriculture in the consumption of mercury as illustrated by temporal trends in mercury uses in the U.S. during the years 1966 to 1970 (*Minerals Yearbook, 1970*, in press). The representative patterns of individual industrial activities may vary

TABLE 3:1

Annual Mercury Consumption by Various Industries in the U. S. During 1966 to 1970. Annual estimates based on consumption data published in *Minerals Yearbook*; *1970* (Cammarota, in press).

	1966 tons/year (%)	1967 tons/year (%)	1968 tons/year (%)	1969 tons/year (%)	1970 tons/year (%)
Electrical apparatus	606.4 (24.6%)	558.3 (23.3%)	676.0 (26.0%)	637.4 (23.9%)	549.1 (25.9%)
Chlorine production	396.9 (16.1%)	493.6 (20.6%)	600.6 (23.1%)	714.7 (26.8%)	517.3 (24.4%)
Paints	308.1 (12.5%)	246.8 (10.3%)	364.0 (14.0%)	336.0 (12.6%)	356.2 (16.8%)
Control instruments	251.4 (10.2%)	256.4 (10.7%)	275.6 (10.6%)	229.4 (8.6%)	167.5 (7.9%)
Dental preparations and pharmaceuticals	81.4 (3.3%)	91.0 (3.8%)	119.6 (4.6%)	122.7 (4.6%)	101.8 (4.8%)
Catalysts and amalgamation	76.4 (3.1%)	93.4 (3.9%)	75.4 (2.9%)	120.0 (4.5%)	84.8 (4.0%)
Agriculture	81.3 (3.3%)	129.4 (5.4%)	117.0 (4.5%)	93.3 (3.5%)	61.5 (2.9%)
Paper and pulp production	22.2 (0.9%)	14.4 (0.6%)	15.6 (0.6%)	18.7 (0.7%)	8.5 (0.4%)
Other uses	640.9 (26.0%)	512.7 (21.4%)	356.2 (13.7%)	394.8 (14.8%)	273.3 (12.9%)
Total consumption	2,465 (100.0%)	2,396 (100.0%)	2,600 (100.0%)	2,667 (100.0%)	2,120 (100.0%)

extensively among different countries, mainly with regard to agricultural uses (Smart, 1968, Gurba, 1971, and Rissanen and Miettinen, to be published), and to paper, pulp, and paint production (Bouveng, 1967, Keckes and Miettinen, 1970, Cooke and Beitel, 1971, and Hanson, 1971).

3.2.1 Industrial Sources

The major part of the mercury produced annually is still consumed by the chlorine-alkali industry to compensate for the losses of mercury in the electrolytic production of chlorine and caustic soda. This type of industry constitutes the largest potential source of mercury released into the atmosphere and surface water. This source of environmental pollution has been identified repeatedly; in many plants all over the world steps have been taken to prevent unnecessary release of mercury. In the U.S. the discharge of mercury from this source was reduced 85% in several large plants by 1970, resulting in a decline of total mercury consumption by the chlorine-alkali industry of 27% (Cammarota, in press). In Sweden the total loss of mercury into the environment was estimated to be between 25 and 38 tons annually, corresponding to 100 to 150 g of mercury lost per ton of produced chlorine. In new plants the losses were reduced to 2 to 3 g Hg/ton (Halldin, 1969, and Hanson, 1971). A similar situation is expected in Canada (Flewelling, 1971) and other countries.

An approximately equal part of the mercury produced annually is used for the production of electric apparatus. Environmental losses connected with the industrial production are considered small (Halldin, 1969). Most disposable equipment (mercury battery cells, fluorescent bulbs, switches, etc.) ends up, however, in landfills, dumps, and incinerators.

Mercury is used extensively as an antifouling and mildew-proofing agent in oil, latex, and ship bottom paints. Nearly 400 tons of mercury are consumed for this purpose yearly in the U.S., and about 10 tons in Canada (Cooke and Beitel, 1971). Only 5 tons were consumed for paint production in Sweden in 1967 (Hanson, 1971).

Annual mercury discharge from the pulp and paper industries in Sweden using phenyl mercury compounds for impregnation of pulp and for slime control between 1940 and 1965 achieved the highest level in 1960, when estimates of the total yearly mercury loss approached 15 tons. After restrictions on the use of mercurials in the pulp production in 1966 to 67, mercury losses from paper and pulp industries declined to less than 1 ton a year (Hanson, 1971). A sharp decline in the use of mercurials also has been reported in the pulp industries in Canada and the U.S. since 1970 (Paavila, 1971, and Cammarota, in press).

3.2.2 Agricultural Sources

Agricultural uses of organomercurial fungicides constituted a considerable portion of mercury production released in the form of highly toxic methyl mercury in past years (Keckes and Miettinen, 1970, Berglund et al., 1971, and Wallace et al., 1971). The legislative elimination of alkyl mercurials from seed treatment and restrictions on the agricultural use of mercury decreased consumption of mercury for these purposes in Sweden by 70% between the years 1964 and 1969 (Lihnell, 1969, and Esbo and Fritz, 1970). Similarly, agricultural uses of mercury in the U.S. decreased by 10% in 1968, 22% in 1969, and 33% in 1970. The annual consumption in 1968 constituted only 48% of the consumption for 1967 (Cammarota, in press). Agricultural uses of mercury in Canada decreased from 18% of total consumption of mercury in 1964 to about 3% in 1970 (Gurba, 1971).

Several estimates have been made on how much the seed treatment by organomercurials contributed to the mercury content in the soil. Methyl and ethyl mercury were used as seed disinfectants in Sweden between 1940 and 1966. Approximately 4,500 kg Hg were consumed yearly for this purpose and a total of 80 tons of mercury was distributed in Sweden during this time (Hanson, 1971). Analysis of cultivated and uncultivated soil in Sweden proved that these seed treatments played only a minor role for the mercury levels in the soil (Andersson and Wiklander, 1965, and Andersson, 1967).

The distribution and biodegradation of residues of mercurial fungicides in the soil also have been studied. Increased levels of mercury after the use of alkyl mercurial fungicides or inorganic mercury were found in the topsoil (Ross and Stewart, 1962, and Andersson, 1967). In contrast, phenyl mercury penetrates easily into the deeper layers. The differences in distribution are explained by different affinities of various mercurials for individual components of the soil (Aomine, Kawasaki, and Inoue, 1967, and Aomine and

Inoue, 1967). Residues of mercurial fungicides are firmly bound in the soil and only small fractions are leached into the surface water (Andersson, 1967, and Saha et al., 1970). Moreover, they can be decomposed and volatilized by microbial action (Kimura and Miller, 1964).

3.2.3 Other Sources

In addition to the losses caused by the intentional industrial use of mercury, release of substantial amounts of mercury also may occur in primary processes of mercury production and in other industrial processes where mercury vapors are generated as a side product. Losses of mercury during mining and smelting of mercury-containing ores were evaluated at 2 to 3% in efficient operations (Cooke and Beitel, 1971). Substantial emissions of mercury during refining of many other metallic ores are suspected, but no estimates have been proposed so far. Procedures are being developed for economically feasible recovery of mercury from gas condensates, waste waters, and slurries (Cammarota, in press).

A recently discovered source of mercury released into the environment is the burning of fossil fuels by unintentional industrial processes. Preliminary information presented at the Environmental Mercury Contamination Conference, 1970, showed that ash from coal burning plants contains negligible amounts of mercury, and that the mercury content of fuels is released completely into the atmosphere. Recently, mercury concentrations in different types of coal were reviewed (Wallace et al., 1971), estimated (Bertine and Goldberg, 1971) and analyzed (Joensuu, 1971, and Ruch, Gluskoter and Kennedy, 1971). The mercury content in 36 American coal samples, determined by mercury vapor detectors after release of mercury by combustion, ranged between 0.07 and 33 ppm, with an average of 3.3 ppm (Joensuu, 1971). Coal samples from Illinois analyzed in another study revealed an average value of 0.18 ppm (Ruch, Gluskoter and Kennedy, 1971). Coal from areas in mercuriferous belts may contain up to 300 ppm of mercury (Wallace et al., 1971). Annual coal production in 1967 was $3 \cdot 10^9$ tons. Joensuu, 1971, estimated that at the conservative figure of 1 ppm for average concentration, the release of mercury from coal burning must be assumed to be in the range of 3,000 tons a year, i.e., much larger than the amount of mercury released by weathering. Similar estimates by Bertine and Goldberg, 1971, were based on assumed lower concentrations of mercury and were approximately 300 times lower.

There is no satisfactory evidence on the mercury content in oil and natural gases used for heating purposes. Preliminary information shows that in mercuriferous belts concentrations of mercury in petroleum can be high and natural gases can be saturated with mercury vapors (White, Hinkle, and Barnes, 1970).

In general, it seems that in recent studies more attention has been focused on the evaluation of potential exposures to secondary sources of mercury through burning of fossil fuels and emissions from refining of ores with mercury impurities than to primary sources of industrial production and consumption of mercury. The role of atmospheric mercury concentrations in the transport of mercury by air masses has not yet been evaluated.

3.3 Possible Routes of Environmental Exposure and Levels of Mercury in the Environment

It can be concluded from the evidence on transport and transformation of mercury in nature as summarized in previous parts of this chapter that all components of the biosphere contain at least minimal traces of mercury and constitute potential sources of exposure for all living organisms, including man.

3.3.1 Possible Routes of Environmental Exposure Through Atmosphere

No satisfactory information exists on the amounts of mercury transferred or accumulated by the atmospheric air masses, and little is known about the abundance and distribution of mercury in the atmosphere. Recent data collected by the U.S. Geological Survey, 1970, proved that mercury concentrations in the atmosphere over nonmineralized land areas range between 3 and 9 ng Hg/m^3. Scattered analyses performed over mineralized areas indicated, in contrast, concentrations of 7 to 53 ng/m^3, and over known mercury mines, 24 to 108 ng/m^3. At the surface levels up to 20,000 were recorded at active mercury mines (McCarthy et al., 1970).

Seasonal, daily and diurnal variations of atmospheric mercury concentrations were recorded. Maximum concentrations were obtained in the middle of the day; levels were lower in the morning and evening, and the minimum concentrations were detected near midnight. Airborne mercury concentrations were inversely related to barometric pressure. Atmospheric concentrations of mercury also change as a function of altitude. At the level of approximately 300 ft a marked drop in mercury concentrations was recorded and similar changes were observed over mineralized areas. Levels at ground surface were 10 to 20 times higher than concentrations 400 ft over the ground (McCarthy et al., 1970).

By older methods levels higher than background were found over urban areas in the U.S., and varied between 10 and 170 ng/m^3 (Cholak, 1952).

Brar et al., 1969, measured 3 to 39 ng Hg bound on particulates in the atmosphere. Dams et al., 1970, found 4.8 ng/m^3 in the atmosphere over an industrial urban area and compared the levels with particulate mercury in the atmosphere over a rural area, i.e., 1.9 ng/m^3. Leites, 1952, observed levels up to 4,000 ng/m^3 in a polluted urban area and 0 to 2,000 ng/m^3 in a suburban area. Goldwater, 1964, reported 0 to 14 ng/m^3 in a metropolitan area, while Saukow, 1953, (quoted by Berglund et al., 1971) gave a value of 20 ng/m^3 for a metropolitan area. No information on the reliability of methods or selection of sampling areas was given.

Similarly, no satisfactory data are available on mercury concentrations in the ambient air in the vicinities of mercury mines and smelting plants, although levels much higher than in control areas must be expected. Fernandez, Catalan, and Murias, 1966, recorded extremely high concentrations up to 800,000 ng/m^3 in 2 localities in residential areas removed approximately 400 meters from the mine and mercury plant at Almadén, Spain, even during winter months. Kournossov, 1962, and Melekhina, 1958, (quoted by Kournossov, 1962) and Vengerskaya, 1952, (quoted by Leites, 1952) observed decreasing mercury levels in the ambient air with increasing distance from a mercury emitting plant, indicating a source of atmospheric contamination. McCarthy et al., 1970, reported airborne concentrations up to 600 ng Hg/m^3 during working hours at a mercury mine in Arizona, in the U.S. These scattered and solitary data obtained by different methods cannot be properly evaluated, but they indicate an urgent need for more detailed studies of atmospheric mercuric profiles in the vicinities of mercury emitting sources.

Concentrations over the ocean are lower than over the ground. Williston, 1968, indicated that winds coming from the sea have lower levels (2 ng Hg/m^3) than winds coming from the industrialized land surface (8 to 20 ng Hg/m^3). These observations confirm land surface as the main source of airborne mercury.

Air concentrations of mercury can be completely washed out by rain, even in polluted areas (McCarthy et al., 1970). Mercury concentrations in the rainfall are, therefore, determined by airborne levels in the area. Eriksson (quoted by Berglund et al., 1971) found, by neutron activation analysis, background levels of mercury in rainfall of about 0.1 ng Hg/g. Higher levels were found in industrialized areas with mercury emissions into the atmosphere. Levels up to 0.2 ng/g were reported in older data (Stock and Cucuel, 1934a). The levels of mercury in the snow ranged between 0.08 and 5 ng/g in a metropolitan area (Stråby, 1968). On the basis of these observations, contributions by rainfall to mercury concentrations in the soil were estimated to be in the range of 0.06 mg/m^2 and higher (Berglund et al., 1971). Westermark and Ljunggren, 1968, found that 0.4 mg Hg/m^2/year was the actual contribution to soil by levels in the rain. Andersson and Wiklander, 1965, estimated the annual contribution at the level of 0.12 mg Hg/m^2.

In conclusion, air over mercury deposits and over industrialized areas with high mercury emissions may accumulate higher concentrations of mercury mainly in zones near the ground. Airborne mercury is being continuously removed from the atmosphere and deposited on the earth surface or water surface by rain or snow, but no data are available on the magnitude of these transfers of mercury in polluted areas. Direct respiratory exposure of populations by inhalation seems to be negligible in nonindustrial areas without natural deposits of mercury. However, there is no information on potential respiratory exposure of population groups living in the nearest vicinity of sources emitting airborne mercury.

3.3.2 Possible Routes of Environmental Exposure Through Hydrosphere

Natural mercury levels in surface water were repeatedly measured in various world localities. The levels in unpolluted rivers reported by Stock and Cucuel, 1934a, Heide, Lerz, and Böhm 1957, and Dall'Aglio, 1968, were all lower than 0.1 ng/g. Ljunggren et al., 1969, found, by neutron activation analysis, concentrations ranging between 0.02 and 0.12 ng/g in Sweden. By the same method, Wiklander, 1968, recorded an average level of 0.05 ng/g.

Recent analyses of surface water by atomic absorption methods in the U.S. indicated nondetectable levels (0.1 ppb) in 34 of 73 samples; 27 samples ranged from 0.1 to 1.0 ng/g, and 10 samples ranged from 1 to 5 ng/g. Only 2 samples were higher than 5 ng/g (Wershaw, 1970). Levels between 0.09 and 0.1 ng/g were reported from atomic absorption analyses in various localities in Canada (Voege, 1971). Samples of drinking water and ground water were analyzed in Sweden. Concentrations varied between 0.02 and 0.12 ng/g with a mean of about 0.05 ng/g (neutron activation analyses; Wiklander, 1968, and Ljunggren et al., 1969).

Surface water draining areas with high natural content or industrial sources of mercury usually have much higher levels. Maximum levels of 0.36 to 0.56 ng/g were found by Hasselrot, 1968, in contaminated areas in Sweden. A single exceptional level of 34 ng/g was recorded in 1969 (Hasselrot). Concentrations up to 136 ng/g were reported in draining areas of rivers with high mercury deposits (Dall'Aglio, 1968), Aidin'yan, 1962, and 1963, (quoted by Wershaw, 1970) found levels between 1 and 3 ng/g in Russian rivers. Zautashvili, 1966, reported levels up to 3.6 ng/g in areas with mercury deposits in Russia.

Wershaw, 1970, analyzed more than 500 samples of industrial effluents in the U.S. Eighty-three percent of all samples were below 5 ng/g. Twelve percent ranged between 5 and 100 ng/g and less than 5% had concentrations higher than 100 ng/g. Only 2 samples revealed concentrations higher than 10,000 ng/g. Cooke and Beitel, 1971, quoted unpublished data by Chou in Canada on mercury concentrations in North American Great Lakes. The levels in Lake Superior were 0.12 ng/g and in Lake Ontario, 0.39 ng/g despite continuing industrial releases of mercury into this system of lakes.

In general, sources of drinking water or even surface water from areas with low levels of natural background do not constitute a primary source of mercury exposure (Jenne, 1970, and 1971). Estimates were made (Berglund et al., 1971) that this type of exposure in man is not higher than 1/20 of his total daily intake of mercury through food and drink.

3.3.3 Possible Routes of Environmental Exposure Through Food Chains
3.3.3.1 Aquatic Food Chains

The prevailing part of mercury wastes reaching water recipients consists of inorganic mercury and phenyl mercury. Larger proportions of methyl mercury, methoxyethyl mercury, or ethyl mercury are exceptional (Jensen and Jernelöv, 1969, and Jernelöv, 1969a and b). All ionized forms of mercury are rapidly bound to organic matter in the water and continue to sediment with the particulate matter. Droplets of metallic mercury sediment by their own weight. Acidity of the surface water is important for the degree of binding of alkyl and aryl mercury; extreme pH values in both directions decrease the adsorption (Miller, Gould, and Polley, 1957). The majority of all forms of mercury accumulate finally in the bottom sediment.

Stock and Cucuel, 1934a, reported concentrations of mercury found in freshwater and seawater fish surprisingly higher than mercury concentrations found in uncontaminated surface waters. Similar levels in fish were later observed also by Raeder and Snedvik, 1941, and 1949. Isolated observations have not attracted any attention until the time of the disaster in Minamata Bay, when high concentrations of mercury were found in shellfish and accumulation of mercury in aquatic organisms was described (Kurland, Faro, and Siedler, 1960).

Systematic studies on fish in Swedish waters were performed in the years 1964 to 67 (Westermark, 1965, Johnels, Olsson, and Westermark, 1967, Johnels et al., 1967, Westöö, 1967b, and Johnels and Westermark, 1969). Levels of several mg of mercury/kg of fish weight were reported from contaminated areas and previous observations were confirmed that levels in fish exceeded considerably the levels of mercury in water recipients from which the samples were obtained. Concentration differences of several orders were established between mercury levels in water and mercury content in fish (Johnels et al., 1967, and Johnels and Westermark, 1969).

A positive correlation was observed between mercury content in the axial muscle and total weight of fish or age of fish (Johnels et al., 1967).

The observed relationship was linear within the weight limits studied. However, variations were higher in areas with extremely high levels of mercury contamination. It was concluded that evidently the degree of exposure is a more influential factor than age or weight. Bache, Gutenmann, and Lisk, 1971, analyzed concentrations of total mercury and methyl mercury in the lake trout (*Salvelinus namaycush*) of Cayuga Lake in New York State and compared them with the precisely known ages of fish ranging from 1 to 12 years. The author confirmed the observations by Johnels et al., 1967, on pike (*Esox lucius*) that the concentrations of both total mercury and methyl mercury increased with the age of the fish.

Subsequently, a survey was conducted on mercury content in fish from water recipients in Sweden. In central Swedish lakes the level of mercury in pike averaged about 0.5 mg/kg. Levels of 1 mg/kg or more were recorded in about 1% of all examined water areas. Only a few localities revealed levels higher than 5 mg/kg (Berglund et al., 1971). The highest levels of mercury in fish ever recorded in Sweden were 17 to 20 mg/kg (Jernelöv, 1969c). The results indicated that mercury content in fish correlated with mercury contamination of the water recipient, although high levels were found in exceptional cases in water without any evidence of contamination by waste waters, and aerial fallout of mercury from dislocated industrial sources had to be suspected as the etiological factor of pollution (Johnels and Westermark, 1969).

Experimental evidence on a direct relationship between mercury concentrations in fish and water contamination was collected by Hasselrot, 1969. Salmon exposed to contaminated water accumulated 20 times higher concentrations of mercury than during the same exposure time in an uncontaminated water stream.

Mercury levels were reported higher than 1 mg/kg in freshwater fish from Finland also (Aho, 1968, Häsänen and Sjöblom, 1968, and Sjöblom and Häsänen, 1969), Norway (Underdal, 1969), Denmark (Dalgaard-Mikkelsen, 1969), and Italy (Ui and Kitamura, 1971). Similar results on mercury contamination of North American wildlife were reported as early as 1968. Levels up to 2.7 mg/kg were observed in fish from Canadian rivers (Fimreite, 1970a) and levels up to several mg/kg in the Great Lakes. Wobeser, 1969, (as quoted by Bligh, 1971) observed levels as high as 10 mg/kg in fish from the Saskatchewan River. Jervis et al., 1970, surveyed by neutron activation analysis the levels of mercury in fish from various localities in Canada. The average concentrations ranged up to 1 mg/kg. Similarly high levels of mercury in fish from the North American Great Lakes were reported in the U.S. at the Environmental Mercury Contamination Conference, 1970. More recent studies on freshwater fish in California, Idaho, Oregon, and Washington indicated maximum levels of 1.9 mg/kg (Buhler, Claeys, and Shanks, 1971). In Idaho, 160 samples of 19 different species from 18 separate areas were analyzed by neutron activation and the highest level was 1.7 mg/kg, recorded in squawfish. More than 19% of all samples exceeded the level of 0.5 mg/kg. Several species of fish were shown to accumulate more mercury than other species from the same water recipients. Catfish, perch, and suckers were representatives of this group; more than 40% of the accumulations analyzed from these species exceeded 0.5 mg/kg (Gebhards, 1971). Values up to a maximum of 1.25 mg/kg were reported by Henderson and Shanks, 1971, from Washington and Oregon, and by Griffith, 1971, from California. Levels of total mercury in the freshwater fish from unpolluted rivers in Japan were found to range from nondetectable levels to 1 mg/kg (Ueda, Aoki, and Nishimura, 1971).

Practically all mercury in fish is in the form of methyl mercury. This has been proved by gas chromatography in Sweden (Westöö, 1966a, 1967a, d, 1968b, c, d, and Westöö and Rydälv, 1969), in the U.S. (Smith et al., 1971), and in Canada (Solomon and Uthe, 1971, and Bligh, 1971), and by mass spectrometry in Sweden (Johansson, Ryhage, and Westöö, 1970). Bache, Gutenmann, and Lisk, 1971, analyzed methyl mercury concentrations in lake trout of precisely known ages and found that total methyl mercury and also relative proportions of methyl mercury to total mercury increased with age. Relative proportions of methyl mercury varied between 30 and 100%. All levels lower than 80% were recorded in the first three years of life. Gas chromatography was used for the identification of methyl mercury and recoveries of an added standard were reported. Studies in Japan using dithizone methods (Ueda, Aoki, and Nishimura, 1971) showed only up to 65% of total mercury in methylated form in fish from freshwater systems. It was further observed that samples with low methyl mercury levels may have up to 49% of the mercury in the form of ethyl mercury. Similar findings of ethyl mercury have not been made in freshwater fish in any other part of the world. The origin of high levels of ethyl mercury in Japan probably can be found in the extensive use of this form of alkyl mercurial for seed dressing and in the wide use of river water for irrigation of rice fields in Japan.

The origin of the methyl mercury concentrations in fish from water recipients where industrial contamination by methyl mercury can be excluded was explained by biological methylation of inorganic mercury by microorganisms or other chemical donors of the methyl group in the bottom mud with mercury sediments (Wood, Kennedy, and Rosen, 1968, Jensen and Jernelöv, 1969, Bertilsson and Neujahr, 1971, Imura et al., 1971, and Landner, 1971). Direct accumulation of methyl mercury by fish from surrounding water has been observed in experimental studies (Hannerz, 1968, and Kitamura, quoted by Tsuchiya, 1969), although the mechanisms by which the fish organism can accumulate methyl mercury have not yet been satisfactorily explained. Biological half-times and excretion of methyl mercury in fish have been explored and proved to be much longer than in mammals (Miettinen et al., 1969a, and Rucker and Amend, 1969). Acute peroral toxicity of methyl mercury in fish is of the same order as in mammals (Keckes and Miettinen, 1970, and Miettinen et al., 1970) and no differences were found between the toxicity of the ionic and protein-bound forms of methyl mercury in pike and rainbow trout (Miettinen et al., 1970).

Species differences in biological half-times of methyl mercury exist even within various fish species living in the same environment (Keckes and Miettinen, 1970). Perch (*Perca fluviatilis*) and pike (*Esox lucius*) represented usually higher levels in nature than any other species (Johnels and Westermark, 1969, and Gebhards, 1971), and their half-lives for methyl mercury were longer than in other fish families (Järvenpää, Tillander and Miettinen, 1970). Biological half-times of inorganic and phenyl mercury generally were shorter than those of methyl mercury in all aquatic species studied (Miettinen et al., 1969b, Miettinen, Heyraud and Keckes, 1970, Ünlü, Heyraud, and Keckes, 1970, and Seymour, 1971).

Reports on the concentrations of mercury in seawater are few. Stock and Cucuel, 1934a, reported 0.03 ng/g, Hamaguchi, Rokuro, and Hosohara, 1961, 0.08 to 0.15 ng/g, Hosohara et al., 1961, 0.15 ng/g, and Hosohara, 1961, 0.27 ng/g in deep seawater. Marine fish accumulate methyl mercury approximately to the same extent as freshwater fish; concentrations of methyl mercury in both types of fish of the same size in unpolluted areas are comparable. Large fish, such as swordfish and tuna, may contain levels up to 1.3 mg/kg and 0.75 mg/kg, respectively, depending upon their size and age (McDuffie, 1971, and in press).

Biological accumulation of short chain alkyl mercurials in fish is of obvious importance to the role in natural food chains. The excretion rate of alkyl mercurials is generally slow also in other animal species (Section 4.4.2.1.1). Consequently, a long-term exposure may lead to the accumulation of mercury in predatory animals fed on fish with higher concentrations of mercury.

Aquatic food chains in predatory animals were studied in Sweden on birds living predominantly or almost exclusively on fish. Extensive studies were performed on osprey (*Pandion haliaëtus*) and great crested grebe (*Podiceps cristatus*) by several authors (Berg et al., 1966, Johnels, Olsson, and Westermark, 1968, Edelstam et al., 1969, and Johnels and Westermark, 1968 and 1969). Further studies were reported on sea eagle (*Haliaëtus albicilla*) (Borg et al., 1966, Henriksson, Karppanen, and Helminen, 1966, and Johnels and Westermark, 1969) and on other sea birds (Borg et al., 1966, 1969a). High levels of mercury were found in tissues and feathers of these predatory birds in coincidence with increasing industrialization and environmental pollution (Johnels et al., 1968, and Johnels and Westermark, 1969). Furthermore, comparative studies on feathers of osprey during their annual migration between Scandinavia and Mediterranean Africa indicated higher levels in feathers acquired in Sweden than in feathers acquired in Africa. Similar results in fish-eating birds are reported from Canada (Keith and Gruchy, 1971, and Fimreite et al., in press).

Aquatic mammals as another type of fish-eating predatory animal were studied in Europe and America. Borg et al., 1969b, found significant levels of mercury in the otter (*Lutra lutra*) and mink (*Mustela vison*) in Sweden. Henriksson, Karppanen, and Helminen, 1969, reported similar results in the northern seal (*Pusa hispida*) in Finland. Average concentrations of mercury in the liver varied in a tenfold range between animals living in the Gulf of Finland and in Finnish lakes, reflecting obvious differences in mercury levels in the aquatic organisms in these two places.

In America, mercury levels increased with age in tissues from fur seals (*Callorhinus ursinus*) on the Pribiloff Islands and in Alaska. Levels of 0.20 mg/kg were found in pups and the concentration range of 10 to 172 mg/kg in the adults (Anas, 1971). Helminen, Karppanen, and Koivisto, 1968, reported 74 to 210 mg Hg/kg of liver tissue in Saimaa seal (*Pusa hispida*). Mercury levels in the whitefish (*Coregonus albula*), which is supposed to be the main component of the food of seals in this region, were only 0.2 mg/kg (Sjöblom and Häsänen, 1969). Explanation was offered by Tillander, Miettinen, and Koivisto, 1970, that the excretory rate of the major part of the methyl mercury in this species is much lower than in other animals. Studies with radioactive methyl mercury revealed a biological half-life of 500 days.

3.3.3.2 Terrestrial Food Chains

Knowledge of the extensive use of methyl mercury for seed treatment in Scandinavia between 1940 and 1966 initiated intensive studies on the transport of mercury through terrestrial food chains. The obvious importance of the problem for wildlife was recognized early and field studies on seed-eating species were started in Sweden in the late 1950's (Borg, 1958).

Pheasants (*Phasianus colchicus*) and goshawks (*Accipiter gentilis*) were selected as typical representatives for the first step in the food chain. Studies were performed on these and other seed-eating species in Scandinavia (Borg, 1958, 1967, Borg et al., 1965, 1966, 1969a, b, Hansen, 1965a, b, Ulfvarson, 1965, Berg et al., 1966, Wanntorp et al., 1967, Johnels et al., 1968, Edelstam et al., 1969, and Johnels and Westermark, 1968 and 1969), in the U.K. (Cowder, 1961), in Ireland (Eades, 1966), and in other European countries (Koeman, Vink, and Goeij, 1969). High tissue levels were uniformly observed. A temporary trend observed in feathers of goshawks, shot in the time period between the start of the 19th century and 1965, showed a sharp increase in the mercury concentrations approximately at the time when seed dressing by methyl mercury started to be widely used in Sweden (Berg et al., 1966, Edelstam et al., 1969, and Johnels and Westermark, 1969). Experimental studies with feeding contaminated food to goshawks confirmed the origin of increased levels of tissue mercury (Borg et al., 1970).

Tissue analyses of upland game birds including also pigeons, waterfowl, and songbirds were recently performed in Canada (Fimreite, 1970a, Fimreite, Fyfe, and Keith, 1970, Wishart, 1970, Keith and Gruchy, 1971) and in the U.S. (Arighi, 1971, Brock, 1971, Buhler, Claeys, and Rayner, 1971, Lauckhart, 1971, and Smith, 1971). Increased levels of mercury were found.

Increased tissue levels of mercury were also found in small seed-eating rodents (Borg et al., 1965, Lihnell and Stenmark, 1967, Fimreite, Fyfe, and Keith, 1970, and Keith and Gruchy, 1971) and other terrestrial mammals (Borg et al., 1965, 1966, and 1969a,b). High concentrations of mercury were observed in tissue of birds of prey (owls, falcons, herons, and hawks); a large number of their eggs also showed high mercury residues (Dustman, Stickel, and Elder, 1970, Fimreite, Fyfe, and Keith, 1970, and Keith and Gruchy, 1971). Elevated levels of mercury in avian species fed by mercury treated seed were observed (Borg et al., 1966, 1969a, b, Tejning, 1967d, Bäckström, 1969a, Fimreite, 1970b, and Norberg, Brock and Shields, 1971) and in tissues of mammals and birds consuming fowl fed by methyl mercury dressed seed (Borg et al., 1970, and Hanko et al., 1970).

In general, hazards to wildlife are obviously involved in the extensive use of alkyl mercurials for seed dressing. Among the injurious effects already observed are increased mortality in many species (Otterlind and Lennerstedt, 1964, Fimreite, 1970b, and in press), reduced hatchability in birds (Borg et al., 1965, 1969a, Kuwahara, 1970a and b, Kiwimäe et al., 1969, Kiwimäe, Swensson, and Ulfvarson, 1970, and Fimreite, Fyfe, and Keith, 1970), and high frequency of fetal malformations (Tejning, 1967d). Recent regulations and restrictions introduced into the seed dressing technology condemned the use of alkyl mercurials (Olsson, 1969, *Minerals Yearbook, 1970,* in press, and Gurba, 1971). As an immediate consequence, tissue levels of mercury in seed-eating birds and their predators decreased substantially (Borg, 1968, and 1969 a, b, Johnels, Olsson, and Westermark, 1968, and Johnels and Westermark, 1969) nearly to the levels observed before the introduction of alkyl mercurials into seed dressing (Berglund et al., 1971). However, even the normal use of mercury fungicides may still cause elevated residues in game birds and, through this step in the food chain, finally reach man.

3.3.3.3 Foodstuffs Other Than Fish

Due to the ubiquity of trace amounts of mercury in nature, all foodstuffs contain low levels of mercury; the exposure of man through the food chain involves primary contamination as well as secondary bioaccumulation processes in the biosphere.

Previous attempts to analyze trace amounts of mercury in individual food components were often limited by detection limits of the analytical methods used (Stock and Zimmermann, 1928a and b, Borinski, 1931a, Stock and Cucuel, 1934a, Gibbs, Pond, and Hansmann, 1941, and Goldwater, 1964.) In recent years, market basket studies for mercury have been repeated in many countries. Smart, 1968, published an extensive review on levels of mercury in foodstuffs from various places in the world. Partial studies were reported from Wales and England (Abbott and Tatton, 1970, and Lee and Roughan, 1970), the U.S. (Corneliussen, 1969) and Canada (Jervis et al., 1970, and Somers, 1971). The most extensive studies were performed in Sweden (Westöö, 1965a, b, c, 1966a, b, c, 1967a and c, 1968a, 1969a, b, c, and 1970, Nordén, Dencker, and Schütz, 1970, and Dencker and Schütz, 1971). These, along with other Scandinavian studies, (Underdal, 1968a and b, 1969, Bonnevie et al., 1969, and Dalgaard-Mikkelsen, 1969) are reviewed by Berglund et al., 1971.

Mercury concentrations in food vary in a wide range. Maximum levels found in Swedish studies were in hog liver (0.18 mg Hg/kg); the mean level in the rest of the foodstuffs investigated was 0.03 mg Hg/kg or less (Berglund et al., 1971).

Average daily intake of total mercury via food was estimated in England to be in the range of 14 μg Hg/day (Abbott and Tatton, 1970) and between 5 to 7 μg Hg/day in the U.S.S.R. (Leites, 1952), about 20 μg/day in the U.S. (Gibbs, Pond, and Hansmann, 1941) and approximately 5 μg/day in Germany (Stock and Cucuel, 1934a). Quantitative studies were performed in Sweden. Total mercury content in 12 analyzed fish-free daily diets in Stockholm varied from 4 to 19 μg Hg/day with a mean value of 10 μg Hg/day (Westöö, 1965c). More recent analyses of 90 duplicate portions of fish-free diets collected from 17 persons in the southern part of Sweden revealed average daily intake of total mercury at the level of 3.6 μg Hg/day (1.0 to 9.3 μg); the mean level in 58 other diets containing fish or fish products was 8.7 μg Hg/day with a range of 1.7 to 30.6 μg Hg (Dencker and Schütz, 1971).

The form in which mercury is present in foodstuffs other than fish also was investigated and varying amounts of methyl mercury were found (Westöö, 1967a, 1968a, 1969a, b, c, and 1970). Maximum relative fractions of methyl mercury (65 to 97% of total mercury) were detected in pork chops, hog liver, hog brain, and reindeer saddle. The lowest levels of methyl mercury were observed in the reindeer kidney and liver.

In conclusion, fish is obviously the most important source of methyl mercury in the food, and the daily intake in fish-eating populations correlates directly with the amounts of fish consumed daily (Berglund et al., 1971, and McDuffie, 1971, and in press). However, relatively high representation of methyl mercury in the foodstuffs other than fish may constitute — in view of the complete absorption of this form of mercury in the gastrointestinal tract (Section 4.1.2.1.2) — an important factor in the quantitative evaluation of the exposure of man to mercury in the contaminated environment, even in populations with low or negligible consumption of fish.

METABOLISM

Gunnar F. Nordberg and Staffan Skerfving

With regard to metabolism and toxicity, it is not enough to consider only the division between inorganic and organic mercury mentioned in the introductory chapter. Elemental mercury differs from inorganic mercury salts and the organic mercury compounds also differ greatly from one another. In this chapter, the description of mercury and its compounds has been disposed accordingly.

4.1 Absorption

Theoretically, mercury and its compounds could enter the human or animal body by the following routes: via the lungs by inhalation, via the gastrointestinal tract by ingestion, via the skin by inunction or accidental exposure, and via the placenta into the fetus. Under exceptional circumstances, direct intravascular injection provides a route of entrance, but the efficiency of different modes of injection will not be dealt with here.

4.1.1 Inorganic Mercury
4.1.1.1 Elemental Mercury
4.1.1.1.1 Respiratory Intake
4.1.1.1.1.1 In Animals

Generally, gases and vapors are deposited in the respiratory tract according to their water solubility. Highly water soluble gases are dissolved in the mucous membrane or fluid of the upper respiratory tract, whereas less water soluble gases and vapors penetrate farther down the bronchial tree and reach the alveoli. Since elemental mercury vapor is only slightly soluble in water, it could be expected to penetrate far down the bronchial tree. This expectation has been fulfilled experimentally. Berlin, Nordberg, and Serenius, 1969, showed in an autoradiographic study that mercury was deposited in similar concentrations in the bronchial tree and the alveoli, with a slight predominance in small bronchioli.

Theoretical considerations on the alveolar transfer of metallic mercury have been presented by Hughes, 1957. He estimated the solubility of elemental mercury in body lipids to be between 0.5 and 2.5 mg/l. Considering that mercury concentration of saturated air can be only 0.06 mg Hg/l. at 40° C, the partition coefficient between air and lipides of alveolar wall and pulmonary blood is approximately 20 in favor of the body. These facts suggest that elemental mercury should pass easily across the alveolar membrane by simple diffusion.

Experimentally, the percentage of inhaled mercury retained by the body has been estimated in a number of animal studies. Hayes and Rothstein, 1962, reported about 100% in rats, and Magos, 1967, calculated 75 to 100% in mice. Gage, 1961a, reported about 50% absorption in rats and earlier reports have stated values down to 25% (Fraser, Melville, and Stehle, 1934, and Shepherd et al., 1941).

In the study by Berlin, Nordberg and Serenius, 1969, it was shown that only about 30% of the whole-body burden of mercury was in the lung after a short (10-min) exposure, meaning that the rest of the mercury had been transferred quickly to the blood via the alveolar membrane. Such diffusion occurs rapidly, as has been shown in rats by Magos, 1968. He found that about 20% of intravenously injected mercury vapor was exhaled after 30 seconds. Part of the mercury (20 to 30% of the whole-body burden) which was originally taken up in the lung was later cleared to the rest of the body with a half-life of 5 to 10 hours in rats and guinea pigs (Hayes and Rothstein, 1962, and Berlin, Nordberg, and Serenius, 1969).

4.1.1.1.1.2 In Human Beings

For man, no direct measurements of the detailed pulmonary deposition of mercury have been reported, but after inhalation of high concentrations of mercury vapor, damage to the lower parts of the bronchial tree and the peripheral lung tissue has been found at autopsy of fatal cases (Matthes et al., 1958, Teng and Brennan, 1959, and Tennant, Johnston, and Wells, 1961). These findings speak in favor of a deposition pattern of mercury in the human lung similar to the one found in animal studies. Matthes et al., 1958, reported high concentrations of mercury in the lungs (6.3 and 9.3 ppm) of 2 infants who died 4 and 7 days after exposure to high concentrations of mercury vapor.

By measurements of the mercury content of inspired and expired air, respectively, Teisinger and Fiserova-Bergerova, 1965, and Nielsen Kudsk, 1965a, found that 75 to 85% of the mercury at concentrations ranging from 50 μg to 350 μg/m^3 of the inspired air was retained in the human body. Nielsen Kudsk, 1965a, also found that the retention fell to 50 to 60% in persons who had consumed moderate amounts of ethyl alcohol. Nielsen Kudsk, 1965a and b, interpreted his results as consistent with a diffusion of mercury vapor into the blood via the alveolar membrane, an opinion further supported by the studies in animals (see Section 4.1.1.1.1.1).

4.1.1.1.2 Gastrointestinal Uptake

Oral intake of liquid elemental mercury was earlier used in the treatment of bowel obstruction (Zwinger, 1776, Ebers, 1829, quoted by Cantor, 1951), without giving rise to mercury poisoning. Mercury has been used in an intestinal decompression tube for the same purpose. Elemental mercury has not infrequently been released into the gastrointestinal tract as a result of its escape from such tubes (Cantor, 1951). No reports of mercury poisoning following such accidents have appeared, and it has long been known that, from a practical point of view, mercury is not absorbed when introduced in the elemental form into the gastrointestinal tract (see, for example, Cantor, 1951). A limited absorption takes place, however, as shown by Suzuki and Tanaka, 1971. The magnitude of this absorption has been illustrated by experimental evidence. Bornmann et al., 1970, administered elemental mercury orally to rats and measured the uptake in the blood and organs. Less than 0.01% of the ingested mercury was absorbed.

4.1.1.1.3 Skin Absorption

Metallic mercury earlier was used widely as a component of ointments in the treatment of syphilis and dermatological disorders. One form of therapy was to cover the patient with the ointment and thereafter to place him in a heated chamber, promoting the uptake of mercury in the body. That mercury was absorbed into the body was obvious from symptoms such as gingivitis, salivation, gastrointestinal disturbances, and tremor, which were more or less obligatory for a successful treatment of the disease (Almkvist, 1928). In this case, however, it is clear that inhalation of mercury vapor could have made an important contribution to mercury absorption.

Cole, Schreiber, and Sollmann, 1930, measured the excretion in urine and feces of patients treated with different kinds of inunctions of mercury ointment. They found that the excretion of mercury was proportional to the concentration of mercury in the ointment. Laug et al., 1947, showed that the ointment base was of importance for absorption of metallic mercury in rats after application onto the skin. Juliusberg, 1901, enclosed the inuncted areas on patients' skin in airtight covers so that no vapors could be inhaled. Nevertheless, he found considerable urinary excretion of mercury. He also made experiments on dogs in which inhalation of vapors from inunctions was prevented completely. Another series of dogs, also treated with inunctions, was allowed to respire the vapors. After 2 to 3 days the dogs were killed and the liver and kidneys were examined for mercury content. In the dogs not inhaling the vapor, an average of 6.2 ppm was found in the kidneys and 1.2 ppm in the livers (4 dogs). In the other series, 12.3 ppm was found in the kidneys and 2.9 ppm in the livers (4 dogs). Schamberg et al., 1918, made experiments with rabbits, in which 1 rabbit inhaled the vapors from the inunction of another rabbit which in turn respired fresh air. In several repetitions of the same procedure, the rabbit which breathed clean air invariably succumbed to mercury poisoning after a brief period, whereas the other rabbit lived throughout the experiment.

All of this evidence shows that a direct penetration of metallic mercury through the skin occurs. However, the investigations mentioned do not include any precise figures for the rate of penetration.

Brown and Kulkarni, 1967, used a report by Forbes and White, 1952, to support the assertion that mercury was absorbed via the skin by police officers working with "grey powder" in the development of finger prints. Though Forbes and White, 1952, made investigations on the possibility of inhalation of an aerosol of mercury droplets, they seem to have overlooked exposure to mercury vapor, the most likely route for absorption of mercury by the police officers, as stated by Rodger and Smith, 1967.

From the conjunctival sac, metallic mercury

can be resorbed to a very limited extent, as shown by Kulczycka, 1965.

4.1.1.1.4 Placental Transfer

No experimental evidence is available on the placental transfer of elemental mercury. Theoretically, it seems possible that this form of mercury penetrates the placental barrier more easily than the poorly penetrating divalent mercuric ions do (Section 4.1.1.2.4). Lomholt, 1928, stated that mercury could be detected in stillborn babies of women treated with mercury inunctions against syphilis.

4.1.1.2 Inorganic Mercury Compounds
4.1.1.2.1 Respiratory Uptake

There are no conclusive data describing the deposition of inorganic mercury compounds in the respiratory tract of animals or man. However, aerosols of mercury compounds are expected to follow general laws governing deposition of particulate matter in the respiratory airways (Task Group on Lung Dynamics, 1966, Air Quality Criteria for Particulate Matter, 1969). Particle size and density are factors of primary importance. In human beings with a respiratory rate of 20 l/min, the deposition in the pulmonary compartment is expected to vary from 10 to 50% depending upon the mass median diameter of aerosol particles from 5 to 0.01 μm, respectively.

Particles deposited on the bronchial mucosa are cleared by means of mucociliary transport within hours, and therefore relatively large particles with a high probability of deposition in the upper airways should be cleared rapidly. For particles deposited in the peripheral lung tissue, however, longer half-lives (from a few days to about 1 year) are expected. The water solubility of the mercury compound is highly important for this part of the clearance as well. Morrow, Gibb, and Johnson, 1964, studied the clearance of highly insoluble mercuric oxide from the lungs of dogs. For an HgO aerosol with a mean diameter of 0.16 μm, they found that 45% of the amount deposited was cleared in less than 24 hours, while the rest was cleared with a half-time of 33 ± 5 days.

Generally, aerosols of inorganic mercury compounds are absorbed via the respiratory system to a lesser degree than mercury vapor. In experiments on rats and mice, Viola and Cassano, 1968, compared the retention of mercury in the organs of rats exposed to mercury vapor with a group exposed to the same concentration of mercury but in the form of a mixture of mercury vapor with an aerosol of mercurous chloride. The retention of mercury in all studied tissues was lower in the aerosol-exposed group. The differences in the brain and heart were especially prominent. These observations are concordant with studies on the bodily distribution of various forms of inorganic mercury (see Section 4.3.1.1). Although the absorption of mercury aerosols is less efficient than that of mercury vapors, cases of poisoning were reported after this type of exposure in man (Kazantzis et al., 1962).

4.1.1.2.2 Gastrointestinal Absorption

Various inorganic mercury compounds have different solubilities in water or gastrointestinal fluids. Differences in physical and chemical qualities of these compounds make the exact evaluation of quantitative aspects of gastrointestinal uptake difficult.

All highly soluble mercuric compounds dissociate easily into mercury ions when dissolved in the gastrointestinal contents and probably have very similar rates of absorption. Prickett, Laug, and Kunze, 1950, found 1.2% in the urine and about 80% in the feces of rats 48 hours after oral dosing (0.5 mg Hg/kg) of mercuric acetate. Ellis and Fang, 1967, gave mercuric acetate to rats by oral tube and found similar values. During the first 48 hours after the dosing (1.3 to 4 mg Hg/kg), they found about 0.5% of the dose in the urine and about 80% in the feces. During the whole period of study, 168 hours, the corresponding figures were 1.5% (urine) and 93% (feces). These studies show that absorption of mercuric acetate is about 20%.

Clarkson, 1971, evaluated the net absorption of mercuric chloride across the gastrointestinal tract in mice to be small, averaging less than 2% of the daily intake when studied by whole body counting. Measurements of the fecal excretion confirmed the conclusions based on the whole body counting. The fecal excretion rate approached nearly 100% of the entire daily dose in food. The dosage schedule was 0.05, 0.5 and 5 ppm of the dry food. The addition of mercuric ion to the food did not influence the physiological status of the experimental animals.

Data on acute cases of poisoning in man from the ingestion of mercuric chloride taken accidentally or with suicidal intent (Sollmann and

Schreiber, 1936) show an important absorption of this form of mercury. Because the patients vomited 20 minutes to 1 hour after taking the poison, it is impossible to calculate any absorption ratio from the data, but the amount found in the body was calculated by Sollmann and Schreiber to be 240 mg as an average for the 3 fatal cases, corresponding to 8% of the dose ingested, a minimum absorption figure. Severe gastroenteritis was present in these cases.

Topical corrosive effects of mercuric chloride are well known to disrupt the permeability barriers in the gastrointestinal tract. As a consequence, the net absorption can vary extremely in both directions depending upon dose and concentrations ingested.

Miettinen (in press) measured the rate of absorption of nontoxic doses of protein-bound mercuric nitrate administered perorally to 7 human volunteers. Only about 15% was retained in the body. The remaining 85% was excreted in the feces during the days immediately following the exposure.

For mercurous compounds which are much less soluble in water, a lower absorption rate could be anticipated. For instance, in the treatment of syphilis (Rosenthal, 1928) 2 doses of calomel (mercurous chloride) 0.6 to 1 g with a 30-minute interval were used. This is an equivalent of the dose of mercury which, according to Sollmann and Schreiber, 1936, produced fatal poisoning when taken as $HgCl_2$. Because the patients treated with calomel for syphilis did not die from mercury poisoning, it can be concluded that the absorption rate was lower than that of mercuric chloride. Calomel in low doses has also been used as a laxative, seldom causing serious symptoms of poisoning. Mercurous compounds, even if not absorbed to a considerable degree, might be partially converted into mercuric ions in the gastrointestinal lumen and, consequently, unknown fractions of bivalent mercury salt may be absorbed. As observed already by Lomholt, 1928, the ultimate absorption rate varies with the time the salt stays in the gastrointestinal tract and with the different contents in the gastrointestinal lumen.

Excessive long-term use of calomel in the treatment of syphilis has caused systemic poisoning with symptoms of stomatitis and salivation (Almkvist, 1928). Calomel in teething powder has given rise to acrodynia in children (Warkany and Hubbard, 1948), probably by gastrointestinal absorption. Mercurous mercury was also used in diuretic therapy. Positive effects seen in the treatment of edema indicate that absorption must have occurred.

Poor absorption of mercurous mercury has been illustrated by autoradiographic studies in mice by Viola and Cassano, 1968.

4.1.1.2.3 Skin Absorption

Soluble mercury compounds have been used extensively for topical application in the treatment of certain dermatological disorders, e.g., psoriasis and seborrheic dermatitis. They have also been used in the prevention of venereal diseases and in the treatment of syphilis. The application of yellow mercuric oxide (HgO) in the conjunctival sac is recommended in the treatment of inflammatory eye disease. Ammoniated mercuric chloride ($HgNH_2Cl$) is still used for dermatological purposes. Young, 1960, and Turk and Baker, 1968, have reported systemic effects after this therapy.

When evaluating exposure via skin application, it is impossible to rule out other routes of exposure. Both inhalation and ingestion can occur, although the direct penetration through the skin is likely to be of relatively greater importance. Because the data concern treatment given to persons with skin diseases, it is difficult to draw conclusions for persons with normal skin. Frithz, 1970, compared the concentrations of mercury in the blood and urine of psoriatic patients and normal volunteers, both treated with ammoniated mercury ointment. He found higher mercury concentrations in both blood and urine from the psoriatic patients than in those from the normal subjects.

Laug et al., 1947, compared the skin penetration of different mercury compounds included in two different ointment bases by measuring the amount of mercury accumulating in the kidneys of rats after application. They found the following average kidney concentrations when a base of 50% lard and 50% lanolin was used as the ointment base for: calomel, 8.8, ammoniated mercury, 19, metallic mercury, 14, and yellow oxide mercury, 23 µg/g. When a base of 50% petrolatum and 50% lanolin was used, the penetration was lower. All ointments contained 25% mercury.

The occurrence of penetration has been further documented both for animal and for human skin. Electron microscopical studies have shown electron dense granules both extracellularly and intracellularly after application of mercuric mercury on the human skin (Frithz and Lagerholm, 1968, and Silberberg, Prutkin and Leider, 1969). Scott, 1959, showed by autoradiography that penetration of the human skin takes about 8 hours. By means of the disappearance technique, Friberg, Skog, and Wahlberg, 1961, showed that mercuric chloride was absorbed to a maximum of 6 percent in guinea pigs in 5 hours at a mercury concentration of 16 mg/ml. At a concentration in the aqueous solution of 48 mg Hg/ml (saturated solution) no resorption could be detected. One ml of solution was applied at a surface of 3.1 cm^2. Further studies with this technique have been reported by Skog and Wahlberg, 1964, and by Wahlberg, 1965a and b, when it was shown that potassium mercuric iodide (K_2HgI_4) was absorbed to a greater extent and exerted a higher percutaneous toxicity than mercuric chloride. If the rates of penetration through the human skin are presumed to be similar to those of guinea pigs, absorption via the skin must be considered as an important route of entry of mercury compounds into the body.

4.1.1.2.4 Placental Transfer

Experimental studies on animals have shown that the placental membrane constitutes an important barrier against the penetration of mercuric ions into the fetus. After injection of high doses of mercuric mercury in the guinea pig, Radaody-Ralarosy, 1938, succeeded in detecting mercury histochemically in the placenta but not in the fetus. Berlin and Ullberg, 1963a, observed by an autoradiographic technique in mice significant accumulation of mercury in the placenta and much lower accumulation in the fetus after intravenous injection of mercuric chloride (0.5 mg Hg/kg). Similar observations were made on rats after intraperitoneal injection by Takahashi et al., 1971. Quantitative determinations were made by Suzuki et al., 1967. The concentration ratios of mercury in maternal blood, placenta, and fetus were 1:19:0.4 after administration of mercuric chloride to mice. For human beings, no conclusive data are available on the transfer of mercuric mercury via the placenta to the fetus.

4.1.2 Organic Mercury Compounds
4.1.2.1 Alkyl Mercury Compounds
4.1.2.1.1 Respiratory Uptake
4.1.2.1.1.1 In Animals

No detailed data concerning uptake and absorption of inhaled vapors or dust of methyl mercury compounds are available.

Several *methyl mercury salts* vaporize relatively easily at room temperature. In some experiments the absorption has been high enough to cause poisoning in monkeys and rats (Hunter, Bomford, and Russell, 1940) and mice (Swensson, 1952, and Hagen, 1955). The salts employed were methyl mercury iodide, chloride, and dicyandiamide.

Östlund, 1969a and b, studied the retention of *di-methyl mercury* after a single inhalation exposure in mice under slight anesthesia. Usually 50 to 80% of offered di-methyl mercury was transferred to the mouse within 45 seconds of exposure. No details on inhaled concentrations were given. Retained amounts corresponded to 5 to 9 mg Hg/kg body weight. The retention course observed in this inhalation experiment did not differ from retention courses obtained after intravenous injection of di-methyl mercury.

No experiments have been published on the respiratory uptake of *ethyl* or *higher alkyl mercury* compounds. Poisoning has been reported after exposure to vapors of ethyl mercury salts (Trachtenberg, 1969).

4.1.2.1.1.2 In Human Beings

There are no experimental data on uptake and absorption of inhaled alkyl mercury compounds in man. Intoxication has been caused by inhalation of vapor or dust of *mono-methyl* (Hunter, Bomford, and Russell, 1940, Herner, 1945, Ahlmark, 1948, Lundgren and Swensson, 1948, 1949, and 1960a and b, and Prick, Sonnen, and Slooff, 1967a and b), *di-methyl* (Edwards, 1865, and 1866), *mono-ethyl* (Höök, Lundgren, and Swensson, 1954, Hay et al., 1963, Schmidt and Harzmann, 1970), and *di-ethyl* (Hill, 1943, and Drogitjina and Karimova, 1956) *mercury* compounds.

4.1.2.1.2 Gastrointestinal Absorption
4.1.2.1.2.1 In Animals

A few experimental observations are available concerning the gastrointestinal uptake of *methyl mercury* compounds. Methyl mercury is stable in acid solutions (Whitmore, 1921, and Mudge and

Weiner, 1958). Studies in rats (Ahlborg et al., to be published), cats (Rissanen, 1969, Albanus et al., to be published), and monkeys (Berlin, Nordberg, and Hellberg, in press) indicate an absorption of more than 90% of the ingested amount of methyl mercury salt or proteinate. Clarkson, 1971, concluded from whole-body counting studies on mice that gastrointestinal absorption of methyl mercury chloride administered in food is practically complete. Fecal radioactivity on the first day of exposure was only 10% of the dose. Investigations on the entero-hepatic circulation of injected methyl mercury salt support this concept (Norseth, 1969b).

Detailed information about the gastrointestinal uptake of *ethyl mercury* is not abundant. However, experimental poisoning occurred after oral administration in several species (Section 8.1.2.2.2.2). This is indirect evidence that considerable absorption occurs in the gastrointestinal tract. From Ulfvarson's (1962) study on rats it seems that ethyl mercury absorption rates are comparable to those of methyl mercury salts. The study on cats performed by Yamashita, 1964, indicates an absorption of more than 90% of the ingested amount.

Itsuno, 1968, gave propyl, butyl, amyl, and hexyl mercury compounds orally to rats. Considerable levels of mercury were found in the organs (Table 4:6).

4.1.2.1.2.2 In Human Beings

Experimental studies on human volunteers indicate an almost complete absorption of *methyl mercury* salt (Ekman et al., 1968a and b, and 1969, Åberg et al., 1969, Falk et al., 1970) and of proteinate (Miettinen et al., 1969b, Miettinen et al., 1971, and Miettinen, in press). The absorption was measured as the difference between the ingested amount and the elimination in feces several days after administration or by whole-body countings. The exposure of the volunteers was low (about 10 and 20 μg), i.e., comparable to "normal" amounts of daily ingested total mercury (Borinski, 1931b, Stock and Cucuel, 1934b, Stock, 1936, Gibbs, Pond, and Hansmann, 1941, Clarkson and Shapiro, 1971, and Dencker and Schütz, 1971).

The high gastrointestinal absorption of methyl mercury compounds has been documented by described cases of poisoning through ingestion of products prepared from contaminated seed (Engleson and Herner, 1952, and Ordonez et al., 1966) or contaminated fish (Niigata Report, 1967, and Minamata Report, 1968).

No experimental studies on the absorption of *ethyl mercury* compounds have been published; however, poisonings have occurred after ingestion of food prepared from treated seed (Jalili and Abbasi, 1961, and Haq, 1963).

4.1.2.1.3 Skin Absorption
4.1.2.1.3.1 In Animals

Friberg, Skog, and Wahlberg, 1961, and Wahlberg, 1965b, showed in guinea pigs that *methyl mercury* dicyandiamide is absorbed from a water solution through intact skin. With various concentrations, a maximum of 6% of the mercury was absorbed in 5 hours. This absorption rate is not too different from the uptake observed with mercuric chloride.

No information is available on *ethyl* or *higher alkyl mercury* compounds.

4.1.2.1.3.2 In Human Beings

Methyl mercury poisoning has been reported in persons who were treated locally with preparations containing methyl mercury thiacetamide (Tsuda, Anzai, and Sekai, 1963, Ukita, Hoshino, and Tanzawa, 1963, Okinaka et al., 1964, and Suzuki and Yoshino, 1969). In these cases, however, the possibility of inhalation exposure cannot be excluded.

No further data are available on skin absorption of *ethyl* or *higher alkyl mercury* compounds. It might be assumed that ethyl mercury, like methyl mercury compounds, can penetrate the skin barrier.

4.1.2.1.4 Placental Transfer
4.1.2.1.4.1 In Animals

After administration of *methyl mercury salts* high levels of mercury have been found in the fetus of mice (Berlin and Ullberg, 1963c, and Suzuki et al., 1967), rats and cats (Moriyama, 1968), and guinea pigs (Trenholm et al., 1971). Östlund, 1969a and b, observed only small amounts of mercury in the fetus after inhalation or intravenous exposure to *di-methyl mercury* in mice.

Mercury levels higher than those of their mothers were demonstrated in the brain of the fetus after injection of *ethyl mercury* phosphate into pregnant mice (Ukita et al., 1967). They used autoradiography.

The placental transfer of methyl and ethyl mercury is also discussed in Section 8.1.1.2.

Placental transfer of *higher alkyl mercury* compounds has not been studied.

4.1.2.1.4.2 In Human Beings

The fact that *methyl mercury* passes the placental barrier in man is documented by the occurrence of prenatal poisoning (Engleson and Herner, 1952, Harada, 1968b, and Snyder, 1971; see also Section 8.1.1.1.1).

Newborn babies of mothers exposed to methyl mercury during pregnancy by consumption of contaminated fish have higher mercury levels in blood cells than their mothers (Skerfving, to be published). Regarding "normal" infants, see Section 6.2.2.

Ethyl mercury has also been stated to have caused fetal poisoning in man (Bakulina, 1968, see Section 8.1.1.1.2). No information is available concerning *higher alkyl mercury* compounds.

4.1.2.2 Aryl Mercury Compounds
4.1.2.2.1 Respiratory Uptake

Hagen, 1955, exposed mice by inhalation to dust of phenyl mercury compounds (see also Section 8.2.2.2). After exposure to phenyl mercury acetate dust with particle size ranging from 2 to 40 μm, no poisoning occurred in 30 hours, while with particle size of 0.6 to 1.2 μm, death occurred after approximately 1 hour.

No further experimental data are available on inhalation exposure to aryl mercury compounds. Data presented in Section 8.2.2.1 show that aerosols of phenyl mercury salts are absorbed by inhalation in man but no quantitative conclusions are possible.

4.1.2.2.2 Gastrointestinal Absorption

Several animal studies have shown that the mercury levels in organs are higher after exposure to phenyl mercury salt than after the same exposure to inorganic mercury salt (e.g. Fitzhugh et al., 1950, Prickett, Laug, and Kunze, 1950, and Swensson, Lundgren, and Lindström, 1959b). This may indicate a better absorption of phenyl mercury from the gastrointestinal tract or a faster elimination of inorganic mercury from the body, or both. Measurements of mercury excreted in the feces during the first days after peroral administration of phenyl or inorganic mercury salts indicated higher absorption of phenyl mercury. Prickett, Laug, and Kunze, 1950, found during 48 hours after single oral administration of 0.5 mg Hg/kg as mercuric acetate to rats, about 80% in the feces, while the corresponding figure for phenyl mercury acetate was 60%. After intravenous administration, 10 and 30%, respectively, were found in the feces (see Section 4.1.1.2.1). Thus it seems that more than half of the phenyl mercury salt was absorbed. Ellis and Fang, 1967, found 50 to 60% of an oral dose of 0.4 to 1.2 mg Hg/kg as phenyl mercury acetate eliminated during 48 hours after administration to rats. Corresponding figures for excreted mercuric acetate were 80 to 90% of the administered dose (see Section 4.1.1.2.1). There are no data on the absorption in man. Tokuomi, 1969, reported that a considerable urinary mercury elimination occurred in a person who had ingested phenyl mercury acetate (Section 8.2.2.1).

4.1.2.2.3 Skin Absorption
4.1.2.2.3.1 In Animals

Goldberg, Shapero, and Wilder, 1950, applied phenyl mercury dinaphthylmethane disulfonate in a buffered water solution on the body surfaces of rabbits and found mercury in the skin, subdermal connective tissue, and muscle. The concentrations in muscles were 3 times higher than that in the solution applied.

Laug and Kunze, 1961, showed that approximately 25% of the mercury applied as phenyl mercury acetate intravaginally in rats 24 hours prior to sacrifice could be recovered in the liver and kidneys.

4.1.2.2.3.2 In Human Beings

Clinical data presented in Section 8.2.2.1 indicate that phenyl mercury acetate is absorbed through the skin. No quantitative conclusions are possible. Intravaginally applied phenyl mercury salts are absorbed to a certain degree (Biskind, 1933, and Eastman and Scott, 1944; see also Section 8.2.2.1), as are phenyl mercury solutions in the conjunctival sac (Abrams and Majzoub, 1970).

4.1.2.2.4 Placental Transfer

After administration of phenyl mercury salts to pregnant mice, mercury accumulation occurs in the placenta but only small amounts of mercury pass to the fetus (Berlin and Ullberg, 1963b,

Suzuki et al., 1967, and Ukita et al., 1967). The mercury levels obtained in fetuses are comparable to those seen after administration of mercuric salts (see Section 4.1.1.2.4) and are considerably lower than after ethyl or methyl mercury salts (see Section 4.1.2.1.4.1).

4.1.2.3 Alkoxyalkyl Mercury Compounds
4.1.2.3.1 Respiratory Uptake

Hagen, 1955, exposed mice to methoxyethyl mercury silicate dust by inhalation. Death occurred after 1.2 to 14 hours under different experimental conditions.

Derobert and Marcus, 1956, have reported 1 case of poisoning after a few hours of inhalation of dust of methoxyethyl silicate. No quantitative conclusions are possible concerning the rate of absorption.

4.1.2.3.2 Gastrointestinal and Skin Absorption

No data are available.

4.1.2.3.3 Placental Transfer

In mice, mercury from methoxyethyl mercury reaches the fetus only to a minor extent, but is accumulated in the placenta and in the fetal membranes in a manner similar to mercuric salts (Berlin and Nordberg, unpublished data). *Substituted alkoxyalkyl mercury* compounds (mercurial diuretics) are discussed in Section 4.1.2.4.

4.1.2.4 Other Organic Mercury Compounds

Mercurial diuretics are active at oral administration, but the absorption is slower and less complete than after parenteral administration. Griffith, Butt, and Walker, 1954, and Leff and Nussbaum, 1957, reported considerable mercury concentrations in kidneys of subjects who had been treated orally with organomercurial diuretics.

Baltrukiewicz, 1969, injected labeled chlormerodrin into female rats early in pregnancy. The description of methods and results is incomplete but the author did state that no radioactivity was present in the litters.

4.1.3 Summary

About 80% of inhaled *elemental mercury vapor* is absorbed in the respiratory system in human beings. Gastrointestinal absorption of elemental mercury is negligible. Skin penetration can take place but the exact rate of this process is not known.

Absorption of aerosols of *inorganic mercury compounds* in the respiratory tract is dependent upon physicochemical characteristics of the aerosol. In exposures to soluble mercury compounds this route of entry can be responsible for the uptake of toxic amounts of mercury in man. Up to 10 to 20% of the ingested amount of easily soluble mercuric salts is absorbed via the human gastrointestinal tract. Animal studies show that high skin absorption of mercuric salts may also occur. Animal experiments have also shown that the placental membrane constitutes a barrier against the penetration of divalent ions into the fetus.

No quantitative data are available on the respiratory uptake of *alkyl mercury* compounds. Poisonings due to inhalation of methyl and ethyl mercury compounds in man and animals indicate a high rate of pulmonary absorption. Methyl mercury compounds are almost completely absorbed in the gastrointestinal tract in animals and man, as is ethyl mercury in animals. Animal experiments indicate high absorption of methyl mercury through the skin. Poisonings in man after cutaneous application of methyl mercury indicate skin absorption, although the possibility of secondary inhalation exposure cannot be excluded. It can be assumed that ethyl mercury is absorbed to an extent similar to methyl mercury, though no experimental data are available. In mice and rats, both mono-methyl mercury and mono-ethyl mercury readily pass the placenta and accumulate in the fetus. Prenatal poisoning by these compounds shows that a similar process occurs in man. Di-methyl mercury reaches the fetus in mice only to a minor extent. No information is available on uptake of higher alkyl mercury compounds.

There are no reliable data on absorption of *aryl mercury* compounds after inhalation. Rapid lethality was reported in mice exposed to fine dust of phenyl mercury. Phenyl mercury acetate is more extensively absorbed from the gastrointestinal tract than mercuric mercury in animal experiments. Available data indicate an absorption of more than half of the ingested amount. Animal data indicate high uptake of phenyl mercury salts through the skin and mucous membranes. After administration of phenyl mercury salts to mice, mercury accumulates in the placenta and only small amounts are found in the fetus. No data are available on the uptake of aryl mercury compounds other than phenyl mercury salts.

Information about respiratory and gastrointestinal absorption of *alkoxyalkyl mercury* compounds is practically nonexistent. Lethal effects were reported in mice exposed by inhalation to fine dust of methoxyethyl mercury. After administration of methoxyethyl mercury to mice, there is little accumulation of mercury in the fetus.

4.2 Biotransformation and Transport
4.2.1 Inorganic Mercury
4.2.1.1 Oxidation Forms of Mercury and Their Interconversions

The toxic effects of all forms of inorganic mercury are ascribed to the action of ionic mercury because elemental mercury (Hg^0) cannot form chemical bonds. Ionic mercury exists in mercurous (Hg_2^{2+}) and mercuric (Hg^{2+}) forms. Oxidation of elemental mercury to mercuric ions occurs according to the reaction: $2 Hg^0 \rightarrow Hg_2^{2+} \rightarrow 2Hg^{2+}$. The mercurous ion is unstable and dissociates further into the mercuric ion (Clarkson, 1968). Ionic mercury forms complexes with SH groups and other ligands in the tissues of the body and only a very small fraction exists in the free form.

In their experimental studies, Clarkson, Gatzy, and Dalton, 1961, found that after exposure of blood to mercury vapor in vitro, 1. no mercury was in the ultrafiltrable fraction after 30 minutes, 2. more mercury was taken up by the blood than could be dissolved physically in the elemental form, and 3. mercury was taken up faster by hemoglobin solution and whole blood than by plasma. These results formed the basis for the conclusion that mercury vapor was oxidized to mercuric mercury in the blood. The authors also concluded that once the mercury reaches the blood, it is quickly oxidized to Hg^{2+} and no differences in distribution or toxicity should exist between inhaled mercury vapor and absorbed mercuric salts.

However, later observations on mice (Berlin and Johansson, 1964, Berlin, Jerksell, and von Ubisch, 1966, and Magos, 1967), on rats, rabbits, and monkeys (Berlin, Fazackerly, and Nordberg, 1969), and on guinea pigs (Nordberg and Serenius, 1969) proved a higher uptake of mercury by the brain, the blood cells, and the myocardium after exposure to mercury vapor than after injection of mercuric salt. This indicates that the chemical state of mercury in blood may vary depending upon the exposure type. Berlin, 1966, observed that mercury in red cells after exposure to mercury vapor was not so firmly bound as after exposure to mercuric salts and proposed that the red blood cells might serve as "accumulators and generators" of metallic mercury capable of interconversion of Hg^0 and Hg^{2+}. The easily diffusible Hg^0 form released from erythrocytes should be responsible for the greater penetration of mercury into the brain after vapor exposure.

Magos, 1967, studied the uptake of mercury in the plasma and blood cells after exposure in vitro to mercury vapor and separated physically dissolved mercury vapor from oxidized mercury by different volatilization rates from the solutions. When samples of diluted blood had been exposed to mercury vapor in vitro at 37°C, 8% of the retained mercury was in the elemental form after 5 minutes of exposure and 4% after 15 minutes of exposure. If these figures are adjusted for a half minute's exposure, it can be extrapolated that nearly all the mercury must exist in the elemental form during this short period. As the total circulation time from the jugular vein to the carotid artery is about 22 seconds in man and shorter in small animals, it can be assumed that a large part of the mercury taken up by the blood in its passage through the lungs still exists in the elemental form at the time at which circulating blood enters the brain vessels.

Autoradiographic observations by Nordberg and Serenius, 1969, of higher concentrations of mercury around small blood vessels in the brain support further the relative ease with which mercury diffuses from cerebral vessels into the brain after exposure to mercury vapors. The studies by Berlin, Fazackerly, and Nordberg, 1969, and by Nordberg and Serenius, 1969, showed high uptake of mercury into the brain after vapor exposure in all of the several mammalian species studied. The general validity of these observations also for man was strongly supported.

Although mercury can temporarily exist in blood in its elemental form, it is ultimately converted to mercuric ions. The exact process of oxidation is unknown, but active participation of enzymatic systems is highly probable (Nielsen Kudsk, 1969a and b). Experimental data seem to show that the reverse process can also occur.

Rothstein and Hayes, 1964, and Clarkson and Rothstein, 1964, reported that after injection of mercuric chloride, approximately 4% of the total excreted amount of mercury was exhaled via the respiratory tract. The chemical form of the exhaled mercury was not identified. Because injected elemental mercury vapor is rapidly exhaled through the lungs (Magos, 1968), it can be assumed that mercury leaves the body through the lungs in the form of such vapor. This explanation is supported by experiments in vitro (Magos, 1967) which proved that about 0.5% of $HgCl_2$ could be volatilized from blood.

4.2.1.2 Transport of Elemental Mercury in Blood and into Tissues

It is evident that during or shortly after exposure to mercury vapor, part of the mercury is transported in the form of elemental mercury in the blood. Differential analyses performed on animals immediately after short exposure to mercury vapor indicated that the erythrocytes contained more mercury than the plasma (Berlin, 1966). Berlin, Fazackerly, and Nordberg, 1969, observed 67 to 84% of the total blood mercury in the blood cells of monkeys and rabbits immediately after exposure to mercury vapor, as compared with 25 to 31% in the blood cells of animals injected intravenously with mercuric ions and sacrificed at the same time as the vapor-exposed animals. The larger amount of mercury in the erythrocytes after exposure to vapors is probably traceable to the dissolved mercury vapor. Hitherto, the reason for the high uptake of elemental mercury by erythrocytes is unknown. Clarkson, 1968, speculated that it may reflect dissolved mercury vapor in the lipid structures of erythrocytes.

Generally, it seems that mercury in the form of elemental mercury vapor dissolves in the body fluids and penetrates biological membranes easily. The distribution seems to be affected by solubility factors, as proposed for the varying red blood cells/plasma ratio. Magos, 1968, injected radioactive metallic mercury vapor intravenously into rats and found that 30 seconds after injection 19% of the mercury had been exhaled through the lungs. The brain contained 0.6% and the blood, 5.9% of the injected dose. For a mercuric chloride injection, the corresponding figures were in exhaled air 1.8%, in the brain 0.3%, and in the blood, 44% of the injected dose. Obviously, diffusion across the alveolar membrane and from blood into tissues is easier for elemental than for ionized mercury. In the tissues metallic mercury is oxidized to mercuric ions, making the re-entry of mercury into the blood more difficult.

4.2.1.3 Transport of Mercuric Mercury in Blood

It follows from the relative rapidity of the oxidation process in the blood and the tissues of Hg^0 to Hg^{++} that after long-term exposure and even shortly after a single exposure, most of the mercury in the body and in the blood is in the form of mercuric ion. Values on the distribution of mercury in blood of workers exposed for a long time to mercury vapor are, therefore, probably a reflection of the dominating amount of mercuric mercury in the blood. Lundgren, Swensson, and Ulfvarson, 1967, found that the ratio between mercury in blood cells and in plasma was about 1:1. Rahola et al., 1971, and Miettinen, in press, recently reported an erythrocyte/plasma ratio of 0.4:1 in the blood of human volunteers who had taken tracer amounts of ^{203}Hg-proteinate by mouth. Suzuki, Miyama, and Katsunuma, 1971a, analyzed both inorganic mercury and total mercury in the blood components of workers exposed to mercury vapor. They found an erythrocyte/plasma ratio of 1.3 for total mercury but only 0.6 for inorganic mercury. They explained the difference by the presence of a certain amount of methyl mercury originating from the general background contamination of foodstuffs with methyl mercury, which is known to accumulate in erythrocytes. Thus it is possible that if the mercuric ion only is considered, the erythrocyte/plasma ratio may be slightly lower than 1 in human beings.

Some animal studies on different inorganic mercury compounds also indicate that mercuric ion is distributed about equally between plasma and erythrocytes (Ulfvarson, 1962, Berlin and Gibson, 1963, Tati, 1964, Suzuki, Miyama, and Katsunuma, 1967, Cember, Gallagher, and Faulkner, 1968, and Takeda et al., 1968a) when equilibrium has been reached. However, the penetration of mercuric mercury into erythrocytes is relatively slow, requiring about 2 hours in the rabbit (Berlin and Gibson, 1963). Takeda et al., 1968a, found that the equilibration took about 4 days after subcutaneous injections in rats (also reported by Ukita et al., 1969).

As chloride is the main plasma anion and

exceeds quantitatively any administered anion, different salts of mercuric compounds are probably handled in the same way by the body once they have been absorbed and dissociated into the body fluids. Thus the transport and excretion of inorganic mercury salts should be the same, regardless of the kind of soluble salt administered. Available experimental evidence is given in Sections 4.3.1 on distribution and 4.4.1 on excretion.

The ultrafiltrable fraction of inorganic mercury in plasma is very small, less than 1% according to Berlin and Gibson, 1963. For further discussion of this value see Section 4.4.1.1.1.2.2. The small percentage is a consequence of the binding of mercury to plasma proteins (Clarkson, Gatzy, and Dalton, 1961, and Cember, Gallagher, and Faulkner, 1968). It is theoretically assumed that diffusible, low molecular weight mercury ligands play a role in the ultrafiltrable fraction of plasma protein (Vostal, 1968, and Rothstein, in press), but the exact nature of this role is still unknown.

At least part of the inorganic mercury in erythrocytes exposed to mercuric chloride in vitro was identified in a protein fraction migrating like hemoglobin on paper electrophoresis (Clarkson, Gatzy, and Dalton, 1961). The distribution of mercury among plasma proteins depends upon the dose of mercury injected into the experimental animal (Cember, Gallagher, and Faulkner, 1968), and is different if mercuric mercury has been added in vitro or in vivo (Farvar and Cember, 1969). Mercury has been found in both albumin and globulin fractions in the plasma (Tati, 1964, Suzuki, Miyama, and Katsunuma, 1967, and Farvar and Cember, 1969). The amount of recovery on the electrophoretic strips was not reported in any of these studies, so it is not possible to judge the representability of the results. It has been shown earlier (Clarkson, Gatzy, and Dalton, 1961) that important amounts of mercury can be lost during electrophoretic procedures and subsequent staining.

4.2.2 Organic Mercury Compounds
4.2.2.1 Alkyl Mercury Compounds
4.2.2.1.1 In Animals
4.2.2.1.1.1 Methyl Mercury Compounds

Ulfvarson, 1962, administered orally or subcutaneously 5 different salts of methyl mercury to rats. There was no obvious difference in metabolism. Irukayama et al., 1965, performed a series of studies on rats, cats, rabbits, and dogs to find out whether there were any differences in metabolism and toxicity after ingestion of methyl mercury chloride and *bis*-methyl mercury sulfide. No definite differences were found. Similar results were obtained by Itsuno, 1968. In cats, no differences in metabolism or toxicity were found among methyl mercury in flesh of contaminated fish, methyl mercury salt added to fish flesh homogenate, or fish liver homogenate incubated with methyl mercury salt (Rissanen, 1969, and Albanus et al., to be published). It seems reasonable to assume that there are no major differences in metabolism or toxicity among different chemical forms of mono-methyl mercury.

4.2.2.1.1.1.1 Transport

After *mono-methyl mercury* compounds are administered, high levels of mercury are found in the blood, mainly in the blood cells and only to a minor extent in plasma. There are some species differences. In mice 75 to 90%, depending on the dose administered, is bound to the blood cells (calculated from data by Östlund, 1969b); in rats about 95% or more (Ulfvarson, 1962, Gage, 1964, Norseth and Clarkson, 1970b); in rabbits (Swensson, Lundgren, and Lindström, 1959b, and Berlin, 1963a) and monkeys (Nordberg, Berlin, and Grant, 1971) about 90%; in pigs about 80% (Bergman, Ekman, and Östlund, to be published); and in cats more than 95% (Albanus et al., 1969, and to be published, Rissanen, 1969, and Albanus, Frankenberg, and Sundwall, 1970). In rabbit plasma at least 99% is bound to plasma proteins (Berlin, 1963a).

Östlund, 1969a and b, made an autoradiographic study on mice exposed to *di-methyl mercury* through inhalation or intravenous injection. In both cases the mercury was readily transported mainly to the fat deposits. The levels in the blood were low.

4.2.2.1.1.1.2 Biotransformation

Theoretically, 2 kinds of biotransformation of *mono-methyl mercury* might occur in the mammal body. The first type includes a metabolic transformation of the methyl group in situ and the second, a breakage of the covalent bond between carbon and mercury. Studies by Östlund, 1969b, and Norseth, 1969b, in mice and rats, respectively, contradict a metabolic transformation of the methyl group.

The slow, even elimination of mercury after

administration of methyl mercury compounds to mice (Östlund, 1969b, Ulfvarson, 1962, 1969a, and 1970, Swensson and Ulfvarson, 1968, and Clarkson, 1971), cats (Albanus et al., to be published), pigs (Bergman, Ekman, and Östlund, to be published), monkeys (Nordberg, Berlin, and Grant, 1971) and man (see Section 4.3.2.1) indicates a rather high stability of the covalent bond. The rather constant distribution of mercury among different organs seen at different times after single administration of methyl mercury to mice (Berlin and Ullberg, 1963c) and rats (Swensson, Lundgren, and Lindström, 1959a, Ulfvarson, 1962, 1969a, and 1970, and Swensson and Ulfvarson, 1968) points in the same direction. Rats fed organs from rats injected with methyl mercury have a distribution similar to that of the injected animals (Ulfvarson, 1969b).

On the other hand, some studies indicate that a breakage of the covalent bond does occur. Östlund, 1969b, found indications of a small breakdown of the carbon-mercury bond in liver and kidneys of mice during a few hours after administration of methyl mercury hydroxide. About 3% of ^{14}C from the labeled methyl group was exhaled as $^{14}CO_2$ during 3 hours after intravenous administration.

Gage and Swan, 1961, and Gage, 1964, investigated the fraction of extractable organic mercury out of total mercury in different rat organs after 6 weeks of injections of methyl mercury dicyandiamide. In liver, spleen, and blood cells, 90% or more was present as organomercury, in plasma and brain, 75%, and in kidney, 55%. See also Section 4.4.2.1.1.1.2.

In ferrets, high fractions of methyl mercury out of total mercury were found in brain while in liver and kidney, they were lower (Hanko et al., 1970). Similar observations have been made in cats (Albanus et al., to be published). In swine, Platonow, 1968a, found high levels of methyl mercury out of total mercury in all organs investigated.

In the studies mentioned above, the fraction of organic mercury out of total mercury was examined. The residue was considered to be inorganic mercury. In studies reported by Norseth, 1969b, and Norseth and Clarkson, 1970a, the fraction present as inorganic mercury was estimated with an isotope exchange technique.

Norseth and Clarkson, 1970b, demonstrated formation of inorganic mercury from methyl mercury in the liver of rats. It is not clear whether transformation also occurred in other organs in the body. Norseth, 1969b, showed that a breakdown occurs in the intestinal lumen. The level of inorganic mercury in the brain was very low, 1 to 4% of the total mercury 28 days after a single administration. In plasma the inorganic fraction was 25%, and in the blood cells, less than 0.2% 10 days after the injection. The absolute levels of inorganic mercury were about equal. In liver and kidney the fractions, rising after injection, reached 50 and 90%, respectively, at 50 days. See also Section 4.4.2.1.1.1.2.

Norseth, 1971, studied the metabolism of methyl mercury in mice. There were some species differences as compared to rats. The main one was a lower fraction of total mercury in the kidney as inorganic mercury in the mice. The relative concentration of inorganic mercury increased gradually and after 22 days was about 30%. The author proposed that inorganic mercury was released from the liver into the bile and into the blood. In the brain the inorganic fraction constituted 2 to 14% of the total mercury and in the blood, 2 to 5%. See also Section 4.4.2.1.1.1.2.

Norseth and Brendeford, 1971, studied the fraction of total mercury present as inorganic mercury in different subcellular rat liver fractions after single injection or long-term oral exposure to methyl mercury dicyandiamide. While the highest total mercury concentrations were found in the microsome fraction (Norseth, 1969a, see Section 4.3.2.1.1.1), the highest levels of inorganic mercury were demonstrated in the lysosomes peroxisomes, which was in accordance with the distribution pattern seen after injection of mercuric chloride (Norseth, 1968, and 1969a).

Östlund, 1969a and b, studied the metabolism of *di-methyl mercury* in the mouse after inhalation and intravenous administration. The major part (80 to 90%) of the mercury was rapidly exhaled and was identified by thin layer chromatography as di-methyl mercury. After 16 hours no di-methyl mercury could be detected in the body. However, within 20 minutes after the administration, a nonvolatile metabolite occurred in the tissues. It was initially seen mainly in the liver, the bronchi, and the nasal mucosa, both in adult mice and in fetuses. After 1 day or more the metabolite had a distribution pattern very similar to that of monomethyl mercury and behaved as such in thin layer chromatography. Thus, the major part of the intact di-methyl mercury behaves like a chemically and physically inert substance while a minor part is metabolized into mono-methyl mercury ion.

4.2.2.1.1.2 Ethyl and Higher Alkyl Mercury Compounds

Ulfvarson, 1962, compared the distribution of mercury in rat organs after oral administration of ethyl mercury cyanide, hydroxide, and propandiolmercaptide. As was the case in the study made by the same author on methyl mercury salts, no definite differences were noted. Takeda et al., 1968a, noted no differences in metabolism between ethyl mercury chloride and ethyl mercury cysteine. It is reasonable to assume that the type of salt is of minor importance for the metabolic fate of ethyl mercury. Platonow, 1969, in a very short communication, stated that mice fed ethyl mercury acetate accumulated less mercury than those given viscera of pigs poisoned by the same compound. Suzuki et al., in press, found no differences in distribution of total mercury or inorganic mercury in rat organs after administration of ethyl mercury chloride or sodium ethyl mercury thiosalicylate.

4.2.2.1.1.2.1 Transport

Ulfvarson, 1962, exposing rats to ethyl mercury salts, showed that mercury accumulated to a considerable degree in the blood cells. Takeda et al., 1968a and b, showed that mercury in blood was almost exclusively present as ethyl mercury bound to a considerable degree to the hemoglobin in the red cells. In vitro studies showed that this binding was firm and that ethyl mercury passed through the stroma only with difficulty. An accumulation of mercury in blood cells after injection of ethyl mercury chloride has also been observed autoradiographically in the cynomolgus monkey and the cat (Ukita et al., 1969, and Takahashi et al., 1971).

4.2.2.1.1.2.2 Biotransformation

Miller et al., 1961, investigated the fraction of mercury present as organic mercury out of total mercury in liver, kidney, and whole blood 2 to 7 days after intramuscular injection of ethyl mercury chloride. The purity of the preparation used was stated to have been 99.5%. The percentage of organic mercury in liver was 89 to 100 and in blood, 67 to 72. In kidney the fraction decreased from 51% after 2 days to 21% after 7 days. See also Section 4.4.2.1.2.1.2.

In a similar study, Takeda and Ukita, 1970, found that in the rat liver 2 and 8 days after administration of ethyl mercury chloride (purity not stated), 94% of the mercury was present in organic and 6% in inorganic form. In the kidney after 2 days, 18% of the mercury was inorganic while after 8 days, the fraction was 34%. The organic mercury was chromatographically identified as ethyl mercury, the major part of which was protein bound. See also Section 4.4.2.1.2.1.2.

Takahashi et al., 1971, studied by thin layer chromatography the fraction present as ethyl mercury in the brain of a cynomolgus monkey 8 days after an intraperitoneal injection of ethyl mercury chloride (stated to be chromatographically pure). Of the total mercury, 96% appeared chromatographically as ethyl mercury while the rest was present as inorganic mercury.

Suzuki et al., in press, studied the fraction of total mercury present as inorganic mercury in brain, liver, and kidney of mice up to 13 days after subcutaneous or intravenous injection of ethyl mercury chloride or sodium ethyl mercury thiosalicylate. In all 3 organs there was an increase of the inorganic mercury fraction with time. At the end of the study about 50% of the mercury in the brain and the kidney was inorganic, while in the liver the corresponding figure was about 30%. In the brain the inorganic mercury concentration increased with time while in the other organs it first increased and then decreased (see also Section 4.4.2.1.2.1.1).

Suzuki et al., in press, quoted Japanese investigations by Sadakane, 1964, and Kitamura et al., 1970, in which it was shown that after administration of different ethyl mercury salts to rats, a considerable fraction of the total mercury in the brain was present as inorganic mercury.

The studies discussed above show that a breakage of the covalent bond between the ethyl group and the mercury occurs in the body and/or in the gastrointestinal tract. Rising fractions of inorganic mercury in an organ indicate either a formation in the organ and an elimination slower than that of the intact organomercurial, or a transformation elsewhere and an accumulation in the organ.

4.2.2.1.2 In Human Beings
4.2.2.1.2.1 Methyl Mercury

In the experiments performed by Åberg, Ekman, Falk, and collaborators (Section

4.1.2.1.2.2), blood cells contained about 10 times more mercury than plasma 24 days after oral administration of methyl mercury. This is in accordance with findings in persons exposed to methyl mercury via fish (Birke et al., 1967, Lundgren, Swensson, and Ulfvarson, 1967, Tejning, 1967b and c and 1968 b, and Sumari et al., 1969).

In an experimental study on metabolism of methyl mercury proteinate ingested with tartar sauce (Miettinen et al., 1969b, and 1971) 5 to 10% of the total body burden was present in the total blood volume during the 86 to 91 days studied. The fraction decreased with time. Five percent in the total blood volume corresponds to about 1% in 1 l. of blood.

Ui and Kitamura, 1971, and Ueda and Aoki (quoted by Ueda, 1969) have reported methyl mercury percentages of 28 to 120% (out of total mercury analyzed by neutron activation and atomic absorption, respectively) in hair from subjects exposed to methyl mercury at various intensities by consumption of contaminated fish (see Section 8.1.2.1.1.2.1). Birke et al., to be published, analyzed total mercury and methyl mercury in blood and hair of subjects exposed to methyl mercury through fish consumption. In whole blood containing 29 and 38 ng Hg/g, 60 and 85%, respectively, were methyl mercury and in 1 sample containing 650 ng/g, 100% was methyl mercury. In hair, 65 to 85% was methyl mercury. Skerfving, 1971, found 40 to 100% in blood cells and 50 to 90% in hair.

Total mercury and methyl mercury levels were studied simultaneously in a few patients from the Minamata and Niigata epidemics of methyl mercury poisoning through fish consumption. Sumino, 1968b (see Section 8.1.2.1.1.2.1) found 13 to 67% methyl mercury (analyzed according to Sumino, 1968a) out of total mercury (dithizone method) in hair. In brain, 60 to 120% of the total mercury (dithizone methods) was present as methyl mercury (Sumino, 1968b, and Tsubaki, personal communication; see Table 8:2). Grant, Moberg, and Westöö, to be published, found only methyl mercury in a brain sample from a patient poisoned by this compound.

4.2.2.1.2.2 Ethyl Mercury

Suzuki et al., in press, studied total mercury and inorganic mercury in blood from persons exposed to sodium ethyl mercury thiosalicylate employed as a preservative in plasma for intravenous use (Section 8.1.2.1.2.2). In 1 sample obtained 5 days after the last infusion from a person suspected to have been poisoned, the ratio between blood cell and plasma levels was about 7. In blood cells, 12% was present as inorganic mercury, while in plasma the corresponding figure was 20%. In brain, about 35% of the total mercury was present as inorganic mercury; in the renal cortex, 69%; in the renal medulla, 51%; in the liver, 31%; and in proximal hair, 5%. In 4 additional subjects the cell/plasma ratios ranged from 2 to 5. The inorganic fractions constituted only a few percent in blood cells while more than half of the mercury in plasma was inorganic. The levels of inorganic mercury in both blood cells and plasma changed little as time elapsed after the last administration. This gave a rising percentage in blood cells and roughly unchanged amount in plasma, in which the total mercury level decreased much more slowly than in blood cells. See also Section 4.4.2.1.2.2.

There is no information on *higher alkyl mercury* compounds. *Substituted alkyl mercury* compounds will be discussed in Section 4.2.2.4 on other organic mercury compounds.

4.2.2.2 Aryl Mercury Compounds

Almost all the work on the metabolism of aryl mercury compounds has been performed with different salts of phenyl mercury. Though no systematical comparative studies have been undertaken, it seems from the published data that there are no major differences in metabolism of different salts of phenyl mercury. In the following descriptions, the metabolism of phenyl mercury will be treated without regard to the type of salt administered. The available data are almost exclusively from animal experiments.

4.2.2.2.1 Transport

High levels of mercury are found in the blood initially after administration of phenyl mercury compounds. At the same dose level the blood concentrations are higher than those found after administration of inorganic mercury salts but lower than after alkyl mercury compounds (Prickett, Laug, and Kunze, 1950, Swensson, Lundgren, and Lindström, 1959b, Ulfvarson, 1962, Berlin and Ullberg, 1963b, Gage, 1964, and Takeda et al., 1968a). The mercury in blood is found mainly in the blood cells (Swensson,

Lundgren, and Lindström, 1959a and b, Ulfvarson, 1962, and Ukita et al., 1969). In rats (Takeda et al., 1968a) and rabbits (Berlin, 1963c) about 90% of the mercury has been found in this fraction. The mercury in the blood cells is bound mainly to the stroma-free hemolyzate (Takeda et al., 1968a). The small fraction of the blood mercury found in the plasma was bound, probably to proteins. Only about 1% passed on ultrafiltration through a cellulose dialysis tube (Berlin, 1963c).

While the initial appearance of mercury in blood after administration of phenyl mercury salts is similar to that of alkyl mercury compounds, the levels and distribution later resemble more what is found after administration of inorganic mercury salts (Ulfvarson, 1962, and Takeda et al., 1968a). In this later phase, the blood mercury levels decrease and a greater fraction of mercury is found in the plasma (Takeda et al., 1968a). As will be discussed in Section 4.2.2.2.2, this is probably to a great extent a result of the breaking of the covalent bond between the mercury atom and the phenyl group. The initial distribution to the blood cells and later to the plasma has been noted also in human beings (Goldwater, Ladd, and Jacobs, 1964).

Shapiro, Kollmann, and Martin, 1968, studied the binding of PCMB to the blood cell surface and entrance into the cells. The PCMB which entered the cell was bound to hemoglobin.

4.2.2.2.2 Biotransformation

In contrast to the considerable stability of the methyl mercury compounds in the organism, a number of studies on different species have supported a fairly rapid breakage of the carbon-mercury bond in phenyl mercury. This is indicated directly by the fraction of organic or inorganic mercury out of total mercury found in organs at different intervals after administration of phenyl mercury compounds, and indirectly through studies of the metabolic fate of the phenyl group as compared to the mercury and through the distribution and excretion patterns of mercury at different times after the administration. The excretion and distribution will be discussed in Section 4.3.2.2 but it should be mentioned here that while the initial patterns of phenyl mercury have certain similarities with those of alkyl mercury compounds, the later patterns are more like those seen after administration of inorganic mercury salts, just as was shown in Section 4.2.2.2.1 in connection with blood mercury level and distribution.

Forty-eight hours after intramuscular injection of phenyl mercury acetate into rats, Miller, Klavano, and Csonka, 1960, found only 20 and 10%, respectively, of the total mercury present as organic mercury in the liver and kidney. In the brain, the levels were so near the analytical zero of the methods used that no conclusions can be drawn. In dogs killed within 24 hours after intravenous administration of phenyl mercury acetate, limited data indicate a low fraction of organic mercury, especially in the kidney. (See also Section 4.4.2.2.1.2). Gage, 1964, repeatedly administered phenyl mercury acetate to rats during 6 weeks. The purity of the phenyl mercury preparation was not stated. Organs were analyzed for total and organic mercury weekly. In the kidney organic mercury made up 1 to 3% of the total mercury throughout the experiment. The information about other organs is less complete, mainly because the levels of organic mercury were low. It seems, however, that in the liver the fraction consisting of organic mercury was less than 20%. (See also Section 4.4.2.2.1.2). In a similar study by Nakamura, 1969, the fraction of phenyl mercury was low.

Daniel and Gage, 1971, (see also Gage, in press) studied the metabolism of ^{14}C-labeled phenyl mercury acetate after subcutaneous administration to rats. About 85% of the radioactivity appeared in the urine within 4 days and about 5% in the breath. Fifty to 60% of the mercury was excreted in the feces and only 12% in the urine. One third of the mercury in the first 24-hour urine was identified as organic mercury. The abundant radioactivity in the urine was associated with sulfonate and glucuronic acid conjugates of phenol. As there was little radioactivity in the expired air, the authors concluded that the breakage of the covalent bond did not result in benzene genesis. Instead, it seemed likely that the breakage occurred after o-hydroxylation (Gage, in press). Hydroxyphenyl mercury salts are instable in vitro in acid cysteine.

Substitutions in the phenyl group affect the fraction of total mercury in the kidney present as inorganic mercury. After administration of p-chloro mercury benzoate (PCMB), almost only inorganic mercury is found in the rat kidney after 4 hours (Clarkson and Greenwood, 1966, and Clarkson, 1969). After administration of p-chloro mercury benzenesulfonate the deposition in the kidney of unmetabolized organomercury is about the same as after PCMB, while the accumulation of inorganic mercury is considerably lower than after PCMB. The data presented by Clarkson, 1969, also indicate that there are species differences in the deposition of inorganic mercury after administration of PCMB and *p*-chloro mercury benzenesulfonate. Vostal, in press, showed a rapid transformation of PCMB into inorganic mercury in the dog kidney.

4.2.2.3 Alkoxyalkyl Mercury Compounds

All the research on the metabolism of alkoxyalkyl mercury compounds has been done with methoxyethyl mercury salts. The data available, from animal experiments only, are far more limited than those regarding the metabolism of alkyl and phenyl mercury compounds.

Substituted alkoxyalkyl mercury compounds (mercurial diuretics) will be discussed in Section 4.2.2.4 on other organic mercury compounds.

4.2.2.3.1 Transport

Though the information is restricted, it seems that initially after administration of methoxyethyl mercury, the mercury in the blood is distributed to a larger extent to the blood cells than to the plasma (Ulfvarson, 1962). The tendency to accumulate in the blood cells is less pronounced than initially at phenyl mercury exposure and far less than at alkyl mercury exposure. In repeated exposure, the distribution between blood cells and plasma later becomes about equal, just as at exposure to inorganic mercury salts or to phenyl mercury (Ulfvarson, 1962).

4.2.2.3.2 Biotransformation

Conclusive evidence shows a fairly rapid breakage of the carbon-mercury covalent bond in methoxyethyl mercury in the rat. Daniel, Gage, and Lefevre, 1971, (see also Gage, in press) administered a single dose of methoxyethyl mercury chloride labeled with ^{14}C to rats. During 24 hours about 50% of the radioactivity appeared in the exhaled air, about 90% of which was incorporated in ethylene (identified by gas chromatography), and about 10%, in carbon dioxide. The radioactivity in the exhaled ethylene corresponded to half of the dose administered. There was an accumulation of mercury in the kidney. A few hours after dosing, inorganic mercury (identified according to Gage and Warren, 1970) made up half of the total mercury in this organ. After 1 day, all the mercury was inorganic. About 25% of the radioactivity was excreted in the urine during 4 days. In the same period about 10% of the mercury was excreted in the urine. The chemical nature of the excreted metabolite(s) of the methoxyethyl group was not clarified. Urinary mercury consisting of appreciable amounts (half) of organic mercury during the first day later became solely inorganic. The liver contained little mercury compared to the kidney. The major part was present as inorganic mercury after 1 day. During the first day there was a considerable excretion of organic mercury in the bile, and later on, some inorganic mercury was excreted through this route. A mercury-free metabolite of methoxyethyl mercury was also excreted in the bile. No radioactivity was found in the feces, which indicates that all of the mercury in the feces was inorganic as a result of degradation in the gut. Organomercury probably was also reabsorbed.

There is also indirect evidence of a breakdown, including some documentation of a change in the mercury distribution pattern with time after administration of methoxyethyl mercury (Ulfvarson, 1969a and 1970). These changes, however, seem to be less pronounced than those after administration of phenyl mercury salts (Ulfvarson, 1962 and 1969a, and Berlin and Nordberg, unpublished data). Mercury from methoxyethyl mercury hydroxide is distributed in the same way as inorganic mercury salt (Ulfvarson, 1962, Berlin and Nordberg, unpublished data). Furthermore, the elimination is similar to that of inorganic mercury (Ulfvarson, 1962, and Berlin and Nordberg, unpublished data). The distribution will be discussed further in Section 4.3.2.3.

4.2.2.4 Other Organic Mercury Compounds

Modern mercury diuretics as a rule have the general formula $R-CH(OY)-CH_2-Hg^+X^-$ (Friedman, 1956, and Mudge, 1970). Y is most often a methyl group. Thus, the compounds might be regarded as substituted methoxyethyl mercury compounds. The metabolism of the mercury diuretics has been extensively studied in man. Some data might be of general interest.

Aside from the use of chlormerodrin as a diuretic, the compound labeled with radiomercury has been used for scanning of brain (Blau and Bender, 1962) and kidney (McAfee and Wagner, 1960). This use has given opportunities for studies of the metabolism in man. It is a pity that the dose of mercury seldom has been stated and that the methods of detection often are described superficially.

Mercury is rapidly cleared out of the blood after intravenous injection in man of chlormerodrin (3-chloromercuri-2-methoxy-propylurea; Neohydrin®) (Blau and Bender, 1962) and meralluride (3-acetomercuri-2-methoxy-succinyl-propylurea; Mercuhydrin®) (Burch et al., 1950). In a few hours the level is only a few percent of the original one. Most of the mercury in plasma is bound to proteins (Milnor, 1950), the fraction bound being dependent on the concentration.

Clarkson, Rothstein, and Sutherland, 1965, Clarkson, 1969, Vostal and Clarkson, 1970, and Vostal, in press, have shown a rapid breakdown of chlormerodrin and mersalyl (3-hydroxymercuri-2-methoxy-1-propylcarbamyl-0-phenoxy-acetate) in the dog kidney. Anghileri, 1964, investigated by paper chromatography the release of mercuric mercury from chlormerodrin injected into rats. At 24 hours after administration the author reported that about 50% of the mercury in the kidney was inorganic. See also Section 4.4.2.4.1.

Kessler, Lozano, and Pitts, 1957, found a high initial accumulation of mercury in the spleen after intravenous administration of the substituted alkyl mercury compound, 2-hydroxypropyl mercury iodide, to dogs. Wagner et al., 1964, introduced the bromide of the compound (1-mercuri-2-hydroxypropanol, MHP) labeled with ^{197}Hg or ^{203}Hg as a diagnostic tool for visualization of the spleen by scanning. (See also Korst et al., 1965, and Loken, Bugby, and Lowman, 1969.)

MHP is bound almost completely to blood cells when added to human blood in vitro (Wagner et al., 1964). It is stated to be bound to the cell surface and to enter the cell, where it is bound to hemoglobin (Shapiro, Kollmann, and Martin, 1968). When added to blood in concentrations of 0.5 to 1 mg Hg/ml it causes splenic sequestration of red blood cells when the blood is reinjected into the circulation (Wagner et al., 1964, and Shapiro, Kollmann, and Martin, 1968). The half-life in the human circulation of 1 to 10 mg Hg as MHP thus injected is 1/2 to 1 1/2 hours (Wagner et al., 1964, Croll et al., 1965, and Fischer, Mundschenk, and Wolf, 1965).

4.2.3 Summary

Elemental mercury introduced into the body is oxidized to mercuric ions. The oxidation occurs at a speed which allows the mercury vapor to exist in the blood during more than 1 circulation through the body. Because of the diffusibility of mercury vapor through biological membranes, a significant part of the transport of mercury from the lungs to the tissues can take place in the form of physically dissolved mercury vapor. After oxidation to mercuric ion has taken place, mercury is distributed about equally between blood cells and plasma or with a slight predominance in plasma. In the plasma, mercuric mercury is bound to different proteins and only a very small fraction is in a "free" ultrafiltrable form.

Data on the transport and biotransformation of *alkyl mercury* compounds are available mainly for methyl and ethyl mercury compounds. In animals and man, exposure to methyl and ethyl compounds gives high levels of mercury in the blood. In human beings exposed to methyl mercury, the ratio between blood cell and plasma mercury levels is about 10. At exposure to ethyl mercury, somewhat lower ratios have been found. The distribution of mercury in the blood is markedly different from that seen at inorganic mercury exposure.

Data on *aryl mercury* compounds are almost completely confined to phenyl mercury compounds. Mercury levels in whole blood are relatively high initially at phenyl mercury exposure in man and animals. More mercury is found in the blood cells than in plasma. At corresponding exposure the blood and blood cell levels are lower than after alkyl mercury compounds. Later, the level and distribution in the blood resemble those seen after exposure to inorganic mercury.

Mercury from methoxyethyl mercury, the only *alkoxyalkyl* mercury compound studied, is distributed similarly to phenyl mercury in the blood, although the initial level might be lower.

Though the covalent carbon-mercury bond in *methyl mercury* has a considerable stability in the body, a certain breakdown has been observed in several animal species. Levels of total mercury higher than the extractable organic mercury have been found. In other studies, considerable amounts of inorganic mercury have been found. Both observations have been made, especially in kidney and liver. It seems that the mercury in the brain is almost completely present as methyl mercury. Inorganic mercury is formed in the liver and in the intestinal lumen of the rat. It is not known whether this also happens within the organs in the body. More limited data on *ethyl mercury* indicate that it is less stable in the body than methyl mercury but still more stable than phenyl mercury. There are no data on higher alkyl mercury compounds.

Phenyl mercury is rapidly transformed into inorganic mercury. This has been shown by studies with compounds labeled in the phenyl group, by analysis of the fraction of organic or inorganic mercury out of total mercury in organs, and is also indicated by a redistribution of the mercury in the body into a pattern similar to that seen after exposure to inorganic mercury salts.

Also, *methoxyethyl mercury* salts are metabolized rapidly in the rat into inorganic mercury, part of which accumulates in the kidney.

4.3 Distribution
4.3.1 Inorganic Mercury

The differences in the transport of the different oxidation forms of mercury in the body have been mentioned in Section 4.2.1. The very initial phase of distribution after exposure to mercury vapor will be dominated by the diffusibility of elemental mercury vapor. As soon as oxidation to mercuric mercury has occurred in the blood and tissues, the oxidized mercury tends to be distributed according to its mercuric form. Therefore, the distribution pattern after exposure to mercury vapor should resemble the pattern after administration of mercuric mercury after prolonged exposure or even relatively soon after a short exposure to Hg^0, with exception for organs protected by barriers especially efficient against the penetration of Hg^{2+}.

4.3.1.1 In Animals
4.3.1.1.1 Mercuric Mercury

The distribution patterns of mercury after administration of mercuric mercury ought to be similar irrespective of the soluble mercuric compound used because mercuric mercury occurs in ionized form as Hg^{2+} in the body fluids. Some experimental support for this can be found in Table 4:1, although the diversity of doses, routes of administration, etc. used by the different investigators make it difficult to see clearly.

The distribution pattern is complicated, changing with time and other factors such as mode of administration, dose, and species. However, the highest concentration of mercury invariably is found in the kidney. Considerable concentrations of mercury also are found in the liver, spleen, and thyroid (Berlin and Ullberg, 1963a, and Suzuki, Miyama and Katsunuma, 1966). Blood contains high concentrations immediately after administration, but mercury is eliminated from blood faster than from most tissues in the body (Friberg, 1956, and Berlin and Ullberg, 1963a, Table 4:1). For the brain the situation is the reverse. Very little penetrates to it, but once mercury has entered, the turnover is very slow (Friberg, 1956, and Berlin and Ullberg, 1963a, Table 4:1). Some data on the distribution among the blood, brain, liver, and kidney for selected time intervals (1 day, 2 weeks, and prolonged exposure) are seen in Table 4:1. It would be desirable also to include in the Table the concentrations found by the different investigators, but many of the most important studies have been performed with radioactive isotope techniques and concentration values have not been expressed on a weight per weight basis. The general trends mentioned above are common in most studies included in the Table, but the numerical values of the ratios between organ concentrations sometimes are considerably different among different investigators even when the dose, species, survival time, etc., are the same. This variation probably is at least partly a reflection of analytical difficulties. The change in distribution with time mentioned above is evident from the Table. Mercury is eliminated from the blood and the liver more rapidly than from the brain and the kidney. This alteration is similar in principle irrespective of species, dose, or route of administration, but the rapidness of the change varies according to species. Numerical values of ratios between different organs also are somewhat different depending upon species and doses. Even strain differences have been shown to influence mercury retention in organs (Miller and Csonka, 1968). For an account of concentration changes among the mentioned organs and other tissues in the body of mice for earlier and intermediate time intervals not given in Table 4:1, the reader is referred to the autoradiographic study by Berlin and Ullberg, 1963a. It is learned from that study that mercury accumulates to a considerable degree in, for example, the colon, bone marrow, and spleen a short time after administration and is retained considerably in the testicles. Suzuki, Miyama, and Katsunuma, 1966, found the thyroid to be one of the sites in the body which accumulated mercury most efficiently.

An account of the detailed distribution among different structures of specific organs follows. The distribution among different components of blood has been dealt with in Section 4.2.1.3. Here, the distribution in the organs "critical" in mercury poisoning, i.e., the kidney and brain, will be dealt with first.

In the kidney, the mercury is not uniformly distributed, as shown already in 1903 by Almkvist, who, by a histochemical method, demonstrated deposits of mercury in the kidney tubules of rabbits given repeated large doses of $HgCl_2$. Friberg, Odeblad, and Forssman, 1957, and Bergstrand, Friberg, and Odeblad, 1958, found that mercury accumulates especially in the distal parts of the proximal tubules, but also in the wide part of Henle's loop and in the collecting ducts. They had given 2 mg Hg/kg (s.c.) in the form of $^{203}HgCl_2$ to rabbits and examined the kidneys by autoradiography 1 and 6 days later.

In the golden hamster, Voigt, 1958, found mercury localized in the proximal convoluted tubules by a histochemical method. He used s.c. (7 to 30 mg Hg/kg) or oral administration of $HgCl_2$ (3700 mg Hg/kg). Reber, 1953, found mercury by means of a histochemical method in the proximal convoluted tubules of mice after high doses of $HgCl_2$. Berlin and Ullberg, 1963a, using lower doses (0.5 mg Hg/kg) and autoradiography observed 2 distribution patterns in the renal cortex of mice, one with a prominent accumulation at the cortico-medullary border and another with equal concentration throughout the cortex.

In the rat, Lippman, Finkle, and Gillette, 1951, saw an especially high concentration at the cortico-medullary border in autoradiographic studies on the kidneys 24 hours after giving 6.6 mg Hg/kg as $^{203}HgCl_2$. Timm and Arnold, 1960, demonstrated mercury histochemically in the proximal convoluted tubules after injection of 4 mg Hg/kg as $HgCl_2$ in the same species.

Taugner, 1966, and Taugner, zum Winkel, and Iravani, 1966, have made extensive studies on the accumulation pattern of mercury in the rat kidney after i.v., i.m., or s.c. injections of $^{197}HgCl_2$ or $^{203}HgCl_2$. They found 2 distribution patterns, one including the whole cortex, corresponding to an accumulation in the middle portion of the proximal convoluted tubule, and another including a prominent accumulation in the cortico-medullary border corresponding to an accumulation in the terminal part of the proximal tubule. The first-named pattern was seen during the first 12 hours after injection and changed to the second pattern after that time if the injected doses were 1 to 10 mg Hg/kg. With lower doses, the second pattern usually was not seen, even at longer post-injection intervals. The development of the second pattern also was partly dependent on the existence of glomerular filtration, as will be discussed in Section 4.4.1.1.1.2.2. The 2 different distribution patterns found in rats are similar to those found in mice by Berlin and Ullberg, 1963a (see above). Further studies in mice (Nordberg, unpublished data) show that the appearance of the 2 patterns is dose- and time-dependent in a similar manner to that just mentioned for rats.

Timm and Arnold, 1960, concluded that their studies indicated a binding of mercury especially to mitochondria in the cells of the proximal tubule. Bergstrand et al., 1959 a, b, carrying out an electron microscopical examination of rats' kidneys after administration of 12 or 25 daily doses of 1 mg Hg/kg body, found an increase in the size of the mitochondria in the proximal convoluted tubules with large amounts of very fine and dense small particles. After fragmentation of the renal tissue and centrifugation at high speed, the radioactivity of ^{203}Hg was found in 2 fractions, corresponding to mitochondria and microsomes. A similar centrifugation study on kidneys from rats receiving a single oral dose of mercuric acetate was reported by Ellis and Fang, 1967. They found the main part of the mercury in the supernatant after centrifugation at 35,000 x g and 12% or less in the mitochondrial fraction, which was always less than what was found in the nuclear fraction (17 to 32%). The microsomal fraction contained between 1 and 12% of the total tissue mercury.

In later electron microscopical studies, changes have been observed in kidney tubule mitochondria (Gritzka and Trump, 1968, Wessel, Georgsson, and Segschneider, 1969), including matrical granular and microcrystalline deposits. Since such changes are also found in other types of irreversible cell injury, they are not necessarily deposits of mercury. Very fine granules found in the cytoplasma after H_2S treatment of the tissue were considered to reflect mercury sulfide in the

TABLE 4:1

Distribution of Mercury in Organs After Administration of Mercuric Mercury to Mammals

Species	No. of animals	Single dose mg Hg/kg	Repeated administration mg Hg/kg/day	Mode of administr.	Compound	Time to sacrifice (single exp.) Exposure time	Blood/ Brain	Blood/ Kidney	Liver/ Brain	Kidney/ Brain	Kidney/ Liver	Reference
Mouse	6	0.005		i.v.	$HgCl_2$	5 min	20	0.55	~32	36		Magos, 1968
	3	0.5		i.v.	$HgCl_2$	1 d	~16		18			Berlin and Ullberg, 1963a[1]
	2	0.01		i.v.	$Hg(NO_3)_2$	1 d				234	13	Berlin, Jerksell and von Ubisch, 1966
	2	0.5		i.v.	$Hg(NO_3)_2$	1 d			24			Berlin, Jerksell and von Ubisch, 1966
	3	0.5		i.v.	$HgCl_2$	16 d			~1	~16	16[2]	Berlin and Ullberg, 1963a[1]
	4	0.01		i.v.	$Hg(NO_3)_2$	16 d			2.0	29	15	Berlin, Jerksell and von Ubisch, 1966
	2	0.5		i.v.	$Hg(NO_3)_2$	16 d			1.8			Berlin, Jerksell and von Ubisch, 1966
Guinea pig	3	0.4		i.v.	$Hg(NO_3)_2$	1 d	3.6	0.011	24.5	328	13	Nordberg and Serenius, 1969
	3	0.4		i.v.	$Hg(NO_3)_2$	16 d	2.1	0.0099	11.3	208	18	Nordberg and Serenius, 1969
Rat	5	0.25		i.v.	$Hg(NO_3)_2$	1 d	5	0.012	13	416	32	Rothstein and Hayes, 1960
	2	0.5		i.v.	$Hg(NO_3)_2$	1 d	5	0.0092	21	525	25	Berlin, Fazackerly and Nordberg, 1969
	3	0.5		i.v.	$Hg(Ac)_2$	1 d					40	Prickett, Laug and Kunze, 1950
	7	0.62		i.v.	$HgCl_2$	1 d					30	Surtshin, 1957[3]
	6	3.0		i.v.	$HgCl_2$	1 d					20	Surtshin, 1957[3]
	1	0.6		or.	$Hg(Ac)_2$	1 d		0.0025			65	Ellis and Fang, 1967
	1	1.2		or.	$Hg(Ac)_2$	1 d	11	0.014	50	800	16	Ellis and Fang, 1967
	3	0.5		or.	$Hg(Ac)_2$	1 d					6	Prickett, Laug and Kunze, 1950
	3	0.01		i.v.	$Hg(NO_3)_2$	12 d	0.4	0.003	1.5	161	104	Ulfvarson, 1969a
	3	0.1		i.v.	$Hg(NO_3)_2$	12 d	0.5	0.0014	2.8	370	132	Ulfvarson, 1969a
	5	0.25		i.v.	$Hg(NO_3)_2$	15 d	1.1	0.0013	2.4	878	366	Rothstein and Hayes, 1960
	2	0.5		i.v.	$Hg(NO_3)_2$	16 d	0.5	0.0005	5.5	1100	200	Berlin, Fazackerly and Nordberg, 1969
	3	1.0		i.v.	$Hg(NO_3)_2$	12 d	0.5	0.0007	3.4	750	22	Ulfvarson, 1969a
	4		0.05	s.c.	$Hg(NO_3)_2$	18 d	1.1	0.0009	10.8	1262	117	Ulfvarson, 1962

TABLE 4:1 (continued)

Species	No. of animals	Single dose mg Hg/kg	Repeated administration mg Hg/kg/day	Mode of administr.	Compound	Time to sacrifice (single exp.) Exposure time	Blood/ Brain	Blood/ Kidney	Liver/ Brain	Kidney/ Brain	Kidney/ Liver	Reference
Rat	9		0.5[4]	s.c.	$HgCl_2$	35 d.[4]	0.5	0.0012	5.4	421	78	Friberg, 1956
	9		0.5[5]	s.c.	$HgCl_2$	35 d.[5]	0.1	0.013	1.3	11	8	Friberg, 1956
	6		0.5[4]	s.c.	$HgCl_2$	39 d.[4]	0.3	0.0009	3.1	291	94	Friberg, 1956
	6		0.5[5]	s.c.	$HgCl_2$	39 d.[5]	0.4	0.036	2.1	13	6	Friberg, 1956
	6		0.5	s.c.	$HgCl_2$	39 d.	1.4	0.064	6.5	222	34	Friberg, 1956
	4		0.5 ppm[6]		$Hg(Ac)_2$	12 mo.					12	Fitzhugh et al., 1950
	4		2.5 ppm[6]		$Hg(Ac)_2$	12 mo.					37	Fitzhugh et al., 1950
	4		10 ppm[6]		$Hg(Ac)_2$	12 mo.					18	Fitzhugh et al., 1950
	4		40 ppm[6]		$Hg(Ac)_2$	12 mo.					48	Fitzhugh et al., 1950
	4		160 ppm[6]		$Hg(Ac)_2$	12 mo.					91	Fitzhugh et al., 1950
Rabbit	3	2.0		s.c.	$HgCl_2$	1 d.	6.5	0.013	50	515	10	Friberg, Odeblad and Forssman, 1957
	1	1.0		s.c.	$HgCl_2$	1 d.	2.6	0.005	10	530	53	Miyama et al., 1968
	1	1.0		i.v.	$HgCl_2$	1 d.	10	0.015	70	665	10	Miyama et al., 1968
	1	0.1		i.v.	$Hg(NO_3)_2$	16 d.		0.006			25	Berlin, Fazackerly, and Nordberg, 1969
	1	0.1		i.v.	$Hg(NO_3)_2$	32 d.					14	Berlin, Fazackerly, and Nordberg, 1969
	2	2.0		s.c.	$HgCl_2$	40 d.	1.3	0.0005	69	2622	38	Friberg, Odeblad, and Forssman, 1957
Monkey	2	0.1		i.v.	$Hg(NO_3)_2$	4 d.	2.5	0.0013	90	1900	21	Berlin, Fazackerly and Nordberg, 1969
	1	0.1		i.v.	$Hg(NO_3)_2$	16 d.	2.0	0.0046	112	429	4	Berlin, Fazackerly and Nordberg, 1969
	1	0.1		i.v.	$Hg(NO_3)_2$	32 d.	1.0	0.0024	53	404	8	Berlin, Fazackerly and Nordberg, 1969

[1] Autoradiographic determination
[2] Kidney cortex vs. liver
[3] Assuming kidney weight = 2.0 g and liver weight = 12 g
[4] Killed 14 days after termination of exposure
[5] After this period exposed to nonradioactive Hg for 14 days, then killed. Organ values represent only radioactive Hg
[6] Concentration in the diet of substance given

electron microscopical study by Wessel, Georgsson, and Segschneider, 1969.

Jakubowski, Piotrowski, and Trojanowska, 1970, found that in rat kidneys a large part of the ^{203}Hg was present in a fraction with a molecular weight of approximately 11,000. To test whether this protein fraction was similar to metallothionein, Wisniewska et al., 1970, injected rats with cadmium chloride and ^{203}HgCl$_2$ and analyzed kidneys by differential centrifugation and gel filtration chromatography. Finding the mercury-containing protein similar to metallothionein, they suggested that mercury might be transported and detoxified in the body by the binding to metallothionein in a manner similar to cadmium (Piscator, 1964, and Friberg, Piscator, and Nordberg, 1971). Studies by Piotrowski et.al., in press, have given further evidence on the binding of mercury in the rat kidney and liver to a small molecular size protein with some characteristics of metallothionein. It was also found that prolonged exposure to mercury gave rise to a larger amount of the mentioned mercury binding protein, especially in the kidney.

As mentioned above, the blood-brain barrier hinders the penetration of mercuric mercury into the brain (Berlin and Ullberg, 1963a, Berlin, Jerksell, and von Ubisch, 1966, Berlin, Fazackerly, and Nordberg, 1969, and Nordberg and Serenius, 1969). After intravenous injection of 0.4 mg Hg/kg as Hg(NO$_3$)$_2$ in guinea pigs, the relatively small amount of the mercury that penetrates into the brain is initially rather uniformly distributed with a predominance in the grey matter compared to the white matter. With time, however, a more differentiated pattern is seen. Generally the concentration in the grey matter diminishes except for remaining high concentrations in certain mesencephalic nuclei on the border line to the rhombencephalon and a few other nuclei, e.g., nucleus dentatus in the cerebellum, which also retain a large amount of mercury for a long time. A prominent concentration was seen in the area postrema and in the plexus chorioideus. In the monkey a similar distribution pattern was seen; in addition to the nucleus dentatus in the cerebellum, the nucleus olivarius inferior and nucleus subthalamicus showed a marked uptake of mercury (Nordberg and Serenius, 1969, and Berlin, Fazackerly, and Nordberg, 1969). The concentration differences among different parts of the brain just described (illustrated by autoradiography) were substantial; up to 250 times higher concentrations of mercury were detected in some brain structures when compared to others. Other investigators (Ulfvarson, 1969a, and Suzuki, Miyama, and Katsunuma, 1971a), measuring mercury concentrations in relatively large parts of the brain, have found much smaller differences among their parts because they could not obtain any similar resolution between brain structures as obtained by autoradiography. Timm, Naundorf, and Kraft, 1966, detected mercury by a histochemical method, especially in the nerve cells of the brain stem, after long oral exposure of rats to mercuric mercury. Fractionation of subcellular particles from brains of animals given mercuric mercury has not been performed, but such studies are available for vapor-exposed animals (see Section 4.3.1.1.3).

Initially, mercury is uniformly distributed in the livers of mice and rabbits but after a few days most of the mercury is in the peripheral parts of the liver lobules (Friberg, Odeblad, and Forssman, 1957, and Berlin and Ullberg, 1963a).

Ellis and Fang, 1967, found mercury in nuclear, mitochondrial and microsomal fractions of rat liver. Similar results were obtained by Norseth, 1968, who used the distribution of marker enzymes for characterization of the fractions. The distribution pattern among the subcellular fractions changed gradually after administration of HgCl$_2$ and especially in the lysosomes an accumulation of mercury was seen with time. Jakubowski, Piotrowski, and Trojanowska, 1970, and Piotrowski et al., in press, made similar studies of the binding of mercury in the liver to those reported above for kidneys. Likewise in the liver, mercury was bound to proteins with both high and low molecular weights. A portion of the mercury which is in fractions corresponding to a molecular weight of about 10,000 has attracted special interest (see above).

Concerning other organs, it is interesting to note the observation by Berlin and Ullberg, 1963a, that mercury accumulates specifically in the wall of the thoracic aorta, but not in the other blood vessels of the mouse. In the same study it was found that mercury was localized specifically in the interstitial cells of the testis and epididymis but was not found in the testicular tubules. A more detailed picture showing this specific distribution in the mouse testicle has been published by Bäckström, 1969b.

4.3.1.1.2 Mercurous Mercury

The distribution of mercury after administration of this form is largely unknown. The earlier mentioned principal considerations on the oxidation of mercury in the body make it probable that relatively soon after administration the absorbed amount will be converted to the mercuric form and distributed accordingly. Data from Lomholt, 1928, on rabbits and from Viola and Cassano, 1968, on mice, support such an assumption.

4.3.1.1.3 Elemental Mercury

The distribution pattern of mercury after exposure to mercury vapor is similar to that seen after administration of a corresponding amount of mercuric mercury, except for higher concentrations in brain, blood, and myocardium. For comparison with mercuric mercury (Table 4:1), the distribution among the blood, brain, liver, and kidney for selected time intervals is shown in Table 4:2. The higher amounts of mercury found in the brain after vapor exposure influence ratios involving this tissue. Unfortunately, there are very few data on blood/brain and blood/kidney ratios, but there is a tendency for these ratios to diminish with time, in principle the same change seen after mercuric mercury exposure.

Concerning the more detailed distribution of mercury in the tissues, there are relatively few studies in which direct comparisons between occurrences after mercuric and elemental mercury exposures have been made. The available data indicate that the distribution is the same in most tissues, in accord with the concept that mercury is rapidly oxidized in the tissues once it has entered them. Since the brain shows the most prominent difference in concentration between vapor exposure and mercuric mercury exposure, it is of interest to examine whether the brain distribution differs between the 2 forms of exposure. Such comparative studies have been performed in guinea pigs by Nordberg and Serenius, 1966, 1969, and in monkeys *(Saimiri Sciureus)* by Berlin, Fazackerly, and Nordberg, 1969. One day and longer after the exposure, the distribution within the brain was essentially the same, irrespective of the mode of administration. The levels after injection of mercuric mercury were generally lower, except for plexus chorioideus and area postrema, structures which are unprotected by the blood-brain barrier.

Details of the distribution pattern have been given in Section 4.3.1.1.1.

Detailed studies of the distribution of ^{203}Hg in the brains of rats and mice after exposure to ^{203}Hg vapors have been performed by Cassano, Amaducci, and Viola, 1966, 1967, and Cassano et al., 1969. They found mercury localized predominantly in the grey matter. In the cells the mercury was localized in the cytoplasm and processes of neurons. They found the greatest concentration of radioactivity in certain nuclei of the mid-brain, pons, medulla, and cerebellum. In the cerebellar cortex the mercury was selectively localized over the Purkinje cells. The distribution of radioactivity in different chemical fractions of nervous tissue was also studied. It was revealed that ^{203}Hg was highly concentrated in the water soluble fraction and in the protein fraction. The lipid soluble fraction contained no detectable radioactivity. The somewhat different distribution in the blood at vapor exposure has been commented upon earlier in the section on Biotransformation and Transport (4.2).

4.3.1.2 In Human Beings

The extremely scarce data on the distribution in man after mercuric mercury exposure are reviewed in Table 4:3. Most of these data are relatively old and analytical errors cannot be excluded. However, when comparing these data with the animal data (Table 4:1), it is seen that the principal distribution pattern is the same, i.e., the highest concentration is found in the kidney, followed by the liver, with low values in the blood and the brain. It is not possible from the limited material to draw any conclusions in regard to eventual differences in the numerical values of the ratios between the different organs in man in relation to animals. The relatively low kidney/liver and kidney/brain ratios seen in Table 4:3 may reflect the pronounced kidney damage with loss of renal tissue which caused the death of persons poisoned with $HgCl_2$. Some data on the distribution in blood in human beings are available (see Section 4.2).

Some more recent data have been reported by Sodee,. 1963, who studied the distribution of ^{197}Hg in the human body by external radioactivity measurements. He found high concentrations of ^{197}Hg in the kidneys and 30% of the injected dose in the liver and spleen. Sodee injected ^{197}HgCl$_2$ intravenously but did not

TABLE 4:2

Distribution of Mercury in Organs After Exposure to Elemental Mercury Vapor in Mammals

Species	No. of animals	Exposure air concentration mg Hg/m³ and time (single exposure)	Time to sacrifice (single exp.) Exposure time	Ratios between concentrations in organs					Reference
				Blood/ Brain	Blood/ Kidney	Liver/ Brain	Kidney/ Brain	Kidney/ Liver	
Mouse	6	i.v. Hg° vapor[1]	5 min	1.24	0.28		4.5		Magos, 1968
	2	4 hrs[2]	1 day			1.31	12.2	9	Berlin, Jerksell, and von Ubisch, 1966
	2	4 hrs[3]	1 day			0.96	10.2	11	Berlin, Jerksell, and von Ubisch, 1966
	4	4 hrs[2]	16 days			0.28	5.4	19	Berlin, Jerksell, and von Ubisch, 1966
	2	4 hrs[3]	16 days			0.14	1.1	8	Berlin, Jerksell, and von Ubisch, 1966
Guinea pig	3	7; 5 hrs	1 day	1.33	0.026	1.28	38.5	30	Nordberg and Serenius, 1969
	3	7; 5 hrs	16 days	0.10	0.002	1.37	40.9	30	Nordberg and Serenius, 1969
Rat	3	1.4; 5 hrs	1 day		0.0006			33	Hayes and Rothstein, 1962
	2	1.0; 4 hrs	1 day	0.38	0.011	1.03	31.4	31	Berlin, Fazackerly, and Nordberg, 1969
	3	1.4; 5 hrs	15 days					320	Hayes and Rothstein, 1962
	2	1.0; 4 hrs	16 days	0.02	0.002	0.16	13.0	78	Berlin, Fazackerly, and Nordberg, 1969
	2	1	6 weeks			0.09	15.5	179	Gage, 1961a[4]
	2	1	4 months			0.02	23.8	1167	Gage, 1961a[4]
	2	0.02-0.03	6.5 months			2.5	5.6	2	Kournossov, 1962

TABLE 4:2 (continued)

Species	No. of animals	Exposure air concentration mg Hg/m³ and time (single exposure)	Time to sacrifice (single exp.) Exposure time	Blood/ Brain	Blood/ Kidney	Liver/ Brain	Kidney/ Brain	Kidney/ Liver	Reference
					Ratios between concentrations in organs				
	3	0.008-0.01	6.5 months			2.8	9.8	3	Kournossov, 1962
	3	0.002-0.005	6.5 months			4.7	8.6	2	Kournossov, 1962
	4	0.1	7-9 weeks					29	Ashe et al., 1953
	2	0.1	3.5 months					75	Ashe et al., 1953
	7	0.1	13-15 months					23	Ashe et al., 1953
	2	0.1	17 months					22	Ashe et al., 1953
Rabbit	2	1;4 hrs	4 days	0.25	0.003	1.8	83	45	Berlin, Fazackerly, and Nordberg, 1969
	1	1; 4 hrs	16 days			3.5	74	21	Berlin, Fazackerly, and Nordberg, 1969
	4	6	6-8 weeks	0.15	0.015	0.36	9.7	27	Ashe et al., 1953[5]
	2	6	10-11 weeks	0.06	0.006	0.50	9.4	19	Ashe et al., 1953[5]
	11	0.9	6-8 weeks	0.30	0.012	2.24	25.9	12	Ashe et al., 1953[5]
	4	0.9	10-12 weeks	0.12	0.004	3.53	27.6	8	Ashe et al., 1953[5]
	1	0.1	8 weeks	0.23	0.015	2.23	31.6	16	Ashe et al., 1953[5]
	2	0.1	3.5 months	0.25	0.007	5.4	30.0	14	Ashe et al., 1953[5]
	4	0.1	10.5 months	0.27	0.003	2.8	7.1	6	Ashe et al., 1953[5]
	2	0.1	19 months		0.04			3	Ashe et al., 1953[5]
Dog	1	0.1	14 months	0.026	0.0013	1.86	20.0	11	Ashe et al., 1953[5]
	1	0.1	19 months	0.012	0.0005	7.37	25.1	3	Ashe et al., 1953[5]
Monkey	2	1; 4 hrs	4 days	0.19	0.011	2.0	17	17	Berlin, Fazackerly and Nordberg, 1969
	1	1,4 hrs	16 days	0.57	0.010	15.9	58	4	Berlin, Fazackerly, and Nordberg, 1969

[1] Single exposure corresponding to 0.005 μg Hg/kg
[2] Single exposure corresponding to 0.01 mg Hg/kg
[3] Single exposure corresponding to 0.5 mg Hg/kg
[4] Brain weight assumed to be 1.8 g, liver weight 12 g, and kidney weight 2.0 g
[5] Concentrations in tissues are given in Table 7:10

report what dose of mercury he injected. Other information on methodological questions is scarce in his report and it is, therefore, difficult to evaluate his data. He suggested the use of ^{197}Hg as a suitable isotope for liver and spleen scanning.

Artagaveytia, Degrossi, and Pecorini, 1970, used ^{197}HgCl$_2$ for differentiation of malignant and nonmalignant thyroid nodules. ^{197}HgCl$_2$ was also used by Rosenthall, Greyson, and Eidinger, 1970, to differentiate malignant and nonmalignant intrathoracic lesions. It was learned in both of the last-mentioned studies that the malignancies accumulated mercury more frequently than did the benign tumors or inflammatory foci.

For mercurous mercury there are no reliable data on the distribution of mercury in man. Some results from old studies have been reviewed by Lomholt, 1928, but it is difficult to draw any conclusions from them.

For mercury vapor exposure, the data are extremely scarce (see, however, Table 4:3). Unfortunately, no blood concentrations are available. Lower kidney/brain ratios in vapor-exposed persons show that the exposure-dependent distribution differences also exist in man. Considerable accumulation of mercury in the brain after exposure to mercury vapor is evident from the data reported by Takahata et al., 1970, and Watanabe, 1971. They studied the distribution of mercury in formalin-treated brain specimens from 2 mercury mine workers who had died from pulmonary tuberculosis. Both workers had been exposed for more than 5 years to high concentrations of mercury vapor and one of them (I) died 6 years after exposure ended. The other man (II) died about 10 years after the end of exposure. In the brains of both men, similarly high concentrations of mercury were found. Especially in the occipital cortex (I:34, II:15 ppm), parietal cortex (I:16, II:17 ppm) and substantia nigra (I:23, II: 18 ppm) high concentrations were found. In other parts, e.g., the caudate nucleus, much smaller concentrations (3 to 4 ppm) were found. By a histochemical method mercury was detected in the lamina III of the cerebral cortex and in the cytoplasm of the Purkinje cells of the cerebellum. Electron microscopical examination of Purkinje cells revealed small electron dense granules in the cytoplasm but not in the nucleus. The data by Takahata et al. show that there are considerable concentration differences among different parts of the human brain and also that the retention even 10 years after termination of exposure can be considerable, thus indicating a very long half-life for mercury in the human brain.

4.3.2 Organic Mercury Compounds
4.3.2.1 Alkyl Mercury Compounds
4.3.2.1.1 In Animals
4.3.2.1.1.1 Methyl Mercury Compounds

Some data on the distribution of mercury after administration of different *mono-methyl mercury* compounds are presented in Table 4:4. Various species, routes of administration, exposures, exposure times, time between end of exposure, and sacrifice and methods of analysis have been used. At repeated exposure, values have been recalculated as daily dose of mercury/kg body weight.

After administration of methyl mercury, the mercury is more evenly distributed among different organs than after administration of inorganic mercury salts. High levels of mercury are obtained in liver, kidney, and blood cells. Within the kidney, higher levels of mercury are found in the cortex than in the medulla (Bergstrand et al., 1959a, Berlin and Ullberg, 1963c, Platonow, 1968a, and Rissanen, 1969). High levels also occur in the spleen and pancreas (Norseth and Clarkson, 1970b).

In relation to liver and kidney, the CNS shows a low mercury level, though the brain mercury level found after administration of methyl mercury is high in comparison with those seen after administration of inorganic mercury salts (e.g., Berlin, 1963b, and Berlin and Ullberg, 1963c). After a single administration of methyl mercury salt to mice (Berlin and Ullberg, 1963c, Suzuki, Miyama, and Katsunuma, 1963) and rats (Swensson and Ulfvarson, 1968, Norseth, 1969b, Ulfvarson, 1969a, 1970, and Norseth and Clarkson, 1970b), the CNS reaches its maximum concentration several days later than the other organs. It seems that the blood-brain barrier delays distribution. The distribution to the brain is accelerated by administration of 2,3-dimercaptopropanol (Berlin and Ullberg, 1963d, and Berlin, Jerksell, and Nordberg, 1965).

Studies on mice (Berlin and Ullberg, 1963a), rats (Friberg, 1959, Swensson and Ulfvarson, 1968, and Ulfvarson, 1969a), cats (Yamashita, 1964), dogs (Yoshino, Mozai, and Nakao, 1966a), pigs (Platonow, 1968a and b, Coldwell and Platonow, 1969, and Bergman, Ekman, and

TABLE 4:3

Distribution of Mercury in Man. Calculations Based on Concentrations in Wet Weight Tissue if Not Otherwise Stated

A. Mercuric mercury

Case No.	Compound	Route of exposure	Blood/Brain	Kidney/Blood	Liver/Brain	Kidney/Liver	Kidney/Brain	Reference
1.	HgCl$_2$	or.		35		1.4		Sollmann and Schreiber, 1936
2.	HgCl$_2$	or.		27		1.3		Sollmann and Schreiber, 1936
3.	HgCl$_2$	or.		158		2.3		Sollmann and Schreiber, 1936
I.	Mercuric benzoate	inj. ?			8.00	5.3	42.1	Lomholt, 1928
II.	HgCl$_2$	or.			18.88	2.2	41.2	Lomholt, 1928
IV.	HgCl$_2$	or.			3.20	5.0	16.0	Lomholt, 1928

B. Mercury vapor

1.	Acute exposure					8.9		Matthes et al., 1958
2.	Acute exposure					6.4		Matthes et al., 1958
I.	Chronic exposure				0.03[1]	21.2[1]	0.5[1]	Watanabe, 1971
II.	Chronic exposure				0.55[2]	3.5[2]	1.9[2]	Watanabe, 1971

[1] Calculated on dry weight values. Values given for cerebellum used for brain and kidney cortex for kidney in calculations.
[2] Calculated on wet weight values. Cerebellum used for brain and kidney cortex for kidney.

TABLE 4:4

Distribution of Mercury in Organs After Administration of Methyl Mercury (MeHg) to Mammals (From Berglund et al., 1971, with some additions.)

Species	No. of animals	Compound/ source	Mercury exposure — Single dose mg Hg/kg body weight	Mercury exposure — Repeated administration, mg Hg/kg body weight/day	Administration route	Exposure time, days	Time single exposure to sacrifice days
Mouse	8[1]	Shellfish		25	or.	11-61	
	5	MeHg acetate		0.5	s.c.	11	
	15	MeHg dicyandiamide		0.05	i.p.	16	
	3	MeHg acetate		2.5	s.c.	10	
	3	MeHg acetate		5	s.c.	10	
	10	MeHgOH	0.03		i.v.		6
	10	MeHgOH	0.3		i.v.		6
	10	MeHgOH	1.0		i.v.		6
	10	MeHgOH	5.0		i.v.		6
	2	MeHgCl	1.0		i.v.		22
Rat	10	MeHgCl	0.1		i.v.		32
	5	MeHgCl		[3]	or.	21	
	5	MeHgCl		[3]	or.	21	
	6	MeHg dicyandiamide		[1]	s.c.	10	
	5	MeHgOH		[3]	or.	21	
	4	MeHgOH		0.05	s.c.	13	
	6	MeHg dicyandiamide		0.65	s.c.	42	
	5	MeHgOH	0.1		i.v.		4
	5	MeHgOH	0.5		i.v.		4
	3	MeHgOH	0.5		i.v.		16
	3	MeHgOH	0.04		i.v.		6
	3	MeHgOH	0.4		i.v.		6
	3	MeHgOH	4.0		i.v.		6
	20	MeHgOH	40		s.c.		3
	20	MeHgOH	40		s.c.		3
	20	MeHgOH		~0.01	or.	180-210	
	20	MeHgOH		~0.06	or.	180-210	
	20	MeHgOH		~0.3	or.	180-210	
Ferret	2[1]	[6]		[5]	or.	35-36[7]	
	2[1]	[6]		[8]	or.	58[9]	
Rabbit	3	MeHg dicyandiamide	1.5		i.v.		11

TABLE 4:4 (continued)

Brain	Liver		Kidney		Whole Blood			References
		liver/		kidney/		blood/	blood/	
μg/g	μg/g	brain	μg/g	brain	μg/g	brain	kidney	
28	72	2.6	64	2.3				Saito et al., 1961
		3.4		8.5´				Suzuki, Miyama, and Katsunuma, 1963
		3.6		14				Berlin, Jerksell, and Nordberg, 1965
4	20	5	40	10	5	1.3	0.13	Suzuki, 1969[2]
6	30	5	60	10	9	1.5	0.15	Suzuki, 1969[2]
0.02	0.08	4.0	0.02	1.0	0.02	1.0	1.0	Östlund, 1969b[2]
0.2	0.6	3.0	0.2	1.0	0.2	1.0	1.0	Östlund, 1969b[2]
0.6	2.1	3.5	0.6	1.0	0.5	0.8	0.83	Östlund, 1969b[2]
3.3	9.8	3.0	4.0	1.2	3.7	1.1	0.93	Östlund, 1969b[2]
0.37	0.7	2	2.7	7	0.5	1.3	0.14	Norseth, 1971
0.02	0.08	4	0.26	13	0.05	3	0.19	Swensson, Lundgren, and Lindström, 1959b
1.6	7.0	4.4	18	11	19	11	1.06	Swensson, Lundgren, and Lindström, 1959b
0.5	2.3	4.6	46	93	9.0	18	0.20	Swensson, Lundgren, and Lindström, 1959b
3.0	14	4.7	52	17	48	16	0.92	Friberg, 1959
1.7	7.2	4.3	24	14	21	12	0.88	Ulfvarson, 1962
0.19	0.92	4.8	4.6	24	3	16	0.65	Ulfvarson, 1962
4	16	4	51	13	~40[4]	20	~0.78	Gage, 1964
0.04	0.17	4.2	0.59	15	0.43	11	0.73	Swensson and Ulfvarson, 1967
0.13	0.52	4.0	1.7	13	2.2	17	1.29	Swensson and Ulfvarson, 1967
0.17	0.48	2.8	2.6	15	1.8	11	0.69	Swensson and Ulfvarson, 1968
0.01	0.04	4	0.49	50	0.14	14	0.29	Ulfvarson, 1969a
0.14	0.41	2.9	2.2	15	1.7	12	0.77	Ulfvarson, 1969a
1.5	4.2	2.8	22	15	19	13	0.86	Ulfvarson, 1969a
	88		140					Ulfvarson, 1969b
21	77	3.7	98	4.7	290	14	2.96	Ulfvarson, 1969b
0.2					1.2	6.0		Ahlborg et al., to be published
1.2					7.8	6.0		Ahlborg et al., to be published
7.0					45	6.5		Ahlborg et al., to be published
37	61	1.6	73	2.0				Hanko et al., 1970
16	47	2.9	65	4.0				Hanko et al., 1970
1.5	2.9	2	2.9	2				Swensson, 1952

TABLE 4:4 (continued)

Species	No. of animals	Compound/ source	Mercury exposure		Administration route	Exposure time, days	Time single exposure to sacrifice days
			Single dose mg Hg/kg body weight	Repeated administration, mg Hg/kg body weight/day			
Cat	3[1]	Shellfish		?	or.	?	
	4[1]	Shellfish		?	or.	?	
	3[1]	MeHgCl		1.4	or.	32	
	2[1]	MeHgI		1.1	or.	33	
	3[1]	Shellfish		?	or.	100	
	3	MeHgSHg		0.9	or.	36	
	4	MeHgSHgMe		1.5	or.	22	
	6[1]	Fish + shellfish		?	or.	?	
	9[1]	Fish + shellfish		?	or.	?	
	1[1]	Fish + shellfish		?	or.	?	
	?[1]	Shellfish		?	or.	?	
Cat	14	Fish + shellfish		?	or.	?	
	11	Fish + shellfish		?	or.	?	
	2[1]	[10]		1	or.	37-45	
	2[1]	MeHgOH		0.2-0.4	or.	76-126	
Dog	3[1]	MeHg thio-acetamide	21		i.v.		5
Pig	1	MeHg acetate	1.0		i.m.		7
	1	MeHg acetate		1.0	i.m.	7	
	1	MeHg acetate	5.0		i.m.		7
	1	MeHg acetate		5.0	i.m.	7	
	2	MeHg acetate	5		or.		7
	2	MeHg acetate		5	or.	7	
	2	MeHg dicyandiamide	1.7		or.		32
	2[1]	MeHg dicyandiamide	27		or.		7
Monkey (*Saimiri sciureus*)	2[1]	MeHgOH		0.3-0.7	or.	35-36[11]	
	1[1]	MeHgOH		0.3-0.7	or.	28[12]	
	1[1]	MeHgOH		0.3-0.7	or.	28[13]	
	2	MeHgOH		0.3-0.7	or.	21[14]	

[1] Symptoms of poisoning in all or some animals in the group.
[2] Values read in diagram in original paper.
[3] Supplied in drinking water 2 mg Hg/1,000 ml.
[4] Level in whole blood calculated from level in blood cells and plasma, with a presumed hematocrit of 50.
[5] Total dose ~35 to 45 mg/kg. Exposure varied between 0 and 1.5 mg Hg/kg/day. Mean exposure ~0.5 mg/kg/day.
[6] Musculature and liver from intoxicated hens whose food was mixed with MeHg dicyandiamide.
[7] Onset of symptoms after about 14 days. Until then mean exposure about 1.4 mg/kg/day.

TABLE 4:4 (continued)

Mercury levels in organs								
Brain	Liver		Kidney		Whole Blood			
µg/g	µg/g	liver/brain	µg/g	kidney/brain	µg/g	blood/brain	blood/kidney	References
9	52	5.8	15	1.7				Takeuchi, 1961
14	82	6.0						Takeuchi, 1961
12	90	7.5	11	0.9				Yamashita, 1964
9	79	8.8	30	3.3				Yamashita, 1964
5	77	15	12	2.4				Yamashita, 1964
15	100	6.7	21	1.4				Yamashita, 1964
26	75	2.8	22	0.8				Yamashita, 1964
9.2	62	6.7	20	2.1	13	1.4	0.65	Kitamura, 1968
13	74	5.8	20	1.6				Kitamura, 1968
10	48	4.8	16	1.6				Kitamura, 1968
19	96	5.0	88	4.6				Kitamura, 1968
2.2	57	26	3.6	1.6				Kitamura, 1968
1.6	26	16	3.6	2.2	1.7	1.1	0.47	Kitamura, 1968
28	100	3.6	62	2.2	61	2.2	0.98	Rissanen, 1969
9	31	3.5	19	2.1	19	2.1	1.0	Albanus et al., 1969
33					12	0.36		Yoshino, Mozai, and Nakao 1966a
0.53	2.2	4.2	1.2	2.3				Platonow, 1968a
3.7	22	6.2	12	3.2	2.0	0.5	0.17	Platonow, 1968a
4.3	17	4.0	11	3.6				Platonow, 1968a
13	50	3.8	57	4.3	7.0	0.5	0.12	Platonow, 1968a
3.9	11	2.8	7.2	1.8				Platonow, 1968b
14	80	5.7	52	3.7				Platonow, 1968b
0.45	2.0			1.6				Piper, Miller, and Dickinson, 1971
23	83		54					Piper, Miller, and Dickinson, 1971
18	9.8	0.5	6.0	0.3	1.5	0.1	0.25	Nordberg, Berlin, and Grant, 1971
14	24	1.7	12	0.9	1.9	0.2	0.16	Nordberg, Berlin, and Grant, 1971
7.3	7.3	1.0	13	1.8				Nordberg, Berlin, and Grant, 1971
2.5	2.7	1.1	4.9	2.0	0.3	0.1	0.06	Nordberg, Berlin, and Grant, 1971

[8] Total dose ~20 to 27 mg/kg. Exposure varied between 0 and 0.6 mg Hg/kg/day. Mean exposure ~0.5 mg/kg/day.
[9] Onset of symptoms after about 21 days. Until then mean exposure about 0.5 mg/kg/day.
[10] Homogenate of liver incubated with MeHgOH.
[11] Symptoms 0 and 6 days, respectively, and killed 1 and 9 days, respectively, after completion of exposure.
[12] Killed 4 days after completion of exposure.
[13] Symptoms 37 days and killed 63 days after completion of exposure.
[14] Killed 85 days after completion of exposure. One animal showed histopathological damage in the CNS.

Östlund, to be published) and monkeys (Nordberg, Berlin, and Grant, 1971) indicate concentration differences among different parts of the CNS. In poisoned dogs, Yoshino, Mozai, and Nakao, 1966a, found higher mercury levels in the calcarine cortex than in other parts of the brain. Studies on repeatedly exposed monkeys (Nordberg, Berlin, and Grant, 1971, and Berlin, Nordberg, and Hellberg, in press) showed an accumulation of mercury in subcortical layers of the cerebellum and the calcarine area. Data on pigs indicate a decreasing mercury level in the nervous system from the cerebral cortex to the peripheral nerves (Platonow, 1968b, and Bergman, Ekman, and Östlund, to be published).

In this connection it may be mentioned that mercury can be demonstrated histochemically in brain tissue from methyl mercury poisoned persons mainly in the glia cells (Oyake et al., 1966, Hiroshi et al., 1967, and Takeuchi, 1968b).

Little is known concerning the distribution of mercury on the subcellular level after administration of methyl mercury. Yoshino, Mozai, and Nakao, 1966b, found almost all mercury in the protein fraction in the rat brain while lipid and nucleic acid fractions contained little mercury. Norseth, 1969a, and Norseth and Brendeford, 1971, studied the distribution in rat liver cells by marker enzyme technique. The highest mercury levels were found in the microsomes; lyzosomes/peroxisomes contained less.

From Table 4:5 it is evident that considerable species-related differences in distribution exist. The ratio between levels in whole blood and brain is 10 to 20 in rats. For other species the more limited data available indicate a ratio of about 1 in mice, 1 to 2 in cats, 0.4 to 0.5 in dogs and pigs, and 0.1 to 0.2 in monkeys. If the considerable variation in the ratio blood cells/plasma among different species (Section 4.2.2.1.1.1.1) is taken into account, it is obvious that the ratio plasma/brain is by far more consistent among different species than the ratio whole blood/brain.

It may be noteworthy that Miller and Csonka, 1968, observed certain differences between 2 different strains of mice in distribution of mercury after administration of methyl mercury.

Studies in mice indicate that the distribution of mercury after a single administration of methyl mercury is dose-dependent (Östlund, 1969b). In rats the distribution is constant at nontoxic doses (Norseth, 1969b, Ulfvarson, 1969a and 1970) but may show a slightly different pattern at high doses (Ulfvarson, 1969b and 1970).

Methyl mercury easily passes through the placental barrier in the species studied (Section 4.1.2). The distribution in the mouse fetus is rather even and comparable to that in the mother (Berlin and Ullberg, 1963c). A characteristic uptake occurs in the fetal lens (Östlund, 1969b).

Östlund, 1969a and b, studied the metabolism of *di-methyl mercury* in mice after inhalation or intravenous administration. The initial distribution was quite different from that of mono-methyl mercury. A rapid distribution occurred, mainly to the fat deposits and to a lesser extent to tissues containing lipophilic cells. The levels in different parts of the CNS were equal to or lower than those in the blood, which was low in concentration. In the liver and kidney, concentrations were moderate and in the adrenal cortex, fairly high. The major part of the di-methyl mercury was rapidly exhaled. After 16 hours, only mono-methyl mercury remained in the body. After 24 hours, mercury was distributed following the characteristic pattern of methyl mercury.

Intact di-methyl mercury in mice did not pass across the placental barrier at all or only to a minor extent (Östlund, 1969b, Section 4.1). The fraction of mono-methyl mercury found in the body after exposure to di-methyl mercury does accumulate in the fetus.

4.3.2.1.1.2 Ethyl and Higher Alkyl Mercury Compounds

Some data on the distribution of mercury after administration of *ethyl mercury* compounds are shown in Table 4:5. The principles used in the compilation were the same as those applied in Table 4:4. Though the data are not as uniform as those on methyl mercury, the general impression is that distribution patterns similar to those after the administration of methyl mercury can be found. This has been shown in studies comparing the distribution of methyl and ethyl mercury (Ulfvarson, 1962, Suzuki, Miyama, and Katsunuma, 1963, and Yamashita, 1964).

Ukita et al., 1969, and Takahashi et al., 1971, studied the distribution in the brain of the cynomolgus monkey up to 8 days after a single intraperitoneal or intravenous injection. Mercury accumulation was observed by whole-body radiography in the cerebral and cerebellar cortices, in the subcortical grey matter, in various nuclei, in the brain stem, and in the grey matter of the spinal

TABLE 4:5
Distribution of Mercury in Organs After Administration of Ethyl Mercury (EtHg) Compounds to Mammals

Species	No. of animals	Compound	Mercury exposure		Adminis-tration route	Exposure time, days	Time exposure to sacrifice days	Mercury levels in organs								
			Single dose mg Hg/kg body weight	Repeated administration, mg Hg/kg body weight/day				Brain		Liver			Kidney			Whole blood
								µg/g		µg/g	liver/ brain	µg/g	kidney/ brain	µg/g	blood/ brain	blood/ kidney
Mouse	5	EtHg acetate[a]	0.5		s.c.	11								44			
Rat	6	EtHgCl[b]	3		i.m.		7	>0.7<	5.3	69	>7.6<		>99<	14	>50<	0.20
	5	EtHgOH[c]			or.	20		0.32	4.5	30	14		94	8.4	26	0.28
	2[1]	EtHg salts[d]		~10	or.	46		22	25	93	1.1		4.2			
	2[1]	(EtHg)₂S		~10	or.			23	100	89	4.4		3.9			
	3	EtHgCl[e]	10		s.c.		8	1.4	11	110	9.2		79	27	19	0.25
	3	EtHg cysteine	10		s.c.		8	1.4	12	95	8.6		68	23	16	0.24
	3	EtHgCl[f]	20		s.c.		8		30	110	11		60			
	?	EtHgCl[g]	1		i.p.		8	0.3	3.3	18						
Cat	3[1]	EtHg[h]		1.0	or.	27		14	200	60	14		4.2			
	5[1]	(EtHg)₂PO₄		1.2	or.	25		10	130	120	13		12			
Calf	1[1]	EtHg-p-toluene sulfonanilide[i]		4.7	or.	38		29	50	60	1.7		2.1	7	0.24	0.12
	1[1]	EtHg-p-toluene sulfonanilide		23	or.	25		18	108	120	6.0		6.7			
	1[1]	EtHg-p-toluene sulfonanilide		47	or.	9		12	58	62	4.8		5.2	23	1.9	0.37
	1[1]	EtHg-p-toluene sulfonanilide	120		or.		3	3.2	46	29	18		9.1	5	1.5	0.17
Monkey	1	EtHgCl[g]	0.8		i.v.		8	1.3	3.0	8.6	2.3		6.6			

The references in this and the following two tables are written as alphabetical footnotes and each small letter refers to the row of data in which it is placed beside the compound as well as to the following rows until a new alphabetical footnote is introduced.

[1] Occurrence of signs of intoxication in some or all animals.
[2] 2mg Hg/1 drinking water.

[a]Suzuki, Miyama, and Katsunuma, 1963
[b]Miller et al., 1961
[c]Ulfvarson, 1962
[d]Itsuno, 1968
[e]Takeda et al., 1968a
[f]Takeda and Ukita, 1970
[g]Takahashi et al., 1971
[h]Yamashita, 1964
[i]Oliver and Platonow, 1960

cord. A pronounced accumulation was observed in the grey matter of the occipital lobe. There were indications that the mercury passed into the brain from the blood and not through the cerebrospinal fluid. A similar distribution was observed in cats by the same technique by Ukita et al., 1969.

In Table 4:6 data have been compiled on the distribution of mercury after administration of *alkyl mercury compounds other than methyl and ethyl mercury*. From the table it is evident that knowledge on higher alkyl mercury compounds is incomplete and inconclusive. In comparative studies in rats and mice the distribution of propyl mercury compounds was similar to that of methyl mercury (Ulfvarson, 1962, Suzuki, Miyama, and Katsunuma, 1963, and Itsuno, 1968). For higher alkyl mercury compounds certain differences have been reported (Suzuki, Miyama, and Katsunuma, 1964, and Takeda et al., 1968b). After *n*-butyl mercury chloride was injected subcutaneously into rats in a single dose of 10 mg/kg, the levels of mercury in the brain were only about half of those found after identical doses of ethyl mercury chloride.

Substituted alkyl mercury compounds will be discussed in Section 4.3.2.4 on other organic mercury compounds.

4.3.2.1.2 In Human Beings
4.3.2.1.2.1 Methyl Mercury Compounds

Of the alkyl mercury compounds, only methyl mercury has been experimentally studied in man. Åberg, Ekman, Falk, and collaborators (Section 4.1.2.1.2.2) in their study on the metabolism of orally administered labeled tracer doses of methyl mercury found that the distribution as measured by whole-body counting was rather constant during the period of 14 days. In 3 subjects, 9 to 11% of the total amount of mercury between the top of the head and the knees was in the head. The major part of this mercury was assumed to be in the brain. In the neck region, 3 to 7% of the radioactivity was found; in the trunk and arms, 51 to 58% (probably a major part in the liver); in the urogenital region, 11 to 14%; and in the thigh region, 16 to 22%.

Based on data from the human tracer dose experiments (Åberg et al., 1969, Miettinen et al., 1969b and 1971) indicating 10% of the total body burden in the head, probably mainly in the brain, and 1% of the total body burden in 1 l. of whole blood, it may be assumed that the ratio between the concentrations in whole blood and brain is 0.1 to 0.2, which agrees well with data on monkeys (Section 4.3.2.1.1.1).

From data on distribution of mercury in cases of methyl mercury poisoning (Section 8.1.2.1.1.3, Tables 8:1, 8:2), it is seen that the levels found in liver and kidney generally were higher than those in the brain. Several authors have analyzed different parts of the brain separately (Ahlmark, 1948, Lundgren and Swensson, 1949, Höök, Lundgren, and Swensson, 1954, Tsuda, Anzai, and Sakai, 1963, Okinaka et al., 1964, Hiroshi et al., 1967, and Tsubaki, 1971, and personal communication). No definite conclusions can be established about differences in concentrations among different anatomical regions or between grey and white matter. The only analysis of peripheral nerve tissue reported so far did not show deviations from levels found in other parts of CNS (Lundgren and Swensson, 1949).

4.3.2.1.2.2 Ethyl Mercury Compounds

Hay et al., 1963, studied the distribution of mercury in a worker who had died of ethyl mercury chloride poisoning. The level in kidney was 82 μg/g, in liver, 17 μg/g, and in different parts of CNS, 1-62 μg/g. Suzuki et al., in press, in a case of suspected ethyl mercury poisoning (Section 8.1.2.1.2.2) found 69 μg/g in the liver, 35 and 43 μg/g in the renal cortex and medulla, respectively, 13-24 μg/g in different parts of the CNS, and 9 μg/g in a peripheral nerve.

There is no information on distribution of alkyl mercury compounds other than methyl and ethyl mercury in man. *Substituted alkyl mercury* compounds will be discussed in Section 4.3.2.4 on other organic mercury compounds.

4.3.2.2 Aryl Mercury Compounds

Only data from animal experiments are available. Almost all studies on the distribution of aryl mercury compounds have been done with phenyl mercury salts.

The distribution pattern of mercury after administration of phenyl mercury compounds is far more complex than that seen after alkyl mercury compounds. Biotransformation of phenyl mercury into inorganic mercury (Section 4.2.2.2) results in redistribution or apparent redistribution. In addition, the distribution pattern is dose-dependent. After a single administration of phenyl mercury or during a repeated exposure there are definite changes in the distribution pattern.

Berlin and Ullberg, 1963b, in an autoradiographic study in mice, initially found a relatively large accumulation in the liver and, later, in the kidney. Similar observations have been made in rats (Ulfvarson, 1962, Takeda et al., 1968a). Ulfvarson, 1969a, showed that in the rat the distribution was completed 30 to 40 days after a single dose. After that time some kind of dynamic equilibrium seems to be established.

While the initial distribution of mercury after administration of phenyl mercury is similar to that seen after administration of short chain alkyl mercury compounds, later on the pattern approaches the distribution of inorganic mercury salts (Swensson, Lundgren, and Lindström, 1959a and b, Ulfvarson, 1962, Berlin, 1963b, Takeda et al., 1968a, and Ukita et al., 1969). There are, however, some definite differences, the main one involving the brain (see below).

Some data on the distribution of mercury after administration of phenyl mercury salts have been compiled in Table 4:7. The data have been selected to show the conditions at single or repeated administration, at high and low exposure, at different times after start or cease of exposure, and in various species. From the table it is obvious that the levels found in the kidney are far higher than levels in other organs. Furthermore, high levels are observed in the liver while the brain levels are low.

Within the kidney, mercury accumulates in the cortex (Bergstrand, Friberg, and Odeblad, 1958, Berlin and Ullberg, 1963b). The distribution is similar to that of inorganic mercury salts. Initially, the distribution in the liver is even, while later, mercury is located in the peripheral parts of the liver lobules (Berlin and Ullberg, 1963b).

Ellis and Fang, 1967, and Massey and Fang, 1968, after administration of phenyl mercury acetate to rats and exposure of tissue slices to phenyl mercury in vitro, studied the incorporation of mercury in different cell fractions from liver and kidney. The highest binding of mercury was observed in the nuclear fraction while the mitochondrial and microsomal fractions contained less. Piotrowski and Bolanowska, 1970, reported that in kidney homogenates from rats exposed to a single dose of phenyl mercury acetate (0.2 to 2 mg Hg/kg), mercury was found in 2 protein classes separated on Sephadex® gel. The one that contained the more mercury (70% after 3 to 7 days) had characteristics of a mercury-metallothionein complex. This complex was found also in liver, serum, and in urine.

As mentioned above, the mercury concentration in the brain is low in relation to those found in kidney and liver. Increased brain concentration can be seen after administration of 2,3-dimercaptopropanol (Berlin and Ullberg, 1963d, Swensson and Ulfvarson, 1967). Furthermore, concentration differences have been noted also within the central nervous system (Friberg, Odeblad, and Forssman, 1957, Berlin and Ullberg, 1963b, Swensson and Ulfvarson, 1968, and Suzuki, Miyama, and Katsunuma, 1971a). The levels found in the brain are comparable to those seen after inorganic mercury salts but much lower than after the corresponding exposure to short chain alkyl mercury compounds (see Tables 4:4 to 4:6).

Miller and Csonka, 1968, proved that the distribution of mercury after administration of a phenyl mercury salt differed between 2 strains of mice.

The distribution pattern of mercury after a single administration of phenyl mercury compounds is dependent upon the dose, indicating a saturation (Cember and Donagi, 1964, Ulfvarson, 1969a, and 1970). In the experiment on chronic exposure performed by Fitzhugh et al., 1950, a saturation at about 40 μg Hg/g kidney seems to have been obtained at high exposure, while in the liver, no such steady level was reached.

In Table 4:7, the ratios of mercury levels between whole blood and brain in the rat range from 0.4 to 12. Though the variation is considerable, the highest ratios generally are seen initially after high exposures, probably mainly due to the initial high blood cell concentrations (Section 4.2.2). With time, the gap between blood and brain levels decreases both after repeated administration and after a single dose. The blood/brain mercury concentration ratio is then lower than after exposure to short chain alkyl mercury compounds (Tables 4:4 to 4:6), and similar to that seen after inorganic mercury salts (Section 4.3.1). The ratio plasma/brain at exposure to phenyl mercury salt is similar to that observed after inorganic mercury salts (Takeda et al., 1968a). As is seen from Table 4:7, the ratios between mercury levels in blood and kidney vary even more than the ratios between blood and brain. The kidney levels

TABLE 4:6

Distribution of Mercury in Organs After Administration of Alkyl Mercury Compounds Other Than Methyl and Ethyl Mercury

Species	No. of animals[1]	Compound[2]	Mercury exposure		Administration route	Exposure time, days	Time exposure to sacrifice, days	Mercury levels in organs								
			Single dose mg Hg/kg body weight	Repeated administration, mg Hg/kg body weight/day				Brain	Liver		Kidney		Whole blood			
								µg/g	µg/g	liver/brain	µg/g	kidney/brain	µg/g	blood/brain	blood/kidney
Mouse	5	n-ProHg acetate[a]	0.5			11										
	3	n-ProHg acetate[b]	3		s.c.		7	0.2	15	5.2	32	13	1.1	5.5	0.03
	3	iso-ProHg acetate	3		s.c.		7	0.14	16	75	22	160	0.33	2.4	0.02
Rat	5	n-ProHgOH[c]		3	or.	21		0.21	6.1	29	21	100	9.4	45	0.45
	1*	n-ProHgBr[d]		~10	or.	46		25	32	1.3	67	2.7			
	1	iso-ProHgBr		~10	or.	10		38	197	5.2	22	0.6			
	2*	(n-ProHg)₂S		~10	or.	21		2-32	28-170	14-5.3	10-110	5-3.5			
	1	(iso-ProHg)₂S		~10	or.	66		15	30	2.0	72	4.8			
Mouse	3	n-BuHg acetate[b]	3		s.c.		7	0.23	11	48	19	83	0.86	3.7	0.05
	3	iso-BuHg acetate	3		s.c.		7	0.24	16	66	21	87	0.80	3.3	0.04
Rat	1	n-BuHgBr[d]		~10	or.	86		8.2	10	1.2	31	3.8			
	1	(n-BuHg)₂S		~10	or.	66		7.2	8.2	1.1	35	4.9			
	1	iso-BuHgBr		~10	or.	88		7.0	7.0	1.0	30	4.3			
Mouse	3	n-AmHg acetate[b]	3		s.c.		7	0.12	6.3	53	17	140	0.46	3.8	0.03
Rat	1	n-AmHgBr[d]		~10	or.	152		2.2	5.0	2.3	37	17			
	1	iso-AmHgBr		~10	or.	152		2.0	4.6	2.3	36	18			
	1	(n-HexHg)₂S		~10	or.	156		4.5	7.2	1.6	43	9.8			

[1] Occurrence of signs of intoxication in some or all animals is indicated with an asterisk.
[2] ProHg = propyl mercury
BuHg = butyl mercury
AmHg = amyl mercury
HexHg = hexyl mercury
[3] 2 mg Hg/l. drinking water

[a] Suzuki, Miyama, and Katsunuma, 1963
[b] Suzuki, Miyama, and Katsunuma, 1964
[c] Ulfvarson, 1962
[d] Itsuno, 1968

TABLE 4:7

Distribution of Mercury in Organs After Administration of Phenyl Mercury (PhHg) Compounds

Species	No. of animals	Compound	Mercury exposure					Mercury levels in organs							
			Single dose mg Hg/kg body weight	Repeated administration, mg Hg/kg body weight/day	Administration route	Exposure time, days	Time exposure to sacrifice, days	Brain μg/g	Liver μg/g	liver/brain	Kidney μg/g	kidney/brain	Whole blood μg/g	blood/brain	blood/kidney
Rat	5	PhHg acetate[a]		1	or.	365			0.05		1.7				
	5[2]	PhHg acetate		3	or.	365			1.5		40				
	5[4]	PhHg acetate		1	or.	540–730			0.25		2.3				
	5[5]	PhHg acetate		3	or.	540–730			3.3		39				
	4	PhHg acetate[b]	0.2		i.v.		1	0.51	0.52	1.0	9.1	18	0.22	0.4	0.02
	8	PhHg acetate	0.5		i.v.		4		0.6		14				
	3	PhHg acetate	0.5		or.		2		0.10		16				
	4	PhHg acetate[c]	3		i.m.		2	1	11	11	79	79			
	4	PhHgOH[d]		0.05	s.c.	18		0.03	0.36	120	33	1100	0.039	1.3	0.001
	6	PhHg acetate[e]		0.65	s.c.	42		<1	1.3	>1.3	90	>100	~0.5	<2	0.006
	2	PhHg acetate[f]	1		s.c.		7			20[6]		1000[6]		5[6]	0.01
	2	PhHg acetate[g]	3		or.		2	0.06	0.58	10	27	45	0.74	12	0.03
	3	PhHgOH[h]	0.05		i.v.		4	0.004	0.081	20	1.7	400	0.006	1.2	0.004
	3	PhHgOH	0.5		i.v.		4	0.017	0.33	20	15	880	0.04	2.4	0.003
	24	PhHg acetate[i]		7	or.	180		0.13	0.52	2.5	16	120	0.22	1.7	0.01
	24	PhHg acetate		8	or.	180		0.29	4.3	15	42	140	0.48	1.7	0.01
	3	PhHgOH[j]	0.5		or.		16	0.018	0.083	4.6	16	890	0.047	2.6	0.003
	3	PhHgCl[k]	10		s.c.		8	0.16	1.3	8.1	18	110	0.29	1.8	0.02
	3	PhHgOH[l]	5		i.v.		3	0.47	11	23	42	89	4.4	9.4	0.1
	20	PhHgOH[m]	25		s.c.		3	0.5	19		52		6	12	0.1

TABLE 4:7 (continued)

Species	No. of animals	Compound	Mercury exposure — Single dose mg Hg/kg body weight	Repeated administration, mg Hg/kg body weight/day	Administration route	Exposure time, days	Time exposure to sacrifice, days	Brain µg/g	Liver µg/g	liver/brain	Kidney µg/g	kidney/brain	Whole blood µg/g	blood/brain	blood/kidney
Guinea pig	2	PhHg dinaphtyl-methane di-sulphonate[n]		0.2	or.	180			3.5		68				
	2	PhHg dinaphtyl-methane di-sulphonate		0.02	or.	180			0.5		7				
Rabbit	2	PhHg acetate[o]	2		s.c.		40	0.010	0.6	60	6	600	0.01	1	0.002
	1	PhHg acetate[f]	0.4		s.c.		7		20[6]		200[6]		1[6]		
Dog	1	PhHg acetate[c]	3		i.v.		1	0	25		100				

[a] Fitzhugh et al., 1950
[b] Prickett, Laug, and Kunze, 1950
[c] Miller, Klavano, and Csonka, 1960
[d] Ulfvarson, 1962
[e] Gage, 1964
[f] Suzuki, Miyama, and Katsunuma, 1966
[g] Ellis and Fang, 1967
[h] Swensson and Ulfvarson, 1967
[i] Piechocka, 1968
[j] Swensson and Ulfvarson, 1968
[k] Takeda et al., 1968a
[l] Ulfvarson, 1969a
[m] Ulfvarson, 1969b
[n] Goldberg and Shapero, 1957
[o] Friberg, Odebad, and Forssman, 1957

[1] 0.1 mg Hg/kg food
[2] Pronounced histological changes in kidney in females, none in males
[3] 10 mg Hg/kg food
[4] Slight histological lesions in females, very slight in males
[5] Pronounced histological changes in females, slight in males
[6] Read from diagram
[7] 1 mg Hg/kg food
[8] 8 mg Hg/kg food

have been found to be 10 to 1,000 times higher than blood levels.

4.3.2.3 Alkoxyalkyl Mercury Compounds

Distribution of simple alkoxyalkyl mercury salts has been studied solely with methoxyethyl mercury salts. Only data from experiments on rats are available.

Published data on the distribution of mercury in rat tissues after administration of methoxyethyl mercury hydroxide are presented in Table 4:8. The comparison is meant to include, as far as possible from the scanty data available, variations in regard to intensity of exposure and to time between exposure and sacrifice. It is obvious that the levels found in the kidney are by far the highest. Fairly high levels have also been found in the liver, while brain concentrations are less prominent.

In comparing the distribution patterns after administration of phenyl mercury compounds (Section 4.3.2.2) and methoxyethyl mercury compounds, some minor differences may be noted. The change in the pattern with time after administration which was prominent in phenyl mercury exposure is less obvious in methoxyethyl mercury exposure. The initial level in the liver seems to be lower than after phenyl mercury (Ulfvarson, 1962). Later, the distribution of mercury is very similar to that seen after administration of phenyl mercury and also of inorganic mercury salts. This may be explained by the more rapid transformation of methoxyethyl mercury into inorganic mercury (Daniel, Gage, and Lefevre, 1971). There is a definite difference when compared to the pattern seen after administration of short chain alkyl mercury compounds (Tables 4:4 to 4:6).

The distribution pattern of mercury after administration of methoxyethyl mercury is dose-dependent. Studies by Ulfvarson, 1969a, indicate that saturation phenomena may occur in some organs.

As shown in Table 4:8, the mercury levels in whole blood were 1 to 75 times higher than the levels in brain. The highest levels were measured shortly after a heavy single dose, while at lower exposures and later after a single high dose, the blood/brain ratio ranged from about 1 to about 4. The kidney levels were 2 to 1,000 times higher than the blood levels.

Norseth, 1967, and 1969a, using marker enzyme techniques, studied the subcellular distribution in the rat liver of mercury administered as methoxyethyl mercury acetate. The distribution of mercury was similar to that seen after inorganic mercury but certain differences were noted in relation to methyl mercury dicyandiamide. *Substituted alkoxyalkyl mercury* compounds (mercurial diuretics) will be discussed in Section 4.3.2.4 on other organic mercury compounds.

4.3.2.4 Other Organic Mercury Compounds

A few notes on the distribution of mercurial diuretics may be of interest. Chlormerodrin, meralluride, mersalyl, and mercaptomerin have been shown to give high levels rapidly after administration in the kidney of rats (Borghgraef and Pitts, 1956, and Anghileri, 1964), rabbits (Aikawa, Blumberg, and Catterson, 1955) and dogs (Kessler, Lozano, and Pitts, 1957, Borghgraef and Pitts, 1956, and Vostal, in press). Within the dog kidney the highest mercury levels occur in the renal cortex (e.g., Greif et al., 1956). The mercury in the cortex has been stated to be located in the convoluted tubules. The levels in other organs are considerably lower than those in the kidney.

Studies in man dosed intravenously with about 0.01 mg Hg/kg body weight as chlormerodrin (Goldman and Freeman, 1971) have been reported. There is an accumulation of mercury in the kidney (McAfee and Wagner, 1960, Blau and Bender, 1962, Kloss, 1962, and Reba, Wagner, and McAfee, 1962). Similarly, renal accumulation has been studied at 400 to 800 times higher doses, the mercury being present mainly in the cortex (Aikawa and Fitz, 1956). Other organs contain little mercury. A considerable part of the mercury in the kidney is rapidly excreted in the urine (see Section 4.4.2.4.2).

The pattern of distribution of mercury after intravenously administered MHP to rats is dose-dependent (Fischer, Mundschenk, and Wolf, 1965). In a short-term study in the dog the highest mercury levels were found in the spleen (Kessler, Lozano, and Pitts, 1957).

The distribution of mercury from labeled MHP administered intravenously directly or after mixing with blood in doses ranging 0.05 to 0.1 mg Hg/kg body weight has been studied in man by radio-activity scanning. There is a rapid increase in the radioactivity over the spleen and over the liver

TABLE 4:8

Distribution of Mercury in Rat Organs After Administration of Methoxyethyl Mercury Hydroxide Compounds

No. of animals	Mercury exposure		Administration route	Exposure time, days	Time exposure to sacrifice, days	Mercury levels in organs								References
	Single dose mg Hg/kg body weight	Repeated administration, mg Hg/kg body weight/day				Brain µg/g	Liver µg/g	liver/brain	Kidney µg/g	kidney/brain	Whole blood µg/g	blood/brain	blood/kidney	
4	0.5		s.c.			0.009	0.25	28	27	3000	0.033	3.7	0.001	Ulfvarson, 1962
4		0.5	i.v.	12	4	0.018	0.30	17	17	940	0.068	3.8	0.004	Swensson and Ulfvarson, 1967
4	0.5		i.v.		4	0.022	0.24	11	12	550	0.054	2.5	0.005	Ulfvarson, 1969a
4	0.05		i.v.		4	0.009	0.061	6.8	2.4	270	0.009	1	0.004	
3	3		i.v.		3	0.15	4.0	27	19	120	1.1	75	0.06	
3	0.3		i.v.		3	0.036	0.17	4	7.6	210	0.082	2.3	0.01	
3	0.03		i.v.		3	0.027	0.047	1.7	1.2	45	0.044	1.6	0.04	
3	3		i.v.		12	0.076	0.51	6.7	0.73	9.6	0.066	0.9	0.09	
3	0.3		i.v.		12	0.007	0.021	3	3.4	490	0.004	0.6	0.001	
3	0.03		i.v.		12	0.002	0.005	2.5	0.42	210	0.001	2	0.002	
20	20		s.c.		3	0.2	44	220	29	140	11	55	0.4	Ulfvarson, 1969b

(Wagner et al., 1964). The maximum spleen level is obtained within a few hours; then there is a rapid decrease in the splenic count and, to a lesser extent, over the liver, and a gradual increase in the kidney (Fischer, Mundschenk, and Wolf, 1965). The kidney has been stated to contain about 75% of the total body burden and the liver, 25% (Croll et al., 1965). The mercury level in the kidney decreases slowly (see Section 4.4.2.4).

4.3.3 Summary

The distribution of *inorganic mercury* is extremely differentiated. The data on distribution are very limited for human beings and the following summary is based mainly on animal data. With the exception of the brain, the distribution is similar after exposure to mercuric mercury and to elemental mercury vapor. The distribution pattern changes so that relatively more mercury is found in the kidneys and the brain with the passage of time after a single exposure. After vapor exposure the concentration in the brain is about 10 times higher than after administration of a corresponding dose of mercuric mercury. Generally the kidney contains the highest concentration of mercury, the liver has the next highest, and thereafter the spleen, the brain, and other organs. The blood contains a relatively large amount of mercury soon after exposure, but the concentration diminishes rapidly with time.

Also, within the various organs a differentiated distribution can be seen. In the kidney mercury is retained predominantly in the tubules. In the brain, the cerebral and cerebellar cortex and certain nuclei take up the mercury.

The distribution of mercury at exposure to *mono-methyl* and *mono-ethyl mercury* salts has been studied in several species, including man. The distribution pattern, far simpler than at exposure to inorganic mercury salts, is relatively unaffected by dose level and time after a single exposure or exposure time. The mercury levels found in different organs differ much less than at exposure to inorganic mercury salts. The highest levels are obtained in liver and kidney. The levels found in CNS are lower but considerably higher than after corresponding exposure to inorganic mercury salts. There is a time lag in brain accumulation of mercury after single administration. High levels are also present in blood cells. In human tracer dose experiments with methyl mercury, about 10% of the total body burden was found in the head, probably mainly in the brain, and 5 to 10% in the blood. In monkeys, accumulation of mercury has been observed in the subcortical layers in the cerebellum and in the calcarine area. Mono-methyl mercury and mono-ethyl mercury pass the placental barrier.

As studied in mice, *di-methyl mercury* was rapidly distributed to the fat deposits and to a lesser extent to the liver and kidney. The main part of the dose was rapidly exhaled in unchanged form while a small part was metabolized into mono-methyl mercury and distributed as such in the manner described above.

The information on the distribution of mercury after *aryl mercury* salts is less complete than that on short chain alkyl mercury compounds. The studies have dealt with phenyl mercury salts almost exclusively. No information is available about the distribution in man. The distribution pattern as seen in animals is far more complex than after short chain alkyl mercury compounds. It is dependent upon the dose level and the exposure time or the time between a single administration and sacrifice. Initially after administration the distribution pattern has certain similarities to that seen after administration of short chain alkyl mercury compounds, while later on it is more similar to that seen after administration of inorganic mercury salts. This is due to breakage of the covalent carbon-mercury bond with a subsequent re-distribution of the inorganic mercury. The level in the kidney is much higher than in any other organ. Within the kidney higher levels are found in the cortex than in the medulla. The levels in CNS are much lower than those seen after corresponding exposure to short chain alkyl mercury compounds but similar to those after inorganic mercury salts.

Information on the distribution of simple *alkoxyalkyl mercury* compounds is confined to methoxyethyl mercury salts in rats and mice. The complex distribution pattern is similar to that seen after exposure to phenyl mercury salts. The pattern is dependent upon dose and time. The highest levels occur in the kidney, while the CNS concentration is low. The redistribution is explained by a rapid breakdown of the carbon-mercury bond.

Mercurial diuretics are distributed to a very large extent to the kidney, mainly to the cortex.

4.4 Retention and Excretion
4.4.1 Inorganic Mercury

From what has been said about the biotransformation and transport of mercury in the body it is conceivable that the retention and excretion of different forms of inorganic mercury approach the excretion of mercuric mercury at the time after the administration at which conversion to this form of mercury has taken place. In accordance with this concept, the conditions for mercuric mercury will be described first and with these as background, the other forms of inorganic mercury will be considered. Data on the usefulness of blood or urine values in prevention and control of occupational exposures will be dealt with mainly in Chapter 7; the theoretical background will be outlined in Section 4.5.

4.4.1.1 Mercuric Mercury
4.4.1.1.1 In Animals
4.4.1.1.1.1 Retention and Risk of Accumulation at Repeated Exposure

The whole-body retention is best illustrated by the biological half-life of mercury. Studies on this entity have been made by daily measurements of total fecal and urinary excretion or by direct whole-body measurements of radioactive mercury in the animal. The work of Prickett, Laug, and Kunze, 1950, and Ulfvarson, 1962, falls within the first category. Their results show that mercuric mercury is eliminated from rats relatively fast in comparison with methyl mercury and at a rate similar to phenyl mercury.

Direct whole-body measurements are the most accurate way of describing the biological half-life. Such measurements have been performed on several species. Rothstein and Hayes, 1960, found an elimination curve for i.v. or i.m. injected rats following 3 consecutive exponential curves with half-lives of 5 days, 25 to 36 days and 90 to 100 days, respectively. Similar elimination curves for rats have been found by Cember, 1969, and Phillips and Cember, 1969. These authors found that the elimination rate was to some extent dose-dependent so that high doses tended to be faster eliminated than low ones. Ulfvarson, 1969a, also reported this effect of different doses.

A whole-body retention curve for mice has been reported by Berlin, Jerksell, and von Ubisch, 1966. From their data a half-life of 2 to 3 days can be calculated for i.v. injected mercuric mercury. Thus the elimination rate is faster for mice than it is for rats. In addition to species differences, strain differences may also influence the retention to a certain extent as shown for 2 strains of mice by Miller and Csonka, 1968. In the monkey *(Saimiri sciureus)* the half-life of i.v. injected $Hg(NO_3)_2$ is similar to or somewhat slower than the one observed for rats (Berlin and Nordberg, unpublished data).

Although the biological half-life of mercury in the whole body gives a general idea about the kinetics of mercury turnover, it is the biological half-life in critical organs that is of importance for the accumulation and risk of intoxication at repeated exposure. The change of the distribution pattern with time after a single injection of mercuric mercury (see Section 4.3.1.1) reflects the varying rates of elimination from different parts of the body. Organs which have uptake and retention conditions favoring a high accumulation of mercury at a particular exposure are the same organs which are critical at that kind of exposure. Thus at acute exposure or even at prolonged exposure to mercuric mercury, the kidney invariably contains the highest concentrations of mercury but at a prolonged exposure certain parts of the brain also are apt to reach high concentration levels. Brain accumulation is particularly prominent at Hg^0-vapor exposure, which will be considered in Section 4.4.1.3.1.1. Retention in, for example, the thyroid and the testicles is appreciable both at Hg^{2+} and Hg^0 exposure.

For a more precise evaluation of the risk of accumulation in different critical organs at different kinds of exposure, a mathematical model for the kinetics of mercury exchange among different body compartments and excretion would be useful. Such a model has proved to be of value for the medical evaluation of methyl mercury toxicity (Figure 4:2 and Section 8.1.2.1.3), but in that case the properties of methyl mercury permitted the use of a 1 compartment system. The different uptake and elimination rates for different organs in the case of mercuric mercury make a multicompartment model necessary. Cember, 1969, postulated a 4 compartment model (kidney, liver, other tissues, and excretion reservoir) and estimated numerical values for the turnover rates among different organs in the rat after a single exposure. With the assumption that Cember's equations are generally applicable, his expression for the quantity of mercury in the kidney = Q_s for a single dose has been used to form an expression for the accumulated amount in the kidney after a certain time of repeated daily exposures. The following expression has been obtained for the accumulated quantity of mercury in the kidney after n days:

$$Q_{s_n} = Q_0 \cdot \frac{f_s K_t}{K_s - K_t} \left(\sum_{\Theta=0}^{\Theta=n} e^{-K_t \Theta} - \sum_{\Theta=0}^{\Theta=n} e^{-K_s \Theta} \right)$$

A key to the symbols and to the numerical values for the rat is

- Θ = Time in days
- Q_0 = Injected dose daily
- K_t = Turnover rate for tissue compartment 0.46/day
- K_s = Turnover rate for the kidney 0.035/day
- $f_s K_t$ = Fraction of mercury transferred per unit time from tissue compartment to kidney
- f_s = 0.45

The corresponding accumulation curve is seen in Figure 4:1. To allow comparisons, experimental data from 2 studies involving repeated exposure are also plotted in Figure 4:1. A clear difference between the theoretical curve and Friberg's (1956) data is seen, whereas Ulfvarson's (1962) data approximately follow the theoretical curve. Friberg's data on excretion (for details, see Section 4.4.1.1.1.2) also showed an earlier equilibrium than predicted by the theoretical accumulation curve. The explanation for that is a shorter half-life during exposure than after exposure (Friberg, 1956). As the theoretical curve was deduced from turnover constants for the kidney obtained from single exposure experiments, these circumstances can be responsible for the difference. The fact that the experimental data by Ulfvarson follow the theoretical curve better may be a question of dosage, because Ulfvarson gave only about 1/10 of the daily dose given by Friberg. As will be seen from data given in Chapter 7, it is impossible to predict a critical level for kidney damage due to inorganic mercury because of the limitations of these available data. It seems, however, in view of the data reported by Fitzhugh et al., 1950, and Ashe et al., 1953, that 200 µg/g in the kidneys of the rats in Friberg's 1956 study is substantially above the critical level. This may be the explanation for the deviation of Friberg's data from the theoretical curve since it has been shown that excretion of renal tubular cells is concomitant with the development of histological changes in the kidney tubules (see Section 7.2.2.1). With all probability such cell excretion also means an increased excretion of mercury. With the assumption of a critical kidney level of 40 µg/g and a kidney weight of 1.5 g in a 200 g rat, an absorbed dose of about 25 µg/kg per day would be necessary to reach the critical level in about 100 days according to Figure 4:1. These values are offered only to exemplify how the given mathematical expression could be used when the critical organ level is known. However, the present uncertainties surrounding both the validity of the theoretical curve and the critical level make definite statements concerning these matters impossible, but it would be of great benefit to get more experimental data.

From the discussion above, it is evident that even for the kidney, for which data are relatively abundant, it is difficult at present to use a mathematical expression for calculation of the critical intake levels at repeated exposure. For other organs which are critical in mercury poisoning, especially the brain, the situation is still more complicated by greatly differing rates of turnover for different parts of the brain. It can be said, however, that the turnover of mercuric mercury in the main parts of the rat brain is as slow as or slower than that in the kidney after single exposure (see Section 4.3 and Table 4:1) so that the accumulation curve will be at least as prolonged as indicated by the theoretical curve in Figure 4:1. For discussion of the brain accumulation at Hg^0-vapor exposure, see Section 4.4.1.3.1.1.

4.4.1.1.1.2 Excretion
4.4.1.1.1.2.1 Urinary and Fecal Excretion

Mercuric mercury (Hg^{2+}) is excreted from the body mainly by the feces and urine but routes such as exhalation, milk, sweat, and hair may also contribute.

Prickett, Laug, and Kunze, 1950, compared the excretion in rats after an oral and after an intravenous dose (0.5 mg Hg/kg) of mercuric acetate. They found 80% of the dose in the feces and 1% in the urine during 48 hours after the oral dose and 10% in the urine and 18% in the feces during 48 hours after the injection. The large part in feces after an oral dose mainly reflects

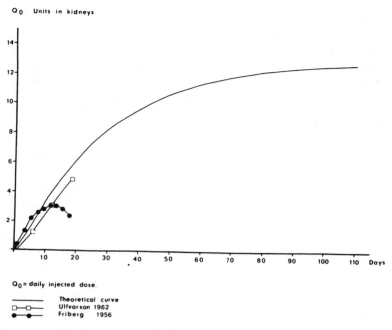

FIGURE 4:1. Accumulation of mercury in rat kidney.

unabsorbed mercury and the 1% in urine comes from the absorbed (see also Section 4.1.1.2.2).

Adam, 1951, found 39% of the dose excreted in the urine and 27% in the feces 14 days after i.v. injection of ^{203}HgCl$_2$ (1 mg Hg/kg) into a rabbit.

Rothstein and Hayes, 1960, also used radioactive mercury ^{203}Hg in their studies in rats on excretion of mercury after a single intravenous injection (0.25 mg Hg/kg) of mercuric nitrate. During the first 9 days the fecal excretion exceeded the urinary excretion, but after that time the urinary excretion prevailed. Because of the large part excreted during the first 9 days, the cumulative excretion in the feces exceeded the cumulative urinary excretion for the whole period of study (54 days). The authors found that intravenously and intramuscularly injected mercuric mercury was excreted at a similar rate. Takeda et al., 1968a, found a high fecal excretion during the first days after a subcutaneous injection of ^{203}HgCl$_2$ (3 mg Hg/kg) in the rat, and later, about equal amounts in feces and urine. The excretion rate in percent of the initial dose was about the same for different doses of mercury (Rothstein and Hayes, 1960) but with higher doses a larger part of the mercury tended to be excreted in the urine compared to the feces (doses 0.025 to 0.25 mg Hg/kg tested).

Cember, 1962, showed that the dose was of importance for the route of elimination of inorganic mercury, thus confirming the above mentioned observation by Rothstein and Hayes, 1960. The same is true of the results reported by Phillips and Cember, 1969, who used intraperitoneal injection. When 0.01 mg/kg was injected, cumulative fecal excretion exceeded cumulative urinary excretion, but when 0.5 mg/kg was given, the opposite happened. Phillips and Cember reported that the total excretion rate was dose-dependent so that an increasing elimination rate was observed with increasing dose. Ulfvarson, 1969a, made a similar observation.

Friberg, 1956, studied the excretion of ^{203}Hg in rats after repeated subcutaneous injection of mercuric chloride at a daily amount of 0.5 mg Hg/kg 7 days of the week. The excretion of mercury in urine and feces reached a relatively constant level after about 2 weeks at which time the output of mercury roughly equalled the administered dose. At this equilibrium stage about 70% was excreted in the urine and about 30% in the feces. Friberg noted a periodic variation in the excretion of mercury in the urine, but no corresponding variability in the fecal excretion.

From the data by Ulfvarson, 1962, on rats s.c. injected with 0.1 mg Hg/kg every second day, it can be calculated that 27% of the total amount injected during the third week was excreted in the urine and 44% via the feces, meaning that the fecal route dominated also at this injection period. As the daily dose given by Ulfvarson, 1962, was only about 10% of the dose given by Friberg, 1956, this is in agreement with studies on single injection, that proportionally more mercury is excreted in the urine with higher doses.

4.4.1.1.1.2.2 Mechanism for Fecal and Urinary Excretion

The mechanism behind fecal excretion of mercury has not been studied as intensely as the urinary one. The limited amount of data available on fecal excretion will be dealt with first in this section.

After injection of ^{203}HgCl$_2$, Berlin and Ullberg, 1963a, showed in the mouse that mercury accumulated in the salivary glands but was cleared from this organ at the same rate as for blood. Mercury appeared in the colonic mucosa very soon after injection and was also found in other mucous membranes of the alimentary tract and in the bile. These data are in favor of a direct transfer of mercury to the contents of the alimentary tract via the mucous membranes of the gastrointestinal tract. However, as mercury gives rise to salivation, the question as to whether mercury enters the gastrointestinal tract mainly by way of the salivation has been discussed since long ago.

Lieb and Goodwin, 1915, found mercury in the gastric contents of rabbits and cats in spite of ligation of the esophagus. They concluded that mercury found in the gastrointestinal tract was not derived from the saliva. Witschi, 1965, studied the mechanisms behind the intestinal excretion of mercury and found that mercury was released through the duodenum, the jejunum, and the colon. Their results indicated that the process of excretion was dependent upon the plasmoenteral circulation.

Friberg, 1956, furnished some data which have a bearing upon the mechanisms of urinary excretion of mercury. After a longer period of repeated injections of radioactive ^{203}Hg, one group of rats was given no further treatment while another group received continued treatment but with nonradioactive mercury. The latter condition increased considerably the excretion of radioactive mercury in urine and diminished the concentration of radioactive mercury, especially in the kidney, in comparison with the control group not receiving any more treatment. These data show that at least a part of the mercury excreted in urine must be derived from the mercury accumulated in kidney tubules.

Berlin and Gibson, 1963, studied rabbits during times up to 5 hours with differing rates of intravenous infusion of mercuric chloride (^{203}Hg, 0.1 to 1 mg Hg per 3 to 4 kg rabbit). The extraction of mercury from the renal arterial blood was found to be 10% or less. About 50% of the total dose infused was taken up by the kidneys but less than 10% was excreted in the urine. Urinary excretions of mercury and blood concentrations were found to be correlated, but there was no correlation between the amount of mercury accumulated in the kidney and the urinary excretion of mercury. In additional experiments, 1 of the ureters was ligated and the accumulation in the kidneys compared after the infusion was completed. About 12% of the total infused dose was found in the kidney with the ligated ureter, and about 15% was in the nonligated kidney. The small difference was explained by decreased blood flow following ligation of the ureter. It would have been much larger if accumulation had been a result of glomerular filtration of mercury. The uptake of mercury in the kidney was, therefore, considered to have occurred directly from the blood. In the same study the ultrafiltrability of mercuric mercury added in vitro was tested. Less than 0.1% of the mercury was found in the ultrafiltrable fraction according to the figures given in the paper, and the authors concluded that the fraction was less than 1%.

Dreisbach and Taugner, 1966, made similar experiments on rats after intramuscular injection of HgCl$_2$. The ultrafiltrability of plasma mercury was found to be 1.13 ± 0.2% (16 determinations on 8 animals). Calculations of the part of mercury which had been filtered through the glomeruli revealed that this was well above the amount excreted in the urine, but even the total nonprotein-bound mercury which had passed the kidneys was much less than the amount accumulated in the kidney tissue. These findings are the same as those by Berlin and Gibson, 1963. In the same work Dreisbach and Taugner studied the uptake of mercury in the kidneys both by measurements and by autoradiography after the ligation of the ureter on 1 side. They found that this treatment decreased the mercury uptake in the ligated kidney to about 30% of the value in the nonligated kidney (they removed the kidneys 10 minutes after the injection of mercury). The autoradiograms showed a difference in the accumulation pattern in the kidney. In the ligated side no mercury was taken up in the corticomedullary border (corresponding to the straight part of the proximal convoluted tubules), whereas a very prominent accumulation was seen in the nonligated kidney. These results oppose those found by Berlin and Gibson, 1963. They are also incompatible with the finding of a filtrable part of the plasma mercury which is by far too small to explain any important part of the kidney accumulation of mercury. Dreisbach and Taugner, admitting difficulties in explaining their findings, concluded that the true ultrafiltrable fraction in blood must be larger than what was obtained by ordinary measurements. Similarly low ultrafiltrable proportions were also found by several other investigators (Kessler, Lozano, and Pitts, 1957, Clarkson, Gatzy, and Dalton, 1961, and Gayer, Graul, and Hundeshagen, 1962) who performed measurements by the same methods. They were also considered unreliable. Dreisbach and Taugner judged that the initial mercury uptake by the proximal convoluted tubule of the renal cortex is not only directly from the blood via the basal membrane, but also by filtration and reabsorption.

Vostal and Heller, 1968, used the avian kidney for an isolated evaluation of tubular mechanisms. By injection of mercuric ^{203}Hg ions in the vena porta renalis connected with the venous system of the leg in birds, they showed that transtubular transfer of Hg^{2+} occurred. No quantitation in relation to glomerular filtration was made.

Gayer, Graul, and Hundeshagen, 1962, used the stop flow technique in studies on dogs, and Mambourg and Raynaud, 1965, used the same technique with rabbits injected intravenously with ^{203}Hg or ^{197}Hg mercuric chloride. The excretion curves obtained had an initial peak and a second ascending component appearing later. The first peak appeared simultaneously with the appearance of radioactive sodium (^{24}Na) injected at the same time as the mercury. This finding shows that there is a tubular transfer of mercury at a zone in the tubule which is also permeable to Na. From a quantitative point of view the initial peak is relatively unimportant. By far the largest part of the urinary excretion of mercury corresponds to the second ascending component which appears at approximately the same volume as inulin but differs from the inulin curve by not reaching a maximum. The authors concluded that mercury was not excreted by glomerular filtration, but probably by a delayed tubular mechanism which made the curve coincide with the inulin curve.

Piotrowski et al., in press, separated urine from ^{203}HgCl$_2$-injected rats by gel chromatography. They found the mercury to be bound mainly to substances with large molecular size.

It is difficult to draw definite conclusions as to the mechanism for renal excretion and accumulation of mercury, because of the variations in the experimental findings reviewed in the foregoing account. Especially the different influences of stoppage of the glomerular filtration by ligation of the ureter must be clarified. At this point the most likely explanation for the differing findings is that different time intervals between stoppage of filtration and kidney removal have been studied. Thus it is possible that during the first minutes after injection glomerular filtration will be of greatest importance for kidney accumulation, but later direct tubular uptake from the blood will dominate. Concerning the mechanism for mercury excretion by the kidneys into the urine, definite conclusions are impossible.

4.4.1.1.1.2.3 Other Routes of Elimination

It was mentioned above that elimination by the fecal and urinary routes roughly equalled the injected amount at the equilibrium stage (e.g., Friberg, 1956), implying that other routes of elimination will be of comparatively minor importance. Clarkson and Rothstein, 1964, have shown that a small part of the mercury injected intracardially as ^{203}Hg(NO$_3$)$_2$ in rats was eliminated from the body in the form of volatilization from the lungs and the body surface. The total amount excreted in this way was about 10% during the first day and later amounted to an average of about 4% of the total excretion from the animals. In the first hours after injection when the blood levels were high, excretion from the lungs was high but thereafter the excretion was about equally divided between the lungs and the body surface.

Berlin and Ullberg, 1963a, showed mercury accumulation in the mammary gland after intravenous injection of HgCl$_2$ into the mouse. They also showed accumulation in the skin. Thus, losses via skin and hair and by lactation will probably make a contribution to the elimination of mercury from the body. Though for certain organic mercury compounds (see Section 4.4.2) the skin and the fur are important routes of elimination, these possibilities are of minor significance for inorganic mercury.

4.4.1.1.2 In Human Beings

Whole-body retention studies of radioactive inorganic mercury in man recently have become available. In addition, we have some knowledge of the whole-body retention and excretion of inorganic mercury resting upon a few incomplete excretion studies.

Rahola et al., 1971, and Miettinen, in press, studied the retention of a single dose of ^{203}Hg after oral ingestion of inorganic mercury in 5 male and 5 female volunteers. Eight subjects consumed the radioactivity bound to calf liver protein and the other two consumed it in the form of unbound ^{203}Hg(NO$_3$)$_2$ in water solution. The amount of radioactivity consumed was 4 to 14 μCi per person corresponding to about 6 μg Hg per person. Whole-body counting was performed during a period of 3 to 4 months. It was found that 85% of the radioactivity passed out with the feces during the first 4 to 5 days, representing mainly unabsorbed mercury (see Section 4.1.1.2.2). The biological half-life of the 15% retained, calculated on the basis of the whole-body measurements, was 42 ± 3 days for the whole group. For the women the average biological half-life was 37 ± 3 days and for the men, 48 ± 5 days. There was no clear difference in biological half-life between protein bound and nonprotein bound mercury. Based on the reported half-life, about 20% of the originally retained dose must still have been in the body after 90 days, when the experiment had ended. Nothing is known about the biological half-life of this 20%. It may be of relevance that the dose of mercury used (about 0.1 μg/kg) is extremely low. In view of the observation in animals (Section 4.4.1.1.1.1 and Section 4.4.1.1.1.2.1) that the elimination rate is to some extent dose-dependent, it would not be unreasonable to expect slightly different elimination rates from human beings exposed to higher doses. Sollmann and Schreiber, 1936, reported a total urinary mercury elimination of 1 to 10 mg during 4 days in 4 orally poisoned persons. Because of the severe kidney damage which even gave rise to anuria, the figures do not tell anything about the rate or percentage of elimination via the urine in moderate dosing.

Some data are recorded on excretion from the past when injections of mercuric mercury were used extensively in the treatment of syphilis. As early as 1886 Welander showed that mercury was present in both urine and feces after such injections. During daily intravenous injections of HgCl$_2$, Bürgi, 1906, (quoted by Lomholt, 1928) found that a steady state of urinary excretion was reached already a couple of days after the initiation of the treatment. When the dose was increased, the urinary excretion of mercury also increased up to daily dose levels of 5 mg HgCl$_2$/day corresponding to a 24-hour urinary excretion of 2 to 2.4 mg. A further increase of the injected doses did not give a corresponding increase of urinary excretion. Excretion studies after intramuscular injection of water soluble mercuric salts were performed by Bürgi, 1906, and Lomholt, 1928. In 2 patients given 5 and 6 intramuscular injections, respectively, of mercuribenzoate, 100 mg once a week, Lomholt, 1928, reported a generally increasing urinary mercury excretion during the first weeks of the study, which included continuous measurements of daily urinary and fecal mercury excretion. Especially during 3 consecutive days following an injection, the urinary mercury excretion was high, reading above 3 mg/day on several occasions. For fecal excretion no clearly corresponding trends appeared. Lomholt found in both cases 3 to 4 times more mercury in the urine than in the feces, as an average for the whole study, and he recovered 35 to 49% of the injected amount in urine and feces. Lomholt, 1928, also included some urinary excretion measurements from Bürgi, 1906, in his report. Two persons were given 10 mg each of HgCl$_2$ daily for 20 and 30 days, respectively. The urinary mercury excretion rose continuously during the experiment and reached about 2.2 mg/day in the 20-day subject and 3 mg/day in the 30-day subject. One of the persons was studied for a week after the termination of the mercury treatment and in that time the mercury excretion diminished to almost half of the maximum value. Twenty-five to twenty-eight percent of the injected mercury was recovered in the urine. These studies tend to show an accumulation of mercury in the body continuing beyond a 30-day period, in consistence with a relatively long biological half-life for the body as a whole. Of course the validity of these old studies may be questioned because of the analytical errors that must have been committed at that time. However, the relation between daily absorbed amount and excreted amount is relatively consistent with more recent studies on human subjects exposed to Hg^{2+} (see above) or exposed to mercury vapor (see Sections 4.4.1.3 and 7.1).

Sodee, 1963, studied the excretion of ^{197}Hg after intravenous injection of 100 μCi. Unfortunately, Sodee did not report the dose of mercury or the method used for obtaining his results, so the validity of his data cannot be judged. The 72-hour urinary excretion was stated to be 75% of the administered dose. These excretion data from intravenously administered mercuric mercury indicate that a large portion of the total excretion takes place via the urine.

In the saliva and the sweat, Lomholt, 1928, found mercury after injection of mercuric mercury but the amounts were small so that it is unlikely that these routes contribute significantly to the elimination of mercury. Of course the material is admittedly limited.

The limited observations on the distribution of mercury in the human body after inorganic mercury exposure (Section 4.3) are the only ones available and from them it is not possible to calculate precisely the accumulation risk for critical organs. Accumulation of mercury, especially in the kidney, has been observed after brief exposure, indicating that the conditions in the human organism do not differ in principle from those more thoroughly documented in animals. If the animal data are also taken into account, it can be said that there is probably a predominant risk of accumulation in the kidney at prolonged exposure to salts of mercuric mercury. This is in accord with the status of the kidney as the critical organ in exposure to inorganic mercury salts.

4.4.1.2 Mercurous Mercury

Almost no reliable quantitative data concerning this form of mercury are available. Lomholt, 1928, found mercury in both the urine and the feces of rabbits injected intramuscularly with suspension of calomel. He made similar studies in human beings injected in the same way as a treatment for syphilis. Results from animals and man indicate a rapid rise in the fecal excretion of mercury initially after the first injection, but a urinary excretion in excess of the fecal one later in the series of injections.

4.4.1.3 Elemental Mercury
4.4.1.3.1 In Animals
4.4.1.3.1.1 Retention and Risk of Accumulation at Repeated exposure

The biological half-life after single exposure to mercury vapor has been followed by Hayes and Rothstein, 1962. They found an elimination curve similar to that for mercuric mercury (see above) with a first exponential curve corresponding to a half-life of 4.5 days and a second one corresponding to about 20 days.

In the mouse, Berlin, Jerksell, and von Ubisch, 1966, found the same rate of whole-body elimination for elemental mercury as for intravenously injected mercuric mercury in one series and a small tendency toward a longer half-life for inhaled mercury vapor in another series. The half-life in this study was about three days, which was the time calculated also from the data given by Magos, 1968. Also in the squirrel monkey, the elimination rate seemed to be roughly equal for inhaled mercury vapor and injected mercuric mercury (Berlin and Nordberg, unpublished data).

The principal considerations in regard to the accumulation risks have been dealt with in the section on mercuric mercury (Section 4.4.1.1.1.1). For the kidney it is probable that the same accumulation curve is valid as for exposure to salts of mercuric mercury. Hg^0 vapor exposure will be given a special section here because of the prominent uptake of mercury in the brain seen after that type of exposure (see Distribution, Section 4.3.1.1.3), which is in concordance with the fact that severe brain damage occurs at prolonged exposure of animals to mercury vapor (Chapter 7). This prominent uptake in combination with the slow rate of turnover of mercury in the brain regardless of whether the exposure is to Hg^{2+} or, Hg^0 vapor, makes it especially susceptible to accumulation at repeated exposure. This slow half-life of mercury in the brain has been illustrated in several studies. Significant are those for the mouse by Berlin and Ullberg, 1963a, and Magos, 1968, and for the rat by Gage, 1961a. In the latter, about 20% of the mercury remained in the brain 6 months after the termination of exposure, whereas the corresponding value for the kidney was only about 1.5% (see also kidney/brain ratios in Table 4:2).

The accumulation curve thus will probably be more prolonged than the kidney curve even when whole brain concentration values are considered.

The markedly differentiated pattern of mercury distribution among different parts of the brain with considerably slower elimination from specific structures compared to others (see Section 4.3.1.1) makes some parts of the brain even more likely to accumulate damaging concentrations of mercury at prolonged exposure than indicated by whole brain concentration measurements. It is not surprising that the brain is the critical organ after chronic exposure to mercury vapor, but the presently available data on retention properties in the specific brain structures are not precise enough to justify their use in calculations of accumulation and critical intake levels.

4.4.1.3.1.2 Excretion

Hayes and Rothstein, 1962, used radioactive ^{203}Hg in their studies on the excretion of mercury in rats after exposure for 5 hours (1.4 mg Hg/m^3). During the first 2 days a fast excretion phase was seen, in which about 4 times more mercury was excreted in the feces than in the urine. Later there was an increase in the urine, but still more than twice as much was excreted in the feces as in the urine. The authors concluded that the excretion after mercury vapor exposure was the same as after injection of mercuric mercury (see above, Section 4.4.1.1.1.2).

Ashe et al., 1953, exposed rabbits to mercury vapor 0.86 mg Hg/m^3 7 hours/day, 5 days/week. The mercury excretion in the urine increased continuously during the first 4 weeks of exposure and then reached an equilibrium, about 0.13 mg/l. This level was maintained until the 12th week, when the exposure was discontinued. Then the mercury concentrations in the urine fell to about 1/3 after 2 weeks and 1/6 6 weeks after exposure.

Gage, 1961a, exposed rats to mercury vapor, 1 mg Hg/m^3, continuously for 28 days. He found a progressive increase in the urinary mercury excretion during the first 10 days. This was followed during the remaining days of exposure by a daily excretion fluctuating around a mean value of about 70 µg Hg/rat/day. The fecal excretion also showed variation around a mean value of about 15 µg/day. The equipment used by Gage does not exclude the possibility of contamination of food or excreta by the mercury vapor. Gage used a dithizone method to analyze mercury in the air and in excreta. Regardless of the possibility of errors, the higher percentage of urinary mercury in the study by Gage in comparison to the one by Rothstein and Hayes mentioned above may be explained otherwise. The explanation may be found in the more prominent excretion of higher doses of mercury in the urine in comparison to the feces observed for mercuric mercury and the increased exchange and excretion of mercuric mercury from the kidneys at repeated exposure described in the section on mercuric mercury.

In conclusion, it is difficult to ascertain whether there is a difference in excretion and whole-body retention of mercury after mercury vapor exposure in comparison to exposure to mercuric mercury. Available data do not speak against the assumption that approximately the same mechanisms and rates govern in both cases, probably because the great majority of the mercury in the body takes the form of mercuric mercury.

4.4.1.3.2 In Human Beings

Intramuscular injections of finely dispersed metallic mercury were used widely earlier in the treatment of syphilis. "Oleum cinerum" contained 40 to 50% finely dispersed (particle size about 7 µm) metallic mercury in oil. Lomholt, 1928, studied the urinary and fecal excretion after injections of oleum cinerum. High amounts of mercury were excreted in the urine (about 4 to 4.5 mg/day) and in the feces (1.2 mg/day). The patient developed stomatitis. Excretion of mercury after inunction of metallic mercury has been reported by Bürgi, 1906, and Lomholt, 1928. The fecal excretion during the first days of the study was usually higher than the urinary, but after the first week the urinary excretion was higher than the fecal.

The most important route for absorption of inorganic mercury in man in industry is the inhalation of mercury vapor. The numerous studies on urinary excretion will be discussed in Chapter 7.

Tejning and Öhman, 1966, performed a careful balance study on 30 workers exposed to Hg^0 vapor in the chloralkali industry in Sweden. Mercury absorption was measured by continuous sampling during 4 days over the workers' breathing zones. In addition, daily excretion in urine and feces was measured and the retention of mercury was calculated for each subject. For 15 workers with a mean mercury exposure of 0.05 to 0.1 mg/m^3 (group 1) a mean urinary excretion of 0.12 mg/day was found. The output in the feces was

0.09 mg/day on the average. In 10 workers with a mercury exposure ranging from 0.11 to 0.2 mg/m^3 (group 2) the mean mercury excretion in the urine was 0.19 and the mean fecal excretion, 0.14 mg/day. The calculated yearly retention of mercury in the body ranged from −54 mg to +47 mg (mean −6 mg) in group 1 and from −32 to +130 mg (mean +51 mg) in group 2. Unfortunately, data on the relation between time of employment and mercury retention were not given in the report by Tejning and Öhman, 1966, so it is not possible to evaluate the time necessary for different individuals to reach a steady state between absorption and excretion. The data as they are presented do show a wide variation in retention among individuals. The marked fluctuation of mercury values from day to day under conditions of relatively constant exposure has been pointed out by several investigators, among them Friberg, 1961, and the commentaries are included in Chapter 7.

The possibility of using salivary excretion as an index of exposure to mercury will be discussed in Section 7.1. In a study of industrial workers by Joselow, Ruiz, and Goldwater, 1968, the low concentrations (about 5 µg/100 ml) found in saliva were only 10% of the concentrations in urine of the same subjects and show that saliva is not a main route of mercury elimination after exposure to mercury vapor.

Precise comparisons between the elimination of mercury after exposure to mercuric salts and after exposure to metallic mercury vapor have not been performed in human beings, but the data at hand in combination with animal data indicate that there are no important differences in excretion. Retention and accumulation conditions for the kidney are probably also similar as judged mainly from animal data. For discussion of data on kidney and whole-body retention and accumulation of mercuric mercury, see above, Section 4.4.1.1.2. For the brain it is probable from studies in several animal species (see Sections 4.2.1 and 4.3.1) that the accumulation risk is more prominent at exposure to mercury vapor, because of the prominent brain uptake that occurs. Animal studies indicate a slow turnover of mercury in the brain. From the data by Takahata et al., 1970, (see Section 4.3.1.2) it is evident that high concentrations of mercury can remain in large parts of the human brain even a long time after exposure has ended. A similar observation has been made by Grant (unpublished data, quoted by Berglund et al., 1971). He found 19 µg Hg/g brain tissue in a dehydrated xylol-extracted specimen from a person who had been exposed to elemental mercury vapor 13 years prior to death. Of the brain concentration only 0.3 µg Hg/g was identified as methyl mercury. The remaining part was probably in inorganic form. Hg^{2+} was identified by thin layer chromatography. The mentioned data indicate that the biological half-life is very long for important parts of the brain. Both Takahata's and Grant's values were derived from tissue specimens which had been prepared in different ways for histological examination. Although it is not probable that such treatments would change the values fundamentally, such an influence could not be excluded with certainty. It would be of great benefit to get data on direct measurements in fresh tissue. Even if one considers the mentioned data it is not possible to make precise evaluations of critical amounts of exposure necessary to reach damaging brain concentrations.

4.4.2 Organic Mercury Compounds
4.4.2.1 Alkyl Mercury Compounds
4.4.2.1.1 Methyl Mercury Compounds
4.4.2.1.1.1 In Animals
4.4.2.1.1.1.1 Retention

The kinetics of metabolism of mercury after administration of methyl mercury compounds have been studied in mice, rats, monkeys and seals. In studies in rats (Ulfvarson, 1962, and Ahlborg et al., to be published) and mice (Clarkson, 1971), whole-body accumulation of mercury at prolonged administration was observed which fit reasonably well with a first degree exponential function.

In seals given a single oral dose of labeled methyl mercury proteinate, a 2-phase whole-body elimination pattern was observed (Tillander, Miettinen, and Koivisto, 1970). In mice (Suzuki, 1969, Östlund, 1969b, Ulfvarson, 1970, and Clarkson, 1971), rats (Swensson and Ulfvarson, 1968, Berglund, 1969, Norseth, 1969b, and Ahlborg et al., to be published), and monkeys (Nordberg, Berlin, and Grant, 1971) elimination patterns for whole-body or for different organs were found which corresponded fairly well with a single phase exponential function, though the elimination rate differed considerably among different species.

In the rat, half of a single dose is eliminated

through feces and urine in about 20 days (Ulfvarson, 1962, and Norseth, 1969b). Longer half-lives have been found at whole-body measurements after administration of labeled methyl mercury (Swensson and Ulfvarson, 1968, and Berglund, 1969), probably because of accumulation of mercury in the fur.

After a single injection of methyl mercury in rats, Swensson and Ulfvarson, 1968, found a slower elimination of mercury from the brain than from other organs during 1 to 3 weeks. Later a dynamic equilibrium seems to have been established among the levels in different organs. Norseth and Clarkson, 1970b, also noted some differences in elimination of mercury from different organs. Elimination from the brain was slower than from the blood.

In the monkey, Nordberg, Berlin, and Grant, 1971, found a biological half-life in blood of 50 to 60 days, while whole-body radioactivity measurements indicated 150 days. The difference was due to accumulation of mercury in fur.

Concerning mice, there is some disagreement as to whether or not the size of a single dose affects the elimination rate. Östlund, 1969b, found a half-life of 3.7 days at administration of 0.03 mg Hg/kg body weight as methyl mercury hydroxide, while 5 mg Hg/g gave a half-life of 12.6 days. Ulfvarson, 1970, in a similar investigation, found a biological half-life of 6 to 7 days irrespective of the dose. Clarkson, 1971, fed mice food containing 0.05 and 0.5 mg labeled mercury as methyl mercury chloride/kg dry weight for 21 days. After stopping exposure the whole-body count decreased with a biological half-life of 8 days. This elimination rate is in accordance with that observed by Suzuki, 1969, in the mouse brain, while elimination from blood was faster with a half-life of about 4 days. Suzuki, Miyama, and Katsunuma, 1971a, reported 6 to 7 days for the mouse brain.

It must be kept in mind that the assumption of a simple, single order elimination pattern may not be strictly valid, since it has been shown (Section 4.2.2.1.1.1) that there is a slow biotransformation of methyl mercury into inorganic mercury which has an elimination pattern quite different from that of methyl mercury (Section 4.4.1.1.1.2).

4.4.2.1.1.1.2 Excretion

The elimination of mercury after administration of mono-methyl mercury occurs mainly via feces, urine, and hair. Some mercury is excreted via the milk.

Urine and feces – In mice, Östlund, 1969b, found that the ratio of excretion in urine/feces was 1/4 during 21 days after a single intravenous injection of *methyl mercury* salt.

Several excretion studies have been performed in rats with different doses in single or repeated administrations and with analyses for varying times. Between 1 to 3 and 10 to 40% of the total elimination has been reported to have occurred through the urine in different studies (Friberg, 1959, Ulfvarson, 1962, Gage, 1964, Swensson and Ulfvarson, 1967, Norseth, 1969b, and Norseth and Clarkson, 1970b).

In pigs 14 days after a single injection of methyl mercury, Platonow, 1968a, found 2% of the dose in the urine and 10% in the feces. Similar results have been reported for cats by Yamashita, 1964. In rats 50 to 90% of the mercury in the urine was organomercury (Gage, 1964, Ahlborg et al., to be published, partly reported by Westöö, 1969b). Norseth, 1969b, and Norseth and Clarkson, 1970b, found 6 to 25% of the total mercury in the urine as inorganic mercury. The fraction was rising during the 24 days studied. In pigs, Platonow, 1968a, identified 20% of the total mercury as methyl mercury.

Swensson, Lundgren, and Lindström, 1959b, and Berlin, 1963c, found a correlation between plasma mercury concentration and urinary mercury elimination in short-term studies on dogs and rabbits, respectively. Norseth, 1969b, and Norseth and Clarkson, 1970b, demonstrated that the fraction of inorganic mercury out of total mercury in the urine was correlated to the fraction in the kidney but not to that in the plasma, indicating a tubular excretion.

In autoradiograms of mice injected with methyl mercury, an accumulation was seen in the bile system and in the mucous membrane of the gastrointestinal tract (Berlin, 1963b). Norseth, 1969b, stated that mercury in the rat bile was probably present as methyl mercury cysteine complex and also smaller fractions as protein-bound methyl mercury and inorganic mercury. A considerable part of the methyl mercury was reabsorbed in the intestine, while some inorganic mercury was formed out of methyl mercury. The resorption of inorganic mercury was small. It was considered likely that mercury was also excreted by routes other than the bile, mainly through shedding of the intestinal epithelium. The author also showed that about 50% of the mercury eliminated through the feces after a single injection was in inorganic form (also Norseth and Clarkson, 1970b). Gage, 1964, found 40 to 50% of the mercury as organomercury after repeated parenteral administration. Takahashi and Hirayama, 1971, showed that most of the mercury in the lumen of the small intestine in rats injected with methyl mercury salt was present as methyl mercury.

Norseth, 1971, studied the elimination of mercury after administration of methyl mercury chloride to mice. The levels of mercury in the bile were higher than those in the blood. By isotope exchange techniques it was shown that only 1 to 7% of the mercury in the bile was inorganic. By separation on a Sephadex® column it was shown that the mercury in the bile was bound to a low molecular compound. On thin layer chromatography the compound moved as methyl mercury cysteine or methyl mercury glutathione, while in paper electrophoresis the characteristics were similar to those of the latter compound. The relative content of inorganic mercury in the feces was 25 to 60%.

In several studies in the rat it has been shown that at corresponding exposures the overall elimination of methyl mercury compounds, at least initially, is considerably slower than that of mercuric mercury salts (e.g. Friberg, 1959, Swensson, Lundgren, and Lindström, 1959a and b, Ulfvarson, 1962, Berlin, 1963b, and Gage, 1964, Section 4.4.2.1.1.1.1). The elimination rate was increased if rats were fed human hair (Takahashi and Hirayama, 1971) or mice fed a thiol containing resin (Clarkson, Small and Norseth, 1971).

Östlund, 1969a and b, investigated the metabolism of *di-methyl mercury* in mice after inhalation or intravenous exposures. The major part of the mercury was rapidly exhaled as di-methyl mercury. After 6 hours 80 to 90% had been eliminated. After 16 hours no di-methyl mercury was detected in the body but a nonvolatile compound remained, chromatographically most probably mono-methyl mercury.

Other routes of elimination – In furred animals a large fraction of the total elimination of mercury occurs through the hair. The elimination in fur has been studied in mice (Östlund, 1969b), rats (Berglund, 1969), cats (Albanus et al., to be published), and monkeys (Nordberg, Berlin, and Grant, 1971). As much as half of the total body burden of mercury might be located in the fur after prolonged administration.

Trenholm et al., 1971, studied the levels of mercury in milk from guinea pigs given single doses of methyl mercury (1 mg Hg/kg) intraperitoneally during pregnancy or gestation. The mercury levels in the milk were generally below 2% of the whole blood levels, which were at or below 2 μg/g. The mercury levels in the milk decreased more rapidly after the injection than those in the blood. Östlund, 1969b, could not demonstrate any exhalation of mercury in intravenously injected mice.

4.4.2.1.1.2 In Human Beings
4.4.2.1.1.2.1 Retention

In the tracer dose experiment performed by Åberg et al., 1969, the daily elimination from the body made up less than 1% of the total body burden. The elimination pattern, estimated from whole-body measurements during 220 to 240 days, was consistent with a first degree exponential function with a biological half-life of 70 to 74 days. The elimination from different regions was measured by repeated scanning. The biological half-life for the head was 64 to 95 days with a mean of 85 days. This should be compared to 60 to 70 days with a mean of 66 days for all the scanned regions together. The authors concluded that the elimination from the head was slower than that from the rest of the body.

Miettinen et al., 1969b, and 1971, in their study on the metabolism of methyl mercury in 15 volunteers, estimated a biological half-life at whole-body measurements of 76 ± 3 (S.E.M.) days after an observation period of 8 months. In 6 subjects the radioactivity was also measured for 91 days in whole blood, blood cells, and plasma (Miettinen et al., 1971). The decay curve of mercury concentration in blood cells showed 2 components, first a rapid decay and then an exponentially slower one. Probably the first part of the decay curve reflected the distribution. For the second part of the curve the biological half-life was 50 ± 7 (S.E.M.) days. The biological half-life of mercury in leg muscle was estimated to be 77 ± 8 days in a group of 5 men and 5 women. Other data may be mentioned on decay of mercury from blood and hair in subjects after an exposure to methyl mercury ceased or diminished. On the basis of data from the Niigata incident, Berglund et al.,

1971, calculated the elimination rates for some patients. In 7 patients the biological half-life in whole blood was 35 to 137 days (median 55 days) and in 8 patients the half-life in hair was 50 to 108 days (median 66 days).

In 2 persons exposed through consumption of methyl mercury contaminated fish but without symptoms of poisoning, Tejning, 1969b, found a biological half-life for mercury in blood cells of 69 and 70 days and in plasma, 76 and 83 days, respectively. In a similar study Birke et al., (to be published) found a half-life in blood cells (corrected for background exposure) in 2 persons of 99 and 120 days, in plasma of the same 2 subjects of 47 and 130 days, and in hair of 5 persons of 33 to 120 days.

A theoretical total body burden accumulation curve for man is shown in Figure 4:2. It has been assumed that the course of elimination is single exponential and that the daily elimination is 1% of the total body burden. About 50% of the steady state level is reached after 2 months and 95% after 1 year. As the elimination from the brain might be slower than that from the rest of the body, it is possible that accumulation takes place during a longer period in this organ.

4.4.2.1.1.2.2 Excretion

Urine and feces — The elimination of mercury in man after a single oral tracer dose of methyl mercury has been investigated by Åberg et al., 1969, and Miettinen et al., 1969b, and 1971. They found the feces to be the main route and only about 10% of the total elimination took place via the urine. Åberg et al., 1969, found during 49 days after administration 3% of the dose in the urine and 34% in the feces. Miettinen et al., 1969b, and 1971, investigated the levels in some 24-hour urinary samples up to 28 weeks after administration. During the first week the amount excreted per day in the urine was about 0.01% of the amount administered while the amount excreted in the feces was about 1.9% per day. This difference decreased as time elapsed. The average elimination per 24 hours during the first month was 0.7 to 0.8% of the total body burden.

Lundgren, Swensson, and Ulfvarson, 1967, found a correlation between blood and urine mercury levels in persons exposed to methyl mercury dicyandiamide. In persons exposed to elemental mercury vapor the urine mercury level was considerably higher at a similar blood level.

Other Routes of Elimination — In the study by Åberg et al., 1969, the hair of the head contained up to 0.12% of the dose/g hair with a maximum after 40 to 50 days. No radioactivity was detected in the semen during 240 days. Miettinen et al., 1971, reported that in 2 hair samples obtained 125 and 274 days after dosing and in 1 sample of beard taken after 108 days, the mercury contents corresponded to 0.05% of the dose/g. From epidemiological studies covering persons exposed to methyl mercury by fish consumption, it is well known that high levels of mercury can accumulate in hair (Section 8.1.2.1.1.2).

Skerfving and Westöö, to be published, analyzed total and methyl mercury in breast milk of heavy fish eaters from Sweden. The total mercury levels in blood cells ranged up to about 100 ng/g. The total mercury levels in milk ranged

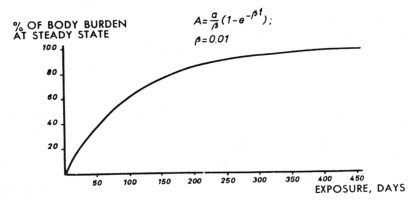

FIGURE 4:2. Theoretical course of accumulation (A) for the total body burden of man at steady state after beginning of exposure to methyl mercury. (From Berglund et al., 1971.) α = dose per unit of time; β = fraction of total body burden excreted per unit of time.

up to about 6 ng/g and were related to the plasma levels. Methyl mercury made up about half of the total mercury in milk or less.

4.4.2.1.2 Ethyl and Higher Alkyl Mercury Compounds
4.4.2.1.2.1. In Animals
4.4.2.1.2.1.1 Retention

From studies in which both methyl and ethyl mercury salts were included, it is evident that the kinetics for the 2 types are similar (Ulfvarson, 1962, Suzuki, Miyama, and Katsunuma, 1963, Yamashita, 1964, and Itsuno, 1968). Takeda et al., 1968a, showed in the rat that the elimination of mercury administered as ethyl mercury salts was slower than that of inorganic mercury salt.

After a single injection of ethyl mercury salt in mice (Suzuki, Miyama, and Katsunuma, 1963) and rats (Takeda et al., 1968a) the decrease of mercury in the brain was slower than in other organs studied. The latter team observed a gradual accumulation of mercury in the rat kidney while the former found no such accumulation in the mouse kidney.

Suzuki et al., in press, studied the elimination of mercury from various organs after a single injection of ethyl mercury salts into mice. In the brain the biological half-life for total mercury was 21 days, while the corresponding figure for the organic fraction was 8 days. The level of inorganic mercury increased during the studied period of 13 days. In the kidney and liver the biological half-life of organomercury was about 4 days, while the level of inorganic mercury decreased very slowly or not at all.

Considerable concentrations of mercury have been found in organs of rats fed propyl, butyl, amyl, and hexyl mercury compounds (Itsuno, 1968, Table 4:6). Takeda et al., 1968a, reported that mercury from *n*-butyl mercury was eliminated more slowly than that from mercuric chloride.

4.4.2.1.2.1.2 Excretion

Urine and feces — Miller et al., 1961, studied the excretion in rats for 7 days after injection of ethyl mercury chloride. The *level* of mercury was 3 to 5 times higher in the feces than in the urine. No quantitative data on elimination were given. In the urine, 56 to 76% of the total mercury was identified as organic mercury. The levels found in the urine and feces were considerably lower than after administration of phenyl mercury chloride (Miller, Klavano, and Csonka, 1960).

Takeda et al., 1968a, studied the elimination of mercury in rats after a single injection of ethyl mercury chloride and ethyl mercury cysteine. On the whole, more mercury was eliminated in the feces than in the urine. In the first period after administration the excretion was 2 or more times higher in the feces than in the urine. After 1 week the eliminations in urine and feces were about equal. Mercuric chloride was eliminated faster than ethyl mercury and to a greater extent in the urine. Takeda and Ukita, 1970, found in a further study that the elimination of mercury during 8 days after a single injection was higher in the urine than in the feces. In the urine about half of the mercury was inorganic while in the feces this fraction was about one third. The organic mercury was identified chromatographically as ethyl mercury. In cats exposed orally to ethyl mercury phosphate, Yamashita, 1964, found elimination of mercury in the feces about 10 times higher than that in the urine.

Excretion data on *higher alkyl mercury* compounds are scanty. Kessler, Lozano, and Pitts, 1957, showed that very little mercury was excreted in the urine during 3 hours after intravenous injection of propyl mercury in dogs. For *n*-butyl mercury chloride in rats, Takeda et al., 1968a, found that more mercury was excreted in the feces than after injection of ethyl mercury. The elimination rate was slower than that for mercuric mercury during the period studied.

Other routes of elimination — In cats, Yamashita, 1964, showed that exposure to different ethyl mercury salts resulted in a considerable accumulation of mercury in hair. Itsuno, 1968, found high mercury levels in hair of rats exposed to ethyl and propyl mercury.

4.4.2.1.2.2 In Human Beings

Suzuki et al., in press, studied the elimination of mercury from blood and excretion in spot samples of urine of persons treated intravenously with a solution containing sodium ethyl mercury thiosalicylate (Section 8.1.2.1.2.2). The first samples were obtained 11 to 22 days after the last administration and repeated sampling was carried out for an additional 7 to 35 days. The biological half-life for mercury in blood cells (almost only organomercury) was about 1 week in 2 subjects. The mercury in plasma (half of it inorganic mercury) was eliminated much more slowly. In 3 persons almost all plasma mercury was inorganic, while in a fourth person, about half of the total mercury was inorganic. There was a relation between the level in total blood or plasma and that in urine.

A few studies have been made on the urinary mercury excretion in workers exposed to ethyl mercury compounds. These studies are discussed in Sections 8.1.2.1.1.1.2, 8.1.2.1.1.3.2 and 8.1.2.1.2.2. No studies on fecal excretion have been published.

Bakulina, 1968, reported that mercury was eliminated via the milk in women earlier poisoned by ethyl mercury phosphate (Section 8.1.1.1.2). The retention and excretion of MHP, a *substituted alkyl mercury* compound, will be discussed in Section 4.4.2.4.2 on other organic mercury compounds.

4.4.2.2 Aryl Mercury Compounds
4.4.2.2.1 In Animals
4.4.2.2.1.1 Retention

The elimination rate of mercury in rats exposed to phenyl mercury salts is dose-dependent; the greater the exposure, the faster the relative elimination (Cember and Donagi, 1964, and Ulfvarson, 1969a). While the elimination pattern in the case of short chain alkyl mercury compounds is close to exponential, the elimination of mercury after administration of phenyl mercury compounds follows a more complicated pattern and changes as time elapses after a single administration. The time required in the rat for half of a given dose to be eliminated has been estimated at 4 to 10 days at repeated administration of 0.05 mg mercury/kg body weight/day (Ulfvarson, 1962). This figure might be compared, however, with the biological half-life for methyl mercury in the rat, 18 days. The risk of accumulation in the body as a whole is thus greater for short chain alkyl mercury compounds. The complicated elimination pattern for phenyl mercury is probably a result mainly of the biotransformation of phenyl mercury into inorganic mercury and the redistribution or apparent redistribution of mercury within the body (see Sections 4.2.2.2 and 4.3.2.2).

The elimination of mercury from the kidney is slower than from the rest of the body (Friberg, Odeblad, and Forssman, 1957, Ellis and Fang, 1967, and Takeda et al., 1968a). The high degree of distribution to and the relatively slow elimination from the kidney mean a definite risk of accumulation in that organ during a continuous exposure. Suzuki, Miyama, and Katsunuma, 1971a, have shown that the elimination of mercury from the mouse brain (half-life 2 to 3 weeks) is slower than from the other organs (half-life about 1 week).

4.4.2.2.1.2 Excretion

The main excretion routes for phenyl mercury compounds are feces and urine. Excretion via hair also occurs (Gage, 1964).

The mercury excretion pattern after administration of phenyl mercury compounds is more complex than that after alkyl mercury compounds. The quantitative aspects of mercury excretion after administration of phenyl mercury have been studied mainly in rats. More mercury is excreted via the feces than via the urine. The fraction of the total combined fecal and urinary elimination appearing in the feces has varied in different studies. The differences probably can be explained to a large extent by differences in the doses administered, in routes of administration, in times of exposure, and in the time spans during which the elimination was observed. In most studies, two thirds or more of the total mercury excretion occurred through the feces (Prickett, Laug, and Kunze, 1950, Ulfvarson, 1962, Cember and Donagi, 1964, Gage, 1964, Ellis and Fang, 1967, Swensson and Ulfvarson, 1967, and Takeda et al., 1968a). In many studies considerable variations in the ratio of mercury excretion in feces and urine have occurred with time after a single dose or in repeated exposure. It is difficult to find a common trend in the reported data.

Cember and Donagi, 1964, showed that the excretion pattern is dose-dependent. As the dose in a single injection was increased from about 0.005 to 0.5 mg Hg/kg the total excretion increased and the ratio of fecal to total excretion increased from 1/8 to 1/2.

Elimination of mercury via the urine starts immediately after injection of phenyl mercury compounds and a correlation between mercury elimination via this route and blood mercury levels has been shown in short-term experiments in dogs (Swensson, Lundgren, and Lindström, 1959a) and in rabbits (Berlin, 1963c). The elimination of mercury from the blood was found to be greater than the mercury excretion in the urine (Berlin, 1963c).

The extent to which an aryl mercury compound is eliminated in the urine seems to be greatly dependent upon the chemical nature. While only about 1% of an intravenous dose of phenyl mercury bromide was excreted during three hours in dogs, as much as about 40% of a dose of p-chloro-mercuric benzoate (PCMB) was eliminated in the same time (Kessler, Lozano and Pitts, 1957).

A considerable fraction of the total mercury elimination in the urine after administration of phenyl mercury acetate does not consist of organic mercury. In rats during 2 days following an intramuscular injection, Miller, Klavano, and Csonka, 1960, identified 40% of the total mercury excretion as organic mercury. During 6 weeks, Gage, 1964, repeatedly injected phenyl mercury acetate into rats. The fraction of the total mercury extractable as organic mercury was about 20% during the first week, and later about 10%. During the first day after a single dose, only organic mercury was found in the urine, while later, only a low percentage was organic. Daniel and Gage, 1971, showed that about 20% of the mercury that occurred in the urine during 4 days after a single injection of phenyl mercury acetate in the rat was in organic form (see also Section 4.2.2.2.2). Neither of the reports stated the purity of the phenyl mercury preparation.

Piotrowski and Bolanowska, 1971, reported that mercury in the urine of rats exposed through a single injection of phenyl mercury acetate was bound partly to proteins of high molecular weight and partly to nonprotein compounds of low molecular weight.

The main route of excretion of mercury is fecal after administration of phenyl mercury. Despite this, fecal elimination has been much less studied than urinary. Prickett, Laug, and Kunze, 1950, after parenteral administration of phenyl mercury acetate, observed an accumulation of mercury in the small intestine of the rat. Berlin and Ullberg, 1963c, by whole-body autoradiography in mice, observed mercury accumulation in the mucous membranes of the gastrointestinal tract, as well as in the liver and in the lumen of the gall bladder. The accumulation was greater after injection of phenyl mercury than after inorganic mercury salt. Gage, 1964, in his 6 weeks' exposure of rats, found 10% or less of the mercury in feces as organic mercury. This was also the case in his single dose experiment. Nakamura, 1969, reported that in rats orally exposed to phenyl mercury acetate for 7 days, the ratio between phenyl mercury and inorganic mercury was about 1 for the content in the stomach, 1/70 in the caecum, and 1/9 in the colon. Daniel and Gage, 1971, and Gage, in press, showed that the excretion of radioactivity in the feces in rats given ^{14}C-phenyl mercury acetate was considerably lower than excretion of mercury, which then was mainly inorganic (see also Section 4.2.2.2.2). The limited data on hand thus indicate that the mercury eliminated through the feces is mainly in inorganic form. It is not known if the breakdown occurs in the gastrointestinal tract or within the body.

4.4.2.2.2 In Human Beings
4.4.2.2.2.1 Retention

Morsy and El-Assaly, 1970, studied by repeated whole-body measurements during 14 days the elimination of ^{203}Hg in a worker accidentally exposed to an unknown amount of labeled *di-phenyl mercury* on an unknown occasion. The elimination revealed a single component exponential pattern with a biological half-life of 14 days.

4.4.2.2.2.2 Excretion

There are no quantitative data on the elimination of aryl mercury compounds in man. The few studies that have been made on the urinary mercury excretion in workers exposed to phenyl mercury salts are discussed in Section 8.2.2.1. In this connection it should be mentioned that in four workers exposed by inhalation to phenyl mercury salt, Massmann, 1957, identified 70 to 90% as organomercury out of a total mercury level of 0.5 to 1.5 mg/l. urine. The author stated, without presenting data, that after exposure had ceased, the fraction of organo-mercury decreased.

4.4.2.3 Alkoxyalkyl Mercury Compounds
4.4.2.3.1 In Animals
4.4.2.3.1.1 Retention

Simple alkoxyalkyl mercury compound studies in animal experiments have involved methoxyethyl mercury salts only. Some data concerning substituted alkoxyalkyl mercury compounds (mercurial diuretics) will be discussed in Section 4.4.2.4.1 on other organic mercury compounds.

The rate of elimination of mercury after administration of methoxyethyl compounds seems to vary at different dose levels. The total elimination from the body is faster at high exposure than at low in rats and mice (Ulfvarson, 1969a, and 1970). At administration of 0.05 mg mercury/kg body weight/day in rats the time required for half of the mercury to leave the body was calculated at 4 to 10 days, in accordance with elimination rates observed at exposure to inorganic and phenyl mercury salts at the same dose level but slower than at exposure to methyl mercury salt (Ulfvarson, 1962). This relation between the elimination rates has been observed also in mice (Berlin and Nordberg, unpublished data).

4.4.2.3.1.2 Excretion

After administration of methoxyethyl mercury compounds, about two thirds of the elimination of mercury occurs through the fecal route and about one third through the urinary (Ulfvarson, 1962, Swensson and Ulfvarson, 1967, and Daniel, Gage, and Lefevre, 1971). The elimination pattern thus is similar to that seen after administration of inorganic and phenyl mercury salts but the fecal elimination is significantly less important than after alkyl mercury compounds.

Daniel, Gage, and Lefevre, 1971, (see also Section 4.2.2.3.2) showed that 1 day after administration of methoxyethyl mercury to rats only organic mercury was excreted in the urine. Later, the urinary excretion consisted of inorganic mercury only. Organic mercury was also excreted in the bile initially. Later some inorganic mercury occurred in the bile. In the feces only inorganic mercury was present, indicating a breakdown of organic mercury in the gut. There were also some indications of a reabsorption of organomercury.

4.4.2.3.2 In Human Beings

There are no data available on retention or excretion of alkoxyalkyl mercury compounds in man aside from the mercurial diuretics which will be discussed in Section 4.4.2.4.2 on other organic mercury compounds.

4.4.2.4 Other Organic Mercury Compounds
4.4.2.4.1 In Animals
4.4.2.4.1.1 Retention

Most studies concerning mercurial diuretics have lasted only a few hours. The retention as indicated by elimination will be discussed in Section 4.4.2.4.1.2.

Anghileri, 1964, showed in rats that the total body burden rapidly decreased during the first week after a single intravenous dose of chlormerodrin. At that time only 5% of the dose remained in the body. Later during the 30 days studied the biological half-life was 8 days. The kidney showed a similar pattern; however, the corresponding biological half-life was 28 days. Miller, Green, and Levine, 1962, found in dogs that the kidney burden was reduced to about 50% in 24 hours and halved once again in 4 to 7 days.

4.4.2.4.1.2 Excretion

Most of the excretion of mercury after administration of organomercurial diuretics occurs in the urine and only a small fraction is eliminated through the stool. Figures of 7 and 10% in the stool have been reported during a short period after injection of chlormerodrin into rats and dogs (Anghileri, 1964, and Miller, Green, and Levine, 1962).

Kessler, Lozano, and Pitts, 1957, showed in dogs that a series of organomercurial compounds (i.e., mersalyl, meralluride, and chlormerodrin) which had a diuretic action were rapidly eliminated in the urine, so that 50 to 75% of the dose was measured there in 3 hours. In contrast, after nondiuretic organomercurials, methyl, propyl, hydroxypropyl (MHP), hydroxyethyl, phenyl, and methoxyethyl mercury salts, only about 1% of the dose appeared in the urine in 3 hours. PCMB was rapidly excreted.

Some species differences have been reported. While in the dog about 50% of a dose is excreted in a few hours (Borghgraef and Pitts, 1956, Borghgraef et al., 1956, and Kessler, Lozano, and Pitts, 1957), in the rat, less than 1% is excreted in the same period (Borghgraef and Pitts, 1956).

Weiner and Müller, 1955, reported that after injection of mersalyl into dogs the mercury in the urine was identified by polarography as a cysteine-like sulfhydryl complex. Anghileri, 1964, stated the presence of inorganic mercury (identified by paper chromatography) in the urine of rats injected with chlormerodrin.

Baltrukiewicz, 1969, studied the mercury levels in suckling newborns of rats which had received injections of chlormerodrin during pregnancy and/or lactation period. About 1% or less of the administered mercury was present in the newborns.

4.4.2.4.2 In Human Beings

4.4.2.4.2.1 Retention

Some studies performed on the retention of mercurial diuretics in man by use of labeled compounds have indicated a complicated pattern including several compartments.

Greenlaw and Quaife, 1962, measured the whole-body activity after single intravenous doses of 0.01 to 0.1 mg Hg as chlormerodrin to 6 volunteers. They found a 2 component elimination curve. About 75% of the dose was eliminated with a biological half-life of about 5 hours and 25% with a biological half-life of 7 days. Blau and Bender, 1962, gave about 10 mg of the same compound. External counts over the kidney showed that the 10% of the dose retained in that organ was eliminated with a biological half-life of about 28 days. The measurements were performed throughout an 80 day period. Bi-exponential elimination patterns of the mercury from the kidney have been reported by Hengst, Ohe, and Kienle, 1967. Baltrukiewicz, 1970, found that about one half of the kidney content of mercury was eliminated in 2 days. The rest was eliminated with a biological half-life of about 60 days. The dose of mercury was not stated. Johnson and Johnson, 1968, who started whole-body measurements 104 days after administration, found a biological half-life of 84 days during 4 months. The dose of mercury was not stated.

Kloss, 1962, using mersalyl (dose of mercury not stated) found a biological half-life for the external radioactivity over the kidney of 10 to 14 days during a 24 day period.

Grossman et al., 1951, during up to 2 months, did not recover in urine and feces all the mercury injected as meralluride. It thus seems that while most of the mercury from mercuric diuretics is rapidly excreted through the urine, a fraction is retained in the body much longer, presumably to a great extent in the kidney. This is also supported by the fact that Butt and Simonsen, 1950, Griffith, Butt, and Walker, 1954, and Leff and Nussbaum, 1957, found considerably higher kidney mercury levels in subjects treated with mercuric diuretics than in persons without known mercury exposure. The rate of elimination is also discussed in Section 4.4.2.4.2.2.

The retention of mercury has been studied by external scanning in persons given single intravenous doses corresponding to 0.05 to 0.1 mg Hg/kg body weight as labeled MHP, in most cases mixed with blood. As was said in Section 4.3.2.4, there is a rapid accumulation of mercury in the spleen, followed soon by a redistribution to kidney and liver, with the greater portion going to the kidney. The maximum kidney retention has been observed to occur 3 to 14 days after dosing (Korst et al., 1965, and Fischer, Mundschenk, and Wolf, 1965). There is then a slow elimination of mercury from the kidney. The biological half-life of the kidney pool has been mentioned briefly to be 45 and 140 days (Croll et al., 1965, and Korst et al., 1965). Measurements in urine and feces have shown a total elimination of 45% of the dose in 27 days (Wagner et al., 1964) and 20% in 20 days (Fischer, Mundschenk, and Wolf, 1965), the former indicating an elimination considerably more rapid than short chain nonsubstituted alkyl mercury compounds, while the latter is similar to those compounds (see Section 4.4.2.1.1.2.1).

4.4.2.4.2.2 Excretion

The elimination of mercury after administration of meralluride to man occurs almost only through the urine, the feces containing only one tenth or less of the total excreted amount (Grossman et al., 1951). Half of a dose is eliminated though the urine in 2 to 3 hours (Burch et al., 1950, and Grossman et al., 1951).

For chlormerodrin, about 50% is eliminated by the urinary route in 8 hours (McAfee and Wagner, 1960, and Blau and Bender, 1962). After this rapid phase the elimination is considerably slower. During 48 hours about 65% of the dose has been recovered in the urine (Blau and Bender, 1962). Wagner et al., 1964, found 3 to 4 times more mercury in the urine than in the feces after single intravenous injection of MHP. On the other hand, Fischer, Mundschenk, and Wolf, 1965, found equal amounts in urine and feces. The fecal excretion rate was close to that reported by Wagner et al., 1964, but the urinary was much lower. A considerably higher urinary excretion rate was reported by Oshiumi, Matsuura, and Komaki, 1969. In that study no fecal analysis was performed. The excretion pattern seems to have been more similar to that of inorganic mercury than to that of short chain alkyl mercury compounds.

4.4.3 Summary

The elimination of *inorganic mercury* from the body is probably similar for exposures to mercuric mercury or elemental mercury. Whole-body measurements in human subjects during 3 to 4 months after a single oral tracer dose indicate a biological half-life of 30 to 60 days. In animals the elimination from the body follows 2 or 3 consecutive exponential curves, with increasing half-lives. The rate of elimination has been shown to be dose-dependent to some extent.

Excretion takes place via the kidneys into the urine and in the feces. The fractions in each of these two are approximately equal, but may fluctuate a little depending upon dosage and route of exposure. It has been shown that a small part of the body burden of mercury can leave the body by volatilization from the lungs and the body surface, by sweat, and by lactation. In man the urinary route is usually somewhat dominant over the fecal route. In spite of considerable efforts, investigators do not yet know in detail the mechanisms for the urinary and fecal elimination of mercury. It seems likely that glomerular filtration and transtubular transport are of importance for urinary elimination. A direct passage over the gastrointestinal mucous membranes is probably of primary importance for the fecal elimination. A mathematical expression for the accumulation of mercury in the rat kidney has been set up to allow exemplification of calculations of critical exposure levels from data on metabolism. Unfortunately data on distribution and half-life of inorganic mercury in critical organs of the human body are insufficient for such calculations. It is evident, however, that high uptake in the kidney, the relatively slow elimination from that organ, and especially the long half-life in certain parts of the brain in combination with a relatively high uptake at Hg^0-vapor exposure, can mean a high accumulation at prolonged exposure. Reported high concentrations of mercury in brains of a few persons exposed to mercury vapor indicate a very slow elimination from some parts of this organ. From these limited data it seems possible that accumulation at repeated exposure can take place over periods of several years.

As regards the *organic mercury* compounds, the elimination pattern is very much dependent upon the rate of degradation into inorganic mercury. The total elimination pattern is a combination of 1 pattern for the intact organomercurial and 1 for the inorganic mercury which is also redistributed in the body after its formation. As was stated in Section 4.2, the rate of breakdown in animal experiments is very different among different organomercurials. It is very slow for methyl mercury. It seems to be faster for ethyl mercury, and is definitely faster for phenyl mercury. For methoxyethyl mercury the rate can be described as rapid. The evaluation of available data is thus difficult.

The elimination of *mono-alkyl mercury* compounds has been studied mainly with methyl mercury. In animal experiments the whole-body elimination of total mercury at exposure to short chain alkyl mercury compounds has been considerably slower than that seen at exposure to mercuric mercury. In human beings exposed to tracer doses of *methyl mercury* the elimination of mercury has followed a single component exponential pattern with a biological half-life of 70 to 90 days, i.e., about 1% of the body burden is eliminated daily. This is not too much in variance with biological half-lives of mercury in hair and blood found in poisoned individuals and in other exposed subjects. There is some evidence that the elimination of mercury might be somewhat slower from the brain than from the rest of the body. The slow elimination of methyl mercury compounds causes a considerable accumulation at continuous exposure. Steady state is not reached until 1 year of exposure has taken place. The distribution pattern favors a high retention in the kidney, the liver, and the brain. Although the information on *ethyl mercury* compounds is less complete, it seems that what was said about methyl mercury is also relevant for ethyl mercury. There are some differences, probably mainly because of the lesser degree of stability in the body. In animal experiments the elimination of total mercury from the brain and the kidney was slower than from other organs. The only *di-alkyl mercury* compound studied is di-methyl mercury. In mice most of a single dose is eliminated in a few hours.

After exposure to *mono-methyl mercury* compounds mercury is excreted mainly via the feces and only to a minor extent via the urine. In furred animals the hair is an important elimination route. In man the fecal elimination is about 10 times the urinary. When methyl mercury has been administered to mice and rats most of the mercury in the feces was inorganic. This is probably the result of reabsorption from and breakdown in the intestine of methyl mercury originating from the bile. In the urine a considerable fraction of organic mercury is found. The available data are far more restricted and less consistent for *ethyl mercury* compounds. In animal experiments the fecal route of elimination seems to be less dominant than at exposure to methyl mercury. In rats and man considerable fractions of inorganic mercury have been found in the urine. In the former species inorganic mercury was also present in the feces. In mice most of a single dose of *di-methyl mercury* is rapidly excreted through exhalation.

Aryl mercury compounds have been studied almost exclusively in animals to which phenyl mercury salts have been administered. The whole-body elimination rate of total mercury after administration of phenyl mercury salts is comparable to that after inorganic mercury salts but considerably faster than that after short chain alkyl mercury compounds. The elimination is dependent upon the dose level and upon the time after a single administration. The pattern thus is far more complicated than after methyl mercury compounds. The elimination rate is slower for the kidney and possibly also for the brain than for the total body. The high uptake of mercury in the kidney and the slow elimination from that organ mean a definite risk of accumulation. At administration of phenyl mercury compounds the ratio between urinary and fecal excretion of mercury is about 2:1 or higher. The main fraction of the mercury found in the urine at phenyl mercury exposure is inorganic. Only a minor proportion of the mercury in feces is organic.

Methoxyethyl mercury salts are the only simple *alkoxyalkyl mercury* compounds that have been studied, and only in animals. The elimination pattern is similar to that of phenyl mercury salts. Thus accumulation should be expected mainly in the kidney. About two thirds of the elimination occur via the feces and about one third via the urine. Most of the mercury in urine is inorganic. In the feces, no organic mercury appears.

The major part of a single dose of mercury administered as *mercurial diuretics* to man is eliminated within a few hours. A fraction of the dose is retained in the kidney and is eliminated considerably more slowly. The excretion occurs almost only through the urine.

4.5 Indices of Exposure and Retention
4.5.1 Inorganic Mercury

The main discussion of these matters with regard to inorganic mercury will be found in Section 7.1. Available data on human beings have been obtained when investigating effects of mercury on industrial workers and have, therefore, been included in Chapter 7. Here only some of the principal considerations and some conclusions which can be drawn from experimental studies will be brought forth. From the fact that the blood/brain and the blood/kidney ratios are not constant, but change with time after an exposure or during a series of exposures (see Section 4.3.1), it follows that blood concentrations will not be useful for indication of the retention in either of these 2 organs which may be critical in specific types of exposure to inorganic mercury. In addition to what has been mentioned earlier in this chapter (Section 4.3.1) some data on the relations among exposure, tissue damage, and concentrations of mercury in organs of experimental animals can be found in Section 7.2.2 and in Table 7:10. In that Table there is a reasonably good correlation among exposure, blood concentration, organ concentration, and organ damage. However, in a special study of blood concentrations in the same rabbits as described in Table 7:10, a comparison was made of blood values before and after a 2 day nonexposure interval. It was seen that the blood values fell to less than half of their value during this period. This confirms the conclusions from Section 4.3.1 that blood values reflect mainly recent exposure, and are not good indicators of accumulations in critical organs if the exposure varies. The mentioned conditions probably provide a main reason for the poor correlation between blood values and signs of intoxication in individual industrial workers (see Section 7.1). However, metabolic factors varying among individuals may also add to the variation. Data on the correlation between urinary and blood concentrations (Figure 7:7) as measured in workers in industry, as well as the correlation between urinary and air concentrations (Figure 7:5) will be discussed further in Chapter 7. Urinary values, which in principle follow blood concentrations, are probably even more dependent upon dosage and metabolic factors than are blood concentrations. The resulting variation from day to day in urinary mercury concentrations is well illustrated by data given in Chapter 7 (Figure 7:6). It has been suggested (e.g., Cember, 1969) that fecal mercury be used in combination with urine values in order to get total excretion values. Both practical and theoretical considerations make such an approach unjustified. Hair and nail samples have been suggested (Berlin, 1963a) as adequate for indication of retention in critical organs, but the difficulties with external contamination in such samples are evident, especially under conditions of industrial exposure. An additional factor which can influence hair concentrations of mercury is the prominent accumulation of methyl mercury in the hair of fish-eating workers. If separate analysis of organic and inorganic mercury is not made, influence from methyl mercury can be of importance for the hair concentration of mercury as shown by Suzuki, Miyama and Katsunuma, 1970. They considered hair analysis useless for evaluation of exposure to metallic mercury. From a theoretical point of view, then, it is difficult to find any index medium suitable for organ retention; even so, the mentioned media (urine, feces, blood) may reflect recent exposure to mercury.

4.5.2 Organic Mercury Compounds
4.5.2.1 Alkyl Mercury Compounds

Most information of use for judging suitable indices of exposure and retention of *mono-alkyl mercury* compounds concerns methyl mercury. The similarities in metabolism between methyl and ethyl mercury salts make similar conclusions valid also for the latter compounds and probably also for propyl mercury.

The slow elimination, the relatively even distribution in the body after administration of methyl mercury, and the stability of the covalent bond between carbon and mercury speak in favor of the assumption that in most mammals, the turnover among different tissues is faster as a rule than the excretion. A turnover as fast or slower than the excretion has been noted only in the CNS.

At least at levels at which no saturation of any tissue or toxic disturbances of the tissues have occurred, the relationship is constant between the mercury levels in different organs and between the levels in different organs and the total body burden. Also, the excretion is related to the body burden. At steady state there is then a constant relationship between the daily dose and the total body burden and the levels in each of the organs.

At exposure to methyl, ethyl, and propyl mercury compounds, the critical part of the body is the nervous system, so indices of the level in the nervous system are of primary interest. Almost all of the data on the metabolism in the nervous system concern the CNS. But considering the simple distribution pattern of mercury at exposure to short chain alkyl mercury compounds, there is probably a constant relationship between levels in CNS and peripheral nerves. There is no information on which is the critical organ at exposure to higher alkyl mercury compounds.

Berglund et al., 1971, on the basis of the available experimental and epidemiological data reviewed in this chapter, concluded that the most reliable index of exposure to methyl mercury and of retention of methyl mercury in the body and in the nervous system is the level of methyl mercury in the blood cells, or, though less reliable, in whole blood. If exposure to other mercury compounds can be excluded, total mercury levels in blood cells are a good index. Supporting evidence for exposure to methyl mercury might be achieved by analysis of total mercury level in plasma, the ratio between levels in blood cells and plasma being about 10 at methyl mercury exposure in man. It must be kept in mind, though, that during exposure to other organic mercury compounds there is a high blood cell/plasma ratio; at exposure to inorganic mercury the ratio is about 1. Total mercury level in whole blood might also be used but it is not possible then to decide the character of the exposure in regard to the mercury compound.

On the basis of data on methyl mercury-exposed but symptom-free subjects (Birke et al., 1967, Tejning, 1967c, and Sumari et al., 1969), Berglund et al., 1971, proposed that there was probably a rectilinear relation between total mercury levels in blood and hair, the hair levels being about 300 times higher than the whole blood levels. The individual variation, however, was considerable. Data from the Niigata epidemic (Tsubaki, personal communications) indicated a relatively higher hair mercury level. Berglund et al., 1971, stated that the discrepancy probably depended upon differences in methods. The mercury levels in the Japanese cases were decaying and probably analyses were made on complete hair tufts. These conditions probably induced relatively too high hair mercury levels in relation to blood. Also, the possibility of analytical errors in blood mercury analyses should be emphasized.

In animal experiments (Swensson, Lundgren, and Lindström, 1959b, and Berlin, 1963c) and in workers exposed to methyl mercury (Lundgren, Swensson, and Ulfvarson, 1967) there is a correlation between levels in plasma and blood, respectively, and urinary levels. The level of mercury in urine at exposure to methyl mercury, however, is low in comparison to levels found at corresponding exposure to inorganic mercury, phenyl mercury salts, or methoxyethyl mercury salts. Due to potential interference from other mercury compounds, urinary mercury levels thus have a limited value as index of exposure to and retention of methyl mercury. A correlation between levels of mercury in plasma and urine has been reported by Suzuki et al., in press, in subjects exposed to ethyl mercury.

In mice exposed to *di-methyl mercury* most of the mercury rapidly left the body in chemically intact form through exhalation (Östlund, 1969b), although a minor fraction was transformed into mono-methyl mercury. It is thus possible that, at least at high exposure, the same indices would be applicable at di-methyl mercury exposure as at mono-methyl mercury exposure. Data are lacking for other di-alkyl mercury compounds.

4.5.2.2 Aryl Mercury Compounds

From animal experiments on metabolism and toxicity of phenyl mercury compounds it has become evident that the levels in the kidney and the nervous system are of primary interest. On the other hand, clinical evidence of kidney and nervous tissue damage at phenyl mercury exposure is scanty (Section 5.2.2.1.2).

As discussed in earlier sections in this chapter, the metabolic pattern at phenyl mercury exposure is complicated. The ratio between levels in blood and kidney is dependent on time after exposure and probably also on the dose level. Thus, the mercury level in blood has limited value as an index of mercury retention in the kidney. The same holds true for the blood level as index of brain concentration.

In view of the rapid transformation of phenyl mercury into inorganic mercury, phenyl mercury analysis most probably would offer no advantage over total mercury analysis.

Whole blood and blood cell mercury levels decrease soon after cessation of exposure. Analysis of total mercury levels in blood cells or whole blood probably could offer some information about recent exposure. It must be realized that the information thus obtained is much more difficult to evaluate than was the case with short chain alkyl mercury compounds (Section 4.5.2.1).

Mercury accumulation in hair has been reported in rats exposed to phenyl mercury salt (Gage, 1964). The available data do not permit conclusions as to whether or not hair mercury levels would be useable as index of exposure and retention. External contamination imposes a problem in the case of hair.

In short-term animal experiments a relationship between blood mercury levels and urine mercury levels has been shown (Swensson, Lundgren and Lindström, 1959b, and Berlin, 1963c). Data from long-term exposure are lacking but as the blood levels were considered to be of limited value as an index of retention, it may be assumed that the same holds true for urinary levels, even if the correlation is present in long- as well as short-term exposure. As was stated in Section 4.4.2.2.1.2, the urinary and fecal excretion patterns are affected by dose level and time after exposure. Thus it is obvious that urinary mercury levels offer limited information as to retention of mercury at phenyl mercury exposure. It is probable, though, that urinary mercury levels can give some information on recent exposure.

There are no data available on aryl mercury compounds other than phenyl mercury salts.

4.5.2.3 Alkoxyalkyl Mercury Compounds

If mercurial diuretics are disregarded, the only alkoxyalkyl mercury compound for which information on metabolism and toxicity is available is methoxyethyl mercury salts. In the case of methoxyethyl mercury no data on hair mercury levels nor on the relation between blood and urinary levels have been reported. Other information needed for judging suitable indices likewise is scanty. It seems, however, that the conclusions made in Section 4.5.2.2 regarding phenyl mercury are valid also for methoxyethyl mercury. Thus neither blood levels nor urinary levels can be considered ideal indices of retention in the whole-body or in organs. They might offer information on recent exposure but caution must be exercised in drawing conclusions.

4.5.3 Summary

The lack of a constant ratio between the mercury concentration in blood and critical organs makes blood hardly suitable as an index medium for the evaluation of retention or risks of intoxication at exposure to different forms of *inorganic mercury*. Because blood concentrations are correlated to both urinary and fecal excretion, the same considerations hold true for these media. Urinary values are influenced by other factors which make them even less suitable than blood values for evaluation of the risks for an individual worker. As an indication of recent exposure they might be useful. On a group basis there is a reasonably good correlation between exposure (probably recent exposure) and urinary or blood values.

The most reliable index of exposure to and retention of *mono-methyl mercury* in the nervous system is analysis of alkyl mercury in blood cells or whole blood. If exposure to other mercury compounds can be excluded, the total mercury level in blood cells or whole blood is a good index. Exposure to organic mercury compounds is indicated by a high blood cell/plasma ratio. If external contamination can be excluded, the alkyl mercury or total mercury level in hair may be used as an index of exposure and retention at the time at which the analyzed part of the hair was formed. At methyl mercury exposure there is a correlation between levels in blood and hair, the hair levels being about 300 times higher than the whole blood levels. Urinary mercury levels are not suitable as an index of exposure and retention because the urinary mercury excretion is low. The information about other mono-alkyl mercury compounds is more incomplete. Considering the similarities in metabolism and toxicity between methyl and ethyl mercury compounds, it is most probable that the same indices may be used. It is reasonable to assume that propyl mercury compounds also may be included in this group. No conclusions are possible about higher alkyl mercury compounds.

Concerning *di-alkyl mercury* compounds the only information available is on the metabolism of di-methyl mercury in mice. Most of a single dose is rapidly exhaled in intact form while a fraction is transformed into mono-methyl mercury. Possibly the same indices would be suitable for di- as for mono-methyl mercury.

Concerning *phenyl* and *methoxyethyl mercury* salts, the only aryl and alkoxyalkyl mercury compounds, respectively, on which information is available, the situation is very similar to that surrounding inorganic mercury. There is no constant relation between levels in blood and either the kidney or the brain. Blood levels thus are unsatisfactory as indices of retention and the same applies to urinary levels. Both blood and urinary mercury levels may give, when cautiously handled, some information on recent exposure.

Chapter 5

SYMPTOMS AND SIGNS OF INTOXICATION

Staffan Skerfving and Jaroslav Vostal

5.1 Inorganic Mercury

5.1.1 Prenatal Intoxication

Although elemental mercury probably penetrates the placental barrier more easily than poorly penetrating mercuric ion (see Section 4.1.1.1.4), no experimental or clinical evidence is available on effects of either elemental or ionized mercury on the fetus. Lomholt, 1928, stated that mercury could be detected in stillborn babies from mothers acutely exposed to mercury inunctions against syphilis, and mercury poisoning has been suggested as causing abortions. However, only a few cases were reported in the old literature (Thompson and Gilman, 1914, quoted by Benning, 1958) and a correct evaluation of the exposure to mercury during pregnancy in sporadic observations published in more recent times (Benning, 1958) is difficult.

5.1.2 Postnatal Intoxication

The fact that postnatal exposure to metallic mercury vapors, fumes, or dust of ionized mercury salts may produce specific symptomatology of mercury poisoning has been known since ancient times and repeatedly described by classic authors. In modern times, poisoning by all forms of inorganic mercury is usually separated into at least 2 clinical entities: 1. acute poisoning caused by inhalation of high concentrations of mercury vapors or caused by accidental ingestion of mercuric salts, usually chloride or cyanide, commonly used as antiseptics in the early decades of this century; 2. chronic poisoning caused exclusively by long-term occupational exposures.

5.1.2.1 Acute Poisoning
5.1.2.1.1 Elemental Mercury Vapor
5.1.2.1.1.1 In Human Beings

Cole, Gericke, and Sollmann, 1922, reported in their studies on the use of mercury inhalations in the treatment of syphilis that single exposures to high concentrations of mercury vapors in the inhaled air cause bronchial irritation and varying degrees of salivation. Since that time only a few cases have been reported in the literature, showing that this form of peracute effects of mercury vapors is rare and usually results from an accident.

Hopmann, 1928, observed 4 persons accidentally exposed to high mercury concentrations in industry. The clinical symptoms mainly involved the respiratory tract and were manifested as coughing, signs of acute bronchial inflammation and chest pain, in addition to excitement and tremor. The symptoms persisted for 2 weeks, followed by a spontaneous complete recovery.

Campbell, 1948, reported dyspnea and cyanosis to be the main symptoms in a 4-month-old infant after massive exposure to mercury vapor. Cough, dyspnea, cyanosis, exudative bronchitis, and vomiting were the symptoms in an adult patient described by King, 1954, without details on the level of exposure.

A detailed analysis of the clinical symptomatology can be found in the descriptions by Matthes et al., 1958, of an accident involving 12-hour exposure to vapors from a space heater freshly painted with a mixture of metallic mercury (65% by volume), aluminum paint, and turpentine. Respiratory difficulties and irritability were the first symptoms in 3 children immediately after the exposure. The course of the disease was characterized by lethargy, followed by restlessness, diarrhea, cough, tachypnea, and respiratory arrest. Necropsy in 3 fatalities from this accident revealed erosive bronchitis and bronchiolitis with interstitial pneumonitis and resulting pneumothorax. Although the participation of turpentine fumes and aluminum products could not be excluded from the pathogenesis of the disease, the authors claimed that the mercury inhalation played a major part in producing the histological changes.

A fatal case in an adult was described by Tennant, Johnston, and Wells, 1961, after 5 hours' exposure to mercury vapor from a ruptured hot mercury vapor boiler. Diffuse pneumonitis with marked interstitial edema and alveolar exudation dominated in the microscopical post-mortem examination. Severe respiratory symptoms or slight symptoms combined with increased urinary excretion of mercury characterized other cases, reported by Haddad and Sternberg, 1963, Hallee, 1969, and Milne, Christophers, and deSilva, 1970.

5.1.2.1.1.2 In Animals

No detailed information exists on similar effects in experimental animals. Microscopical evidence of mild damage to the brain, kidney, heart, and lungs was found in rabbits exposed to mercury vapor at 29 mg Hg/m^3 for only 1 hour and severe changes were induced after a period of 4 hours or more (Ashe et al., 1953).

5.1.2.1.2 Inorganic Mercury Salts
5.1.2.1.2.1 In Human Beings

Highly dissociated inorganic salts of bivalent mercury have an intense local corrosive action. Ingestion of these salts or of concentrated solutions of them causes extensive precipitation of proteins at contact with mucous membranes of the gastrointestinal tract and is immediately followed by a characteristic symptomatology: local pain and gray appearance of oral and pharyngeal mucosa, gastric pain, and vomiting. If the ingested amount is minimal and/or the first reactive vomiting effective enough to empty the stomach, the symptomatology is restricted to the proximal parts of the gastrointestinal tract. If larger amounts of dissociated salts are ingested, high concentrations of ionized mercury occur in the small intestine, causing another symptomatology: abdominal pain and severe protrusive bloody diarrhea, containing necrotic parts of the intestinal mucosa. A profound circulatory collapse and sudden death may occur.

The most characteristic organ change in acute mercury poisoning is acute renal failure (Smith, 1951) including oliguria or complete anuria with azotemia and retention of metabolic waste products in the body. Prior to the treatment by artificial kidney, mortality was high. Hull and Monte, 1934, studied a group of 300 intoxications by mercuric chloride. About 2% of all patients died in early traumatic shock, 9% had oliguria or transient anuria (mortality was 55%), and in 13% anuria lasted more than 24 hours (mortality 92%).

Another group of 46 intoxications with acute renal failure after ingestion of mercuric chloride was reported by Valek, 1965. Inflammatory changes in the oral cavity were registered in 23 patients (50%), epigastric pain and vomiting in 44 patients (96%), and hematuria in 27 patients (58%). Thirty-three patients had severe diarrhea and 21 had blood in feces. Anuria developed in all cases within 24 hours after intoxication. Its duration (4 to 29 days) was not related to the ingested amount of mercury (0.1 to 8.0 g). Patients were treated by extracorporeal hemodialysis and BAL. Mortality was approximately 20%, but none of the patients died of uremia.

Microscopically, necrosis of the proximal tubular epithelium was seen in the first days (Stejskal, 1965). During the second week of therapy desquamation of the necrotic epithelia, with transport into the more distal parts of the nephron and the first signs of regenerative processes in the form of flat cells underneath the necrotic masses were observed. In the third to fifth weeks regeneration progressed with unequal rates in individual cases depending upon the success of treatment.

Cell necrosis of the renal tubular epithelium may develop either by direct toxic action of mercuric ions on the cell proteins or by disturbances in the renal circulation. Since disturbances of circulation in peritubular capillaries, caused by vascular spasms, develop immediately after the poisoning (Oliver, MacDowell, and Tracy, 1951) experimental evidence is available for both mechanisms. Intestinal changes probably are caused by similar mechanisms occurring in the capillaries of intestinal villi mucosa and in submucosal vessels (Schimmert and Wanadsin, 1950).

5.1.2.1.2.2 In Animals

Edwards, J. G., 1942, administered mercuric chloride to rabbits, guinea pigs, and frogs, and described development of acute total or segmental necrosis in the terminal portions of renal tubules. Mustakallio and Telkkä, 1955, observed cellular changes in the straight terminal portion of Henle's loop 24 hours after subcutaneous injection of mercuric ions into rats. At higher doses, the initial and middle portions of the proximal tubule were also affected and changes in the succinic dehydrogenase activity were observed. Bergstrand et al., 1959a, observed electron microscopical changes in the mitochondria of the proximal tubule in the rat kidney after repeated subcutaneous administration of mercuric chloride.

Gritzka and Trump, 1968, observed renal tubular lesions in rat kidney by electron microscopy 3 to 6 hours after subcutaneous injection of 4 mg $HgCl_2$/kg body weight. Rodin and Crowson, 1962, found similar histological changes in rats. Taylor, 1965, in studies on the time course of the development of renal damage, reported changes in the lower segment of the proximal tubule 24 and

36 hours after intramuscular injection of 1.25 mg Hg as mercuric chloride/kg body weight in female rats. With a higher dose, 5 mg Hg/kg, changes were observed already after 6 hours and more prominently after 12 hours.

Oliver, MacDowell, and Tracy, 1951, studied renal effects of inorganic mercury. In the dog, 5 hours after intravenous injection of 24 mg $HgCl_2$/kg body weight, typical signs of cortical ischemia were revealed by fluorescence techniques. The results were similar in rabbits 18 hours after administration of 15 mg $HgCl_2$/kg body weight. The patchiness of the cortical ischemia after nephrotoxic doses of mercuric salts suggests that the effects are caused more by local disturbances of blood flow within the cortical tissue than by overall reduction of the circulation in the entire kidney.

Flanigan and Oken, 1965, postulated that acute renal failure and anuria in rats 24 hours after injection of 18 mg $HgCl_2$/kg resulted from a primary decrease in glomerular filtration rate due to afferent arteriolar constriction. On the other hand, Bank, Mutz, and Aynedjian, 1967, stated that anuria after 4 mg Hg/kg body weight occurred in the presence of normal glomerular filtration rate and was the result of a complete absorption of the filtrate through an excessively permeable damaged tubular epithelium.

Biber et al., 1968, combined microdissection techniques with studies of functional changes in the damaged nephron by micropuncture 72 hours after administration of 5 mg Hg as mercuric chloride/kg in rat subcutaneously. This type of administration permitted slow development of renal damage without complicating vascular disturbances. Simple necrosis of tubular epithelium limited to lower portions of the proximal tubules was the principal microscopic finding. Microdissection techniques localized the necrotic changes into the distal three fourths of the length of the proximal tubule. The authors did not observe anuria or oliguria in the experimental animals. Therefore, it was concluded that anuria or decreased inulin clearance with or without oliguria can be a result of several possible mechanisms, including increased tubular leakage and reabsorption of inulin through an abnormally permeable tubular epithelium when the tubular lumen is completely obstructed by necrotic cellular masses. Or, primary reduction of glomerular filtration rate may be caused by preglomerular vasoconstriction or decrease in arterial pressure. Relative importance of the individual types of mechanisms probably varies with the severity of anatomic damage produced.

5.1.2.2 Chronic Poisoning

Chronic poisoning is caused almost exclusively by occupational inhalation exposures. Usually, various combinations of mercury vapors and dust of inorganic salts or elemental mercury alone are the sources of mercury exposure. Occupational exposures to mercuric dust alone are uncommon. The classic symptomatology of chronic mercury poisoning is reported in the literature without any distinction as far as the form of inhaled mercury is concerned. No attempt will be made for the purposes of this section to separate the chronic effects of elemental mercury vapor from the effects of inorganic mercury dust.

5.1.2.2.1 Nonspecific Signs and Symptoms

Weakness, fatigue, anorexia, loss of weight, and disturbances of gastrointestinal functions have always been associated with fully developed clinical forms of chronic poisoning following long-term exposures to inorganic mercury vapors and dusts. The specific symptomatology usually dominated these subtle signs of mercury exposure (e.g., Neal et al., 1937, 1941, and Neal and Jones, 1938). In modern times, industrial mercury exposure has declined to substantially lower levels and the importance of subtle signs is emphasized, since they precede specific symptoms of mercury poisoning. A thorough analysis of clinical symptomatology reported by Smith et al., 1970, clearly documents this conclusion. Loss of appetite and loss of weight were the most predominant symptoms in exposed groups and correlated well with the exposures. Gastrointestinal symptoms were reported more frequently in the exposed group than in controls, but they did not reveal a direct dose-response relationship with the exposure.

Biochemical effects of mercury also have been studied. Webb, 1966, has published an extensive review of studies on biochemical inhibitory effects in vitro and in vivo. Rentos and Seligman, 1968, did not find any relation between red blood cell glutathione levels or plasma alkaline phosphatase and exposure. On the other hand, clinical evidence has been reported for changes in serum enzyme activity with respect to lactic dehydrogenase

(LDH), alkaline phosphatase, and cholinesterase (Kosmider, 1964). Singerman and Catalina, 1970, examined 154 mercury miners and controls in Spain to detect enzymatic alterations attributable to mercury exposure. Changes of LDH isoenzymograms in serum, inhibition of LDH isoenzymes in urine, and inhibition of Na/K ATPase in erythrocyte membranes were found among exposed people. The significance of these biochemical indices for the diagnosis of mercury poisoning has not yet been shown.

5.1.2.2.2 Oropharyngeal Symptoms

Changes and symptoms in the oral cavity often have been prominent and early (Hamilton, 1925). Among the most frequent oral symptoms reported by the classic authorities in chronic exposure to mercury were queer metallic taste, sensation of heat in the oral mucosa, swollen, painful, readily bleeding gums, and increased flow of saliva. However, these observations were mainly dated to the time at which industrial exposures were often high and standards of hygiene low. Inflammatory changes of the gum with swollen and bleeding margins were usually not easily distinguishable from the pyorrhea of neglected oral hygiene. The degree of ptyalism varied. In extreme cases several liters of saliva were collected per day, while in other cases salivation was not observed.

West and Lim, 1968, in their study of high exposures in mercury mines and mills, observed soreness of the mouth with spongy gums as the most common symptom among the exposed workers. Other oral signs were loose teeth, bleeding gums, sore throats, dry mouth or salivation, and black lines on the gums. Smith et al., 1970, in their study of people chronically exposed to mercury vapors, did not find any objective abnormalities of teeth and gums related to exposure. In fact, the controls showed a higher incidence of abnormal teeth than the exposed workers.

5.1.2.2.3 Symptoms Related to Central Nervous System

The first scientific description of these symptoms was written by Kussmaul, 1861, in Germany. Today, the following are the most common manifestations: 1. asthenic-vegetative syndrome known as micromercurialism, 2. characteristic mercurial tremor involving the hands and subsequently other parts of the body, and 3. personality changes known as erethism (erethismus mercurialis Kussmaul).

5.1.2.2.3.1 Asthenic-vegetative Syndrome

The asthenic-vegetative syndrome, or micromercurialism, originally described by Stock, 1926, was based on the observation of psychological changes in persons exposed for long periods of time to low concentrations of atmospheric mercury. The symptomatology was later characterized by decreased productivity, increased fatigue and nervous irritability, loss of memory, loss of self-confidence, and ultimately, by miniature symptomatology of classical mercurialism: muscular weakness, vivid dreams, pronounced decrease of productivity, depressions, etc. Lvov, 1939 (quoted by Trachtenberg, 1969), mentioned that in many cases the symptomatology of micromercurialism may be falsely diagnosed as neurasthenic syndrome or hysteria, etc. Matusevic and Frumina, 1934 (quoted by Trachtenberg, 1969), showed that the chronic effects of low concentrations of mercury may be manifested by functional changes in the vegetative nervous system. Trachtenberg, 1969, stated that the clinical picture of micromercurialism is based not only on a minor intensity of classic symptoms of chronic mercury poisoning, but also might have its own characteristic symptomatology originating from disturbances in the cortical centers of the central nervous system and manifested by functional changes of organs of the cardiovascular, urogenital, or endocrine systems. Details concerning this syndrome, now used in the diagnosis of micromercurialism by Russian authors, are discussed in Section 7.1.2.1.2.

5.1.2.2.3.2 Mercurial Tremor

Mercurial tremor usually is preceded by other minor nervous symptoms of mercurialism such as insomnia and irritability. However, the presence of tremor in clinical symptomatology is one of the most characteristic features of mercurialism. With the continuation of exposure to mercury vapors or dust, the tremor develops gradually in the form of fine trembling of the muscles interrupted by coarse shaking movements every few minutes. The tremor usually begins in the fingers but it might just as well be seen on the closed eyelids, lips, and on protruded tongue. The frequency of the tremor in mercury intoxication cannot be generalized. Taylor, 1901, traced a frequency of 8 cycles per second and Kazantzis, 1968, found around 5 cycles per second. The tremor is intentional. It stops during sleep even in extreme cases. Psychotherapy cannot cure the tremor; it disappears gradually after cessation of exposure.

Classical authors reported that in progressive cases of mercury intoxication with continuing exposure, the tremor could spread throughout the limbs and they might jerk and jump. In extreme cases there might be a generalized tremor involving the whole body, even in the form of chronic spasms which could not be stopped by the strength of several men. Hamilton, 1925, described a patient with violent clonic muscular spasms throughout the entire leg, uncontrollable by morphine. The spasms stopped only after surgical anesthesia was applied. Similar symptomatology was recently described by Pieter Kark et al., 1971.

The mercurial tremor is central in its origin. Already the classical authorities emphasized the systemic character of the symptoms and localized the lesions to the cerebellum or cerebellar tracts (Guillain and La Roche, 1913, quoted by Hamilton, 1925). Beliles, Clarke, and Yuile, 1968, have observed fine tremor and perivascular histological changes in the medulla oblongata in rats chronically exposed to mercury vapor. In a recent experimental study, 2 of 6 male rabbits, exposed intermittently to mercury vapors 4 mg Hg/m^3 for 13 successive weeks, developed fine tremor and clonus in the fore and hind legs. Concentrations of mercury in the cerebellum and thalamus were determined by neutron activation analysis at 1.8 to 3.0 μg Hg/g wet tissue and were higher than in the remaining structures in the central nervous system (Fukuda, 1971).

Recently, Wood and Weiss, 1971, analyzed the tremor induced by industrial exposure to inorganic mercury by measuring the discriminative motor control. They showed that the tremor decreased in magnitude parallel with a fall of blood levels of mercury and with cessation of exposure. Neither therapy with N-acetyl-penicillamine nor propranolol administration produced marked improvement in performance.

5.1.2.2.3.3 Mercurial Erethism

The classic syndrome of erethism described originally by Kussmaul, 1861, is characterized by changes in behavior and personality, symptoms which appear late in long-term exposure to high concentrations of mercury. Increased excitability as well as depressive symptoms was reported. The final clinical symptomatology might depend upon the personality of the exposed worker. Loss of memory, insomnia, lack of self-control, irritability and excitability and/or timidity, anxiety, loss of self-confidence, drowsiness, and depressions constitute only a part of the rich spectrum of symptoms. Delirium with hallucinations, suicidal melancholia, or even manic-depressive psychosis were described in the most severe cases with excessive exposure (Hamilton, 1925). In recent times only minor psychic disturbances, e.g., insomnia, shyness, nervousness and dizziness are clinically observed in workers with high mercury exposures (Smith et al., 1970). Major disturbances were not reported even in severe exposures with mg/l. amounts of mercury excreted in the urine (West and Lim, 1968).

5.1.2.2.4 Renal Effects

Although acute renal effects of mercuric ion are well known, long-term effects of small doses of mercury on renal function and morphology have been denied for many years. In early descriptions, proteinuria in workers with massive exposures to mercury was usually ascribed to be a consequence of an acute phase of chronic mercury intoxication and not to be a specific feature of the clinical picture of chronic intoxication. Classic descriptions of chronic mercury poisoning do not usually mention proteinuria at all.

Later, transient kidney injury, with much better prognosis than acute renal failure, was included in descriptions of chronic exposures to mercury vapors or dust. Neal and Jones, 1938, and Neal et al., 1941, found increased incidence of proteinuria in workers exposed to mercury vapor in concentrations up to 0.7 mg Hg/m^3. Riva, 1945, and Jordi, 1947, mentioned for the first time investigations on the clinical symptomatology of the nephrotic syndrome (massive albuminuria, hypoproteinemia, edema) in workers exposed to mercury vapors in an ammunition factory and explained the syndrome as manifestations of hypersensitivity to mercury. Similar findings were published by Ledergerber, 1949, in 8 workers with high levels of mercury in the urine.

Friberg, Hammarström, and Nyström, 1953, reviewed the literature and published clinical details of another 2 cases of a transient nephrotic syndrome among workers exposed to elemental mercury vapors in the chlorine-alkali industry.

Normal blood pressure, normal glomerular filtration rate, edema, and high proteinuria (1 to 2 g/100 ml urine) with low levels of plasma albumin dominated the clinical description.

In 1960, Smith and Wells described 3 cases of proteinuria among workers exposed to mercury vapors with urinary levels of mercury between 180 and 1,000 μg Hg/l. Protein concentrations in the urine were 0.03 to 0.13 g/100 ml. The patients had no other signs of mercurialism. Removal from exposure resulted in cessation of proteinuria. Urinary proteins were separated by electrophoresis; albumin, α-2, and β-globulin with normal amino acid composition were highly prevalent. No attempts were made to analyze serum proteins or lipids. Earlier, Goldwater, 1953, reported 2 similar cases with reduced plasma proteins, disturbed albumin/globulin ratio, and elevated plasma cholesterol.

Kazantzis et al., 1962, analyzed the clinical status and occupational history of 3 workers exposed to mercury vapors and inorganic mercury salts. Their urinary levels of mercury were about 1 mg Hg/l. The syndrome was characterized by edema and ascites, high urinary losses of protein, hyaline casts in urinary sediment, and normal glomerular filtration rate. Plasma albumin levels were low (1.2 to 1.3 g%) and serum cholesterol was higher than 500 mg%. Electrophoresis of urinary proteins revealed the presence of albumin and an electrophoretic pattern unlike that found in tubular defects. Percutaneous renal biopsy was performed in 2 cases and showed abundance of lipids and vacuolization in the epithelium of proximal tubules. No abnormalities were seen in the glomeruli. Renal concentrations of mercury ranged between 10 to 20 μg/g fresh tissue.

Clennar and Lederer, 1958, reported similar cases. They analyzed the renal tissue for mercury and found 15 μg Hg/g fresh tissue. Burston, Darmady, and Stranack, 1958, supplemented the clinical picture with the microscopic finding of fatty degeneration and necrotic changes in tubular epithelium in a similar patient. The evaluation of these cases is difficult because congestive cardiac failure may be responsible for a similar morphological picture.

No satisfactory evidence on the long-term effects of mercuric salts on the renal function is available for animals. Morphological findings in the kidney tissue, associated with edema and ascites, were observed in hamsters after repeated subcutaneous injections of mercurous chloride (Zollinger, 1955).

In conclusion, the exposure to inorganic mercury may produce proteinuria in exceptional cases. At continued exposure, the proteinuria can lead to excessive losses of serum albumin with the development of a nephrotic syndrome (Squire, Blainey, and Hardwicke, 1957). The mercury-induced nephrotic syndrome has several specific characteristics. It does not occur in all members of exposed populations and is not directly dose-related (see Section 7.1.2.2). As repeatedly shown, mercury-induced proteinuria is transient and the prognosis of the nephrotic syndrome is good. With cessation of exposure or specific therapy the recovery usually is complete. The proteinuria is characterized by massive losses of albumin and α-2-globulin with electrophoretic mobilities identical with those of corresponding serum proteins. However, in microscopic examination the glomeruli do not show any abnormalities and tubular origin of protein is improbable. The mechanisms of this proteinuria are completely unknown. It has been suggested that a hypersensitivity or idiosyncracy to mercury, similar to that suspected in cases of acrodynia (see Section 5.1.2.3), exists in persons showing this symptomatology.

5.1.2.2.5 Ocular Symptomatology (Mercurialentis)

Atkinson, 1943, examined a group of 70 workers with different types of exposure to inorganic mercury and in all 37 workers with exposure times longer than 5 years, he observed a colored reflex from the anterior capsule of the lens. This grayish-brown or even yellow reflex was not easily seen except in the slit lamp beam. It was most prominent in the center of the lens and faded toward the periphery with small cracks and defects. The reflex was present in all patients who revealed other symptoms of mercury poisoning but was also present in persons without symptoms of mercury poisoning. The author concluded that the occurrence of this symptom is an indication of a hazardous mercury exposure. Later, Atkinson and von Sallman, 1946, expressed the belief that the reflex is caused by an actual deposition of mercury on the lens. Abramowicz, 1946, reported similar observations after local application of mercurial ointment to the lids over long periods of time. In 1950 Rosen introduced a new term for this reflex, mercurialentis.

Among repairers of direct-current meters, Locket and Nazroo, 1952, showed that the brown reflex was not related to age. The reflex did not cause any visual symptoms or other ocular disturbances. In accidental peroral intoxication by inorganic mercury, no reflex was observed in spite of diplopia, nystagmus, and retinal edema.

Burns, 1962, examined 70 workers in a thermometer factory where 57 persons were constantly exposed to mercury vapors for periods ranging between 1 and 48 years. Definite mercurialentis was present in 56 of them. The author proposed that mercury is absorbed from the atmosphere or from local applications, through the cornea, and accumulates over the years on the anterior surface of the lens in the pupillary area until a visible, permanent deposit is formed. He also considered mercurialentis as a symptom of exposure rather than a symptom of poisoning.

5.1.2.3 Hypersensitivity or Idiosyncracy

Individual variability in the tolerance to exposure to mercury has been observed repeatedly in occupationally exposed adults. In the first decades of this century more attention was directed toward the suspected hypersensitivity to mercury in relation to the extensive use of mercury compounds in therapy, especially in children.

In 1947 Fanconi, Botsztejn, and Schenker reviewed 56 cases in which people had reacted to trace amounts of mercury with pathological manifestations such as skin rash, stomatitis, and gangrene of mucous membranes in the mouth. They concluded that some persons might have a lower tolerance to mercury than others. Idiosyncracy to mercury, in contrast to classic symptoms of mercury poisoning, was reported mainly in connection with locally applied preparations of mercury. The first reports of cases have been ascribed to Deakinn, 1883, and Green, 1884 (quoted by Gibel and Kramer, 1943), who observed idiosyncracies in adults exposed to 5 to 10% preparations of ammoniated mercury ointment. Idiosyncracy in children was described in the 1930's (Harper, 1934) and 1940's (Bass, 1943). Gibel and Kramer, 1943, collected reports of 14 cases published between 1883 and 1942 of idiosyncratic reactions following the use of mercurial ointments (1 to 15% mercury), mercurous chloride, metallic mercury, and diaper rinses containing dilute mercuric chloride. Reactions varied from mild erythema to morbilliform rash and severe papulovesicular eruptions covering the entire body, followed by scaling and exfoliative dermatitis. Cutaneous reaction appeared 1 to 11 days after the application was started.

A more specific form of disease, described as an untoward systemic reaction to mercury, is *acrodynia*, variously known as pink disease, erythredema, Feer's vegetative neurosis, and erythredema polyneuritis. Selter, 1903, was probably the first to describe the characteristic symptomatology. The disease affects only children between the ages of 4 months and 4 years. The name "pink disease" is derived from the rash, the color of raw beef. Other symptoms are coldness, swelling, and irritation of the hands, feet, cheeks and nose, usually followed by desquamation, loss of hair, and ulceration. The onset of the disease is characterized by gradually increasing irritability, photophobia, sleeplessness, and profuse perspiration, particularly in the extremities, leading rapidly to signs of general dehydration. Neurological symptoms include tremor, decreased tendon reflexes, marked hypotonia, muscle weakness, and ligament relaxation, permitting the typical "salaam position" of the child in bed. The profuse perspiration is accompanied by enormously dilated and enlarged sweat glands and desquamation of the soles and palms. Fingers and toes are edematous because of hyperplasia and hyperkeratosis of the skin, and the pain leads the child to rub his hands and feet in a characteristic fashion. The prognosis is usually good.

In several cases of acrodynia, changes in the peripheral nerves characterized by demyelinization of fine nerve bundles were described (Patterson and Greenfield, 1923). Secondary changes found in anterior horn cells were interpreted as a retrograde reaction to the degeneration of motor nerves (Orton and Bender, 1931).

The pathogenesis of acrodynia was originally ascribed to different toxins. Mercury has been suspected as a causative agent since 1922 (Zahorsky). Subsequently, the disease was suggested to be an allergic reaction to mercury. The importance of mercury as a cause of the disease received strong support from Warkany and Hubbard, 1948, who found increased concentrations of mercury in urine in one of their patients during an investigation of the presence of toxic metals in the urine. The authors reviewed 20 cases of florid acrodynia and proved that in 18 patients the concentration of mercury in the urine was higher than 50 µg/l. There was no detectable mercury in the urine in 40 out of 49 controls. One control child excreted 50 to 100 µg/l. This child had been treated earlier with calomel tablets. Three other mercury-excreting controls had considerably lower levels in subsequent samples. Active search for acrodynic children during the following years revealed that 64% of 189 urine samples from children with acrodynia contained over 50 µg Hg/l., whereas only 2 samples with mercury concentrations above 50 µg/l. were found among 87 control children. Moreover, previous exposure to mercury was also discovered in the few cases of acrodynia in which the urinary spot samples were free of mercury (Warkany and Hubbard, 1951, and 1953).

In England, Holzel and James, 1952, compared 2 areas, Manchester and Salford counties, where acrodynia was relatively common, and Warwick, where the frequency of the disease was low. A positive correlation was found between the use of a mercury-containing teething powder and the frequency of the disease. Skin sensitivity tests performed on 10 patients with florid acrodynia were positive in only 1 case. One case of skin hypersensitivity was registered also in the control group of 30 children. The results were considered as evidence against an allergic character of the disease. Urinary levels of mercury were considered abnormal in only 65% of 94 patients examined. The authors regarded increased urinary excretion of mercury as proof that mercury is an etiological agent in infants with temporarily decreased tolerance due to an unknown factor. This factor was assumed to be independent of exposure to mercury and contributive to the provocation of the disease.

Several authors concentrated their attention on the symptoms of increased sympathetic activity. Cheek and Hicks, 1950, found hemoconcentration and low levels of plasma sodium in children with acrodynia and revived an older theory that acrodynia is a disorder of the vegetative nervous system. Later, Cheek, Bondy, and Johnson, 1959, suggested that mercury potentiates the action of epinephrine in the body and that coexistence of sympathetic stress and exposure to mercury can give rise to the specific symptomatology. Experimental results of Axelrod and Tomschick, 1958, gave evidence that methyl transferase, active in biotransformation of epinephrine, can be blocked by the mercuric ion. Over-activity of the sympathetic nervous system in acrodynia also was confirmed by the studies of Farquahar, 1953, and Farquahar, Crawford, and Law, 1956. These authors reported higher urinary excretion of catecholamines in acrodynic children than in normal control infants subjected to stress during the time of urine collection.

The mercury theory received support from Bivings and Lewis, 1948, who used dimercaptopropanol for the treatment of 1 patient who showed remarkable improvement after several days. The same treatment was used successfully in a number of other cases by Bivings, 1949. However, results of other studies were not always equally favorable (Fanconi and von Murait, 1953, and Baumann, 1954), and BAL was replaced by N-acetyl-penicillamine as the therapeutic agent of choice in later studies (Hirschman, Feingold and Boylen, 1963, and Bureau et al., 1970).

In conclusion, there is no doubt today that exposure to mercury was related in some degree to the increased incidence of acrodynia, especially since the disease was almost eradicated after withdrawal of mercury from the common therapeutic agents used in children. Today, acrodynia is reported only sporadically (Hirschman, Feingold, and Boylen, 1963, and Bureau et al., 1970).

The disease almost disappeared, without proper analysis and without elucidation of the mechanisms that induced it (Warkany, 1966). The disease was never produced in animals. Simultaneous administration of mercurous chloride and sympathetic stimulation in rats only potentiated the effects of epinephrine on insensible perspiration, hemoconcentration, and sustained hypertension. The animals developed weakness and coldness of the extremities but did not exhibit the full symptomatology of the disease (Cheek, Bondy, and Johnson, 1959).

5.1.3 Summary

Acute intoxication by inorganic mercury can be provoked by: 1. accidental inhalation of high concentrations of elemental mercury vapors, causing bronchial irritation, erosive bronchitis, and diffuse interstitial pneumonitis, and 2. ingestion of dissociable inorganic salts of bivalent mercury that can produce local necrotic changes in the gastrointestinal tract, circulatory collapse, or acute renal failure with oliguria or anuria.

Early stages of chronic poisoning by inorganic mercury, usually by industrial exposure to elemental mercury vapor alone or in combination with dust of mercuric salts, are characterized by anorexia, loss of weight, and minor symptomatology of the central nervous system (the asthenic-vegetative syndrome; micromercurialism). The symptoms are increased irritability, loss of memory, loss of self-confidence, and insomnia. Later phases are characterized by mercurial tremor, psychic disturbances, and changes in personality (erethismus mercurialis).

In exceptional cases, chronic exposure to inorganic mercury may produce transient proteinuria and a benign form of the nephrotic syndrome. Deposition of mercury on the anterior surface of the eye lens (mercurialentis) is only a sign of exposure, not a symptom of chronic mercurialism.

Idiosyncracy to trace amounts of inorganic mercury was reported in the older literature, mainly in connection with local application of mercury preparations. A specific form of systemic reaction to mercury, acrodynia (pink disease, erythredema) has been described. There is no doubt that this systemic reaction was related in some degree to exposure to mercury, but because the disease was almost eradicated before a proper analysis of the mechanisms inducing it had been completed, a definite relation to mercury exposure has never been established.

5.2 Organic Mercury Compounds
5.2.1 Alkyl Mercury Compounds

Exposure to alkyl mercury compounds may occur within the uterus (prenatal intoxication), or it may occur after birth (postnatal intoxication). The symptoms and signs in victims of prenatal intoxication show certain dissimilarities to those present in intoxicated adults.

5.2.1.1 Prenatal Intoxication
5.2.1.1.1 In Human Beings

From Minamata, 22 cases of prenatal *methyl mercury* intoxication have been described (Harada, 1968b). In connection with the Niigata incident, no proven case was found but one suspect case was noted (Tsubaki, 1971). Engleson and Herner, 1952, reported on one case in a newborn whose mother during pregnancy had eaten porridge containing seed dressed with methyl mercury. Snyder, 1971, described a case of prenatal poisoning in an infant whose mother had consumed during pregnancy meat from hogs which had been fed seed grain treated with methyl mercury.

The prevalence of symptoms in prenatal intoxication in Minamata is shown in Table 5:1. The clinical picture was that of an unspecific infantile cerebral palsy. All patients had motor disturbances, mainly ataxic, and mental symptoms. Murakami, 1971, reported that 14 patients had malocclusion and 2 had other congenital malformations. The prognosis is poor. Two of the Minamata patients died. Medical and physical treatment had only a slight effect (Tokunaga, 1966, and Kitagawa, 1968).

Post-mortem pathological findings have been reported for 2 cases from Minamata (Matsumoto, Koya, and Takeuchi, 1965, and Takeuchi, 1968a). The brains were hypoplastic with a symmetrical atrophia of cerebrum and cerebellum involving both cortex and subcortical white matter. Microscopically, a decreased number of neurons and distortion of the cytoarchitecture were noted in the total neocortex. In cerebellum the cell loss was seen mainly in the granular cell layer. The pathological changes observed cannot be distinguished from those often seen in cerebral palsy of unknown etiology.

Ten cases of prenatal intoxication with *ethyl*

TABLE 5:1

Prevalence of Symptoms in 22 Cases of Prenatal Methyl Mercury Intoxication in Minamata. (From Harada, 1968b.)[x]

Symptoms	Prevalence %
Mental disturbance	100
Ataxia	100
Impairment of gait	100
Disturbance in speech	100
Disturbance in chewing and swallowing	100
Brisk and increased tendon reflex	82
Pathological reflexes	54
Involuntary movement	73
Salivation	77
Forced laughing	27

[x]Visual fields and hearing not examined.

mercury have been reported by Bakulina, 1968. The mothers had shown symptoms of poisoning up to 3 years before pregnancy. In 3 of the prenatal cases, severe mental and neurological symptoms in accordance with those seen in Minamata were described. The symptomatology in the other 7 children was only briefly mentioned. In some of them, born one and one half years or more after the onset of symptoms in the mother, decreased weight and floppiness at birth were the only symptoms reported. Mercury was found in the breast milk of 3 of the mothers (see Section 8.1.1.1). The author emphasized the possible importance of the postnatal exposure from this source.

5.2.1.1.2 In Animals

Spyker and Sparber, 1971, have reported behavioral disturbances (low activity, backing, inappropriate gait, and difficulties in swimming) in 30-day-old offspring of rats injected with *methyl mercury* during pregnancy. Neurological signs with disappearance of righting reflexes occurred 2 and one half months later.

Okada and Oharazawa, 1967, found a decrease in weight in litters of mice given *ethyl mercury* phosphate during pregnancy. A high frequency of cleft palate in the litter was reported by Oharazawa, 1968 (quoted by Clegg, 1971). Ataxia was observed by Morikawa, 1961b, in a newborn kitten whose mother had been given *bis-ethyl mercury* sulphide during pregnancy. No study allows definite conclusions about the clinical picture.

A few papers have been published on the morphological changes in prenatal *methyl mercury* intoxication in mammals (Matsumoto et al., 1967, Moriyama, 1968, and Nonaka, 1969). None of these reports allows for definite conclusions regarding the morphology.

Morikawa, 1961b, and Takeuchi, 1968b, have described the neuropathology in 1 kitten of a cat poisoned with *bis-ethyl mercury* sulfide. The main finding was cerebellar hypoplasia of granular cell type. Seven additional kittens in the litter were said to have had similar brain damage.

5.2.1.2 Postnatal Intoxication
5.2.1.2.1 In Human Beings
5.2.1.2.1.1 Local Effects

Dermatitis and eczema were reported after cutaneous contact with *methyl mercury* (Dillon, Weston, and Booer, 1935, Lundgren and Swensson, 1949, and Berkhout et al., 1961), *ethyl mercury* (Dillon, Weston, and Booer, 1935, Vintinner, 1940, Goldblatt, 1945, Ritter and Nussbaum, 1945, Schulte, 1946, Ellis, 1947, and Cohen, 1958), and *tolyl mercury* (Goldblatt, 1945) compounds.

At inhalation exposure to alkyl mercury compounds, irritation of the mucous membranes of the nose, mouth, and throat occurs (Koelsch, 1937, and Lundgren and Swensson, 1949). The symptoms start after a short exposure and usually disappear soon after the termination of exposure.

Sodium ethyl mercury salicylate (Merthiolate®, Thimerosal), formerly used extensively as a topically applied antimicrobial agent, caused hypersensitivity reactions (Ellis and Robinson, 1942, Lane, 1945, and Underwood et al., 1946). The compound is used widely at present as a preservative in solutions for parenteral injection. Cutaneous tests in persons with (Epstein, Rees, and Maibach, 1968) or without (Hansson and Möller, 1970 and 1971a and b) skin disorders have revealed 7 to 16% positives, with higher percentages in young people than in old. The reaction is neither clearly allergic nor clearly irritant (Hansson and Möller, 1970). Positive reactions have been stated not to be caused by earlier exposure by injection (Hansson and Möller, 1971a), and generalized reactions do not occur after subcutaneous injections in subjects with positive skin tests (Hansson and Möller, 1971a). When incorporated in test solutions for intracutaneous use (e.g., tuberculin), it may give false positive reactions (Hansson and Möller, 1971b). The frequency of cross-sensitivity with inorganic mercury or other organic mercury compounds is low (Fregert and Hjorth, 1969). The sensitizing properties have been claimed to be due to the salicylic acid part of the molecule (Ellis, 1947, and Gaul, 1958).

5.2.1.2.1.2 Systemic Effects

The description of *methyl mercury* intoxication in this report is based mainly on observations from the epidemics in Minamata 1953 to 1960 and in Niigata 1964 to 1965. Methyl mercury poisoning is often referred to as the Minamata disease. In Minamata a total of 98 patients were diagnosed (Tokuomi et al., 1961, Harada, 1968a, Tokuomi, 1968, and Takeuchi, 1970) and in Niigata, 48 patients (Tsubaki, 1971).

About 80 more cases of intoxication with methyl mercury have been reported in subjects exposed occupationally (Hunter, Bomford, and Russell, 1940, Herner, 1945, Ahlmark, 1948, Ahlborg and Ahlmark, 1949, Ahlmark and Tunblad, 1951, Lundgren and Swensson, 1948, and 1949, Koelsch, quoted by Zeyer, 1952, Höök, Lundgren, and Swensson, 1954, Prick, Sonnen, and Slooff, 1967a and b), exposed through medical treatment (Tsuda, Anzai, and Sakai, 1963, Ukita, Hoshino, and Tanzawa, 1963, Okinaka et al., 1964, and Suzuki and Yoshino, 1969), exposed through consumption of methyl mercury dressed seed (Engleson and Herner, 1952, and Ordonez et al., 1966), or exposed through meat from swine fed such seed (Storrs et al., 1970a and b, Center for Disease Control, 1971, and Curley et al., 1971).

Systemic intoxication may occur after a short or long exposure time. In regard to the clinical picture, there does not seem to be any clear difference between acute and chronic poisoning. A characteristic feature is the latency period of weeks to months between a single dose exposure and the onset of symptoms (Höök, Lundgren, and Swensson, 1954).

When fully developed, the clinical picture contains 3 main symptoms: 1. sensory disturbances in the distal parts of the extremities, in the tongue, and around the lips, 2. ataxia, and 3. concentric constriction of the visual fields. Hearing loss, symptoms from the autonomic and extrapyramidal nervous systems, and mental disturbances also occur.

The relative frequency of symptoms reported in some adult patients from Minamata (Tokuomi, 1968) and from Niigata (Tsubaki, 1971) is given in Table 5:2.

Morphological changes in methyl mercury poisoning have been reported by Hunter and Russell, 1954, Tsuda, Anzai, and Sakai, 1963, Okinaka et al., 1964, Oyake et al., 1966, Hiroshi et al., 1967, Prick, Sonnen, and Slooff, 1967b, and Takeuchi, 1968a. A review of the findings was made recently (Berglund et al., 1971). A neuron degeneration and loss with gliosis occur mainly in the cerebral cortex in the calcarine area and in the precentral and postcentral areas, superior frontal gyrus and frontal areas. In cerebellum there is a granular cell loss leading to atrophy. Takeuchi, 1970, mentioned damage in peripheral nerves. Other reports have denied such changes (Hunter and Russell, 1954, and Prick, Sonnen, and Slooff, 1967b).

Extensive laboratory examinations gave few positive findings. There were unspecific changes in the electroencephalograms in most of the examined patients (Harada et al., 1968). Some cases were reported to have changes in electromyography and electroneurography (Murone et al., 1967, and Tsubaki, 1969). Proteinuria (e.g., Ordonez et al., 1966) and porphyrinuria (Matsuoka et al., 1967, and Tokuomi, 1968) also have been noted. Taylor, Guirgis, and Stewart, 1969, have found an inhibition of serum phos-

TABLE 5:2

Prevalence of Symptoms in Adult Cases of Methyl Mercury Intoxication from Minamata (34 cases) and Niigata (40 cases). (From Tokuomi, 1968, and Tsubaki, 1971.)

Symptom	% of cases	
	Minamata	Niigata
Constriction of visual fields	100	74
Hearing impairment	85	68
Disturbance of sensation		95
Superficial	100	
Deep	100	
Ataxia		
Adiadochokinesis	94	72
Dysgraphia	94	
In buttoning	94	
In finger-finger, finger-nose tests	81	
Impairment of gait	82	
Dysarthria	88	
Romberg's sign	43	
Tremor	76	
Extrapyramidal symptoms		15
Muscular rigidity	21	
Ballism	15	
Chorea	15	
Athetosis	9	
Contractures	9	
Tendon reflexes		
Exaggerated	38	8
Weak	9	50
Pathologic reflexes	12	14
Hemiplegia	3	
Salivation	24	
Sweating	24	
Slight mental disturbance	71	43

phoglucose isomerase and an increased urinary protein excretion in subjects occupationally exposed to one or more of the compounds used for seed dressing, such as methyl, phenyl, alkoxyethyl or tolyl mercury, though there was no clinical evidence of poisoning.

In severe cases, the prognosis was bad. Some regression of the sensory symptoms occurred and physical treatment relieved some of the motor disturbances (Tokunaga, 1966, and Kitagawa, 1968), but in most cases the symptoms remained. In Minamata 43 out of 98 patients had died in 1968 (Takeuchi, 1970) and in Niigata, 6 out of 48 had died in 1971 (Tsubaki, 1971).

Some 250 cases of poisoning by *ethyl mercury* compounds have been reported. Most of them have not been described in detail. The majority of patients had eaten seed dressed with various ethyl mercury compounds (Jalili and Abbasi, 1961, Kantarjian, 1961, Haq, 1963, and Dahhan and Orfaly, 1964). A few cases due to occupational exposure by inhalation or due to medical treatment with mercuric ointments also have been reported (Prumers, 1870 [quoted by Swensson, 1952], Veilchenblau, 1932, Merewether, 1946, Pentschew, 1958, Höök, Lundgren, and Swensson, 1954, Drogitjina and Karimova, 1956, Saito et al., 1959, Hay et al., 1963, Katsunuma et al., 1963, and Schmidt and Harzmann, 1970). Suzuki et al., in press, mentioned, without describing the symptomatology, a case of poisoning by infusion of plasma solution containing sodium ethyl mercury salicylate as a preservative.

The clinical picture in ethyl mercury poisoning has shown a similarity to that seen in methyl mercury poisoning with some possible differences. Motor weakness of the extremities with progressive muscular atrophy and widespread fasciculations, a syndrome similar to that of amyotrophic lateral sclerosis, was reported by Kantarjian, 1961, in persons poisoned by ethyl mercury p-toluene sulfonanilide-treated seed. A cardiac affection with changes in the electrocardiogram was described in some cases (Welter, 1949, quoted by Schmidt and Harzmann, 1970, Jalili and Abbasi, 1961, Dahhan and Orfaly, 1964, and Mnatsakonov et al., 1968). Such changes were associated with hypokalemia (Mnatsakonov et al., 1968).

Ethyl mercury exposure has been associated with symptoms from the gastrointestinal tract, including abdominal pain, vomiting, and diarrhea (e.g., Veilchenblau, 1932, and Jalili and Abbasi, 1961). It is not clear whether these symptoms are local or systemic. In some of the reports on ethyl mercury poisoning, albuminuria has been mentioned (Jalili and Abbasi, 1961, Hay et al., 1963, Haq, 1963, and Dahhan and Orfaly, 1964).

There have been a few reports on the morphological changes in ethyl mercury poisoning (Hay et al., 1963, and Schmidt and Harzmann, 1970). The findings are in accordance with those described in methyl mercury poisoning.

A few cases of *di-ethyl mercury* intoxication have been reported (Hill, 1943, Welter, 1949, quoted by Tornow, 1953, and Ashbel, 1959, quoted by Meshkov, Glezer, and Panov, 1963). From the scanty descriptions of symptoms available, there does not seem to be any clear difference between mono-ethyl mercury poisoning and di-ethyl mercury intoxication.

5.2.1.2.2 In Animals

Symptoms which might be of local origin were noted after inhalation or oral exposure to *methyl mercury* (Hunter, Bomford, and Russell, 1940, Hagen, 1955, Saito et al., 1961, and Takeuchi et al., 1962) or *ethyl mercury* (Oliver and Platonow, 1960, and Palmer, 1963) compounds.

Systemic intoxication by *methyl mercury* compounds in mammals has been described in mice (Hagen, 1955, Gage and Swan, 1961, Saito et al., 1961, Takeuchi et al., 1962), rats (Hunter, Bomford, and Russell, 1940, Swensson, 1952, Hagen, 1955, Kai, 1963, Takeuchi, 1968b, Berglund, 1969), ferrets (Hanko et al., 1970), rabbits (Swensson, 1952, Kai, 1963, Irukayama et al., 1965), cats (Takeuchi, 1961, Kai, 1963, Yamashita, 1964, Pekkanen, 1969, Albanus, Frankenberg, and Sundwall, 1970), dogs (Kai, 1963, Irukayama et al., 1965), swine (Storrs et al., 1970a, Piper, Miller, and Dickinson, 1971), and monkeys (Hunter, Bomford, and Russell, 1940, Nordberg, Berlin, and Grant, 1971, and Berlin, Nordberg, and Hellberg, in press).

In most species, anorexia and weight loss were the first signs of intoxication. However, neurological symptoms dominated the clinical picture in all of the species studied. The symptoms have shown a definite similarity. The most prominent sign first seen was an ataxia, including weakness and clumsiness in the hind legs. In swine (Storrs et al., 1970a) and monkeys (Nordberg, Berlin, and

Grant, 1971, and Berlin, Nordberg, and Hellberg, in press) blindness was noted.

Morphological changes in methyl mercury intoxication were reported in mice, rats, ferrets, cats, and monkeys. A review of the findings was given recently (Berglund et al., 1971). Damage was noted in cerebral and cerebellar cortex and in peripheral nerves with their dorsal roots separately or in combination. The pattern is similar to that observed in human beings. Peripheral nerve damage has been reported to be the first lesion present in rats (Miyakawa et al., 1970). Grant (in press) observed peripheral nerve damage in rats without neurological signs of poisoning. Grant also reported clinically silent CNS damage in monkeys. Miyakawa et al., 1971a, gave some evidence of regeneration of peripheral nerve lesions in rats. The same authors (1971b) observed slight muscle lesions in rats. Fowler (1972) reported tubular kidney damage in rats.

Ethyl mercury poisoning has been described in rats (Takeuchi, 1968b, Itsuno, 1968, and Krylova et al., 1970), rabbits (Schmidt and Harzmann, 1970), cats (Morikawa, 1961a, Yamashita, 1964, and Takeuchi, 1968b), sheep (Palmer, 1963), swine (Taylor, 1947), and calves (Oliver and Platonow, 1960).

The symptoms were similar to those in methyl mercury poisoning. Oliver and Platonow, 1960, reported that heavy oral exposure produced predominantly gastrointestinal disturbances in calves, while prolonged administration of lower doses gave rise to mainly neurological symptoms. They also found electrocardiographic disturbances and changes in the serum protein pattern. Kidney damage was reported in rabbits (Schmidt and Harzmann, 1970), and calves (Oliver and Platonow, 1960). Cardiac lesions were noted in mice, rats, and rabbits (Trachtenberg, Goncharuk, and Balashov, 1966, Krylova et al., 1970, Schmidt and Harzmann, 1970, and Verich, 1971). Blindness may occur in swine (Taylor, 1947).

Morphological changes in CNS in cats poisoned by 6 different ethyl mercury compounds were reported by Morikawa, 1961a. The general pattern was in accordance with that seen in methyl mercury poisoning. Powell and Jamieson, 1931, and Schmidt and Harzmann, 1970, found kidney damage in rabbits exposed to ethyl mercury compounds.

Symptoms of *di-ethyl mercury* intoxication in dogs have been described by Hepp, 1887. Of the other alkyl mercury compounds, only *n-propyl mercury* has been associated with this kind of intoxication (Itsuno, 1968, and Takeuchi, 1968b).

5.2.2 Aryl Mercury Compounds

Murakami, Kameyama and Kato, 1956, found malformation in litters of mice exposed to *phenyl mercury* acetate during pregnancy. No other study has indicated prenatal effects.

Because of the instability of aryl mercury compounds, there is an obvious accompanying risk of exposure to inorganic mercury.

5.2.2.1 In Human Beings
5.2.2.1.1 Local Effects

Phenyl mercury compounds are well known to cause dermatitis in skin exposure (Levine, 1933, Dillon Weston and Booer, 1935, Gross, 1938, Wilson, 1939, Vintinner, 1940, McCord, Meek, and Neal, 1941, Goldblatt, 1945, Cotter, 1947, Rost, 1953, Sunderman, Hawthorne, and Baker, 1956, Massmann, 1957, Morris, 1960, Ladd, Goldwater, and Jacobs, 1964, and Hartung, 1965). Some of those reactions have been claimed to be allergic (Mathews and Pan, 1968).

5.2.2.1.2 Systemic Effects

Only a few poisonings have been blamed on exposure to aryl mercury compounds. The clinical picture is not at all as uniform as in alkyl mercury poisoning and the causal relationship to the exposure to mercury is often doubtful.

Birkhaug, 1933, reported slight abdominal pain and diarrhea after ingestion of about 100 mg of mercury as *phenyl mercury* nitrate during 24 hours. No albuminuria was noted. Koelsch, 1937, reported in 3 persons who had inhaled phenyl mercury unspecific symptoms such as fatigue, dyspnea, edematous inflammation of mucous membranes, increased body temperature, and pain in the chest and in the extremities. In one case, there might have been kidney damage with proteinuria and edema. Bonnin, 1951, observed headache, vomiting, abdominal pain, signs of meningism, and paresthesia in 1 person exposed through inhalation of phenyl mercury and methoxyethyl mercury compounds. The symptoms remained for several days.

Cotter, 1947, described the symptoms of 10 subjects said to have been exposed to various phenyl mercury salts. The exposure was not very well defined; it was not clear whether exposure to other substances had also occurred. Most of the patients showed evidence of liver damage. Anemia and other blood changes were also present.

A case of nervous system involvement was reported by Brown, 1954. The patient had been exposed to phenyl mercury acetate through inhalation. The symptoms consisted of gingivitis, with a possible mercury line and dysphagia, dysarthria, motor weakness in arms and legs, abnormal tendon reflexes, and muscular fasciculations. There were no sensory disturbances (see also Section 8.2.2.1).

Goldwater et al., 1964, found a transient albuminuria without other symptoms in a subject exposed heavily by having phenyl mercury acetate sprayed on his skin.

5.2.2.1.3 Hypersensitivity or Idiosyncracy

In a child exposed to mercury from bedroom walls painted with a paint containing *phenyl mercury* propionate, Hirschman, Feingold, and Boylen, 1963, described the syndrome of pink disease (acrodynia). Mercury vapor detector measurement showed that elemental mercury was emitted from the freshly painted wall. It is not known whether exposure to phenyl mercury had also occurred. Goscinska, 1965, reported a similar case in which a child had been sprayed on the face, lips, hands, and clothes with phenyl mercury acetate. The symptoms, which occurred 2 months after the exposure, were convulsions, tremor, ataxia, visual and aural impairment, abnormal electroencephalogram, mental retardation, acrodynia, and aminoaciduria.

Mathews and Pan, 1968, reported a case of severe asthma and urticaria probably caused by exposure to phenyl mercury propionate in hospital linens. There was a positive skin test reaction to phenyl mercury compounds. No skin irritation or sensitization or other clinical manifestations were observed in 1,500 hospitalized patients who had come in contact with phenyl mercury propionate-treated fabrics over a period of 6 months (Linfield et al., 1960).

5.2.2.2 In Animals

Only a few reports have described the symptoms of *phenyl mercury* intoxication and a typical symptomatology cannot be stated. Hagen, 1955, found pulmonary symptoms in mice exposed through inhalation to a dust of phenyl mercury salts. Renal damage was observed in mice, rats, and rabbits given phenyl mercury salts intraperitoneally or intravenously (Weed and Ecker, 1933).

Fitzhugh et al., 1950, found a decreased weight gain, histopathological kidney damage and reduced survival period in rats exposed to phenyl mercury acetate in the diet at varying levels for 12 to 24 months. After oral administration, Tryphonas and Nielsen, 1970, observed anorexia, diarrhea, weight loss, and renal damage in swine. At repeated administration of phenyl mercury acetate to mice, Gage (in press) did not observe paralysis of the type seen at methyl mercury administration.

Morphological investigations in phenyl mercury poisoning are incomplete. Kidney damage has been reported in mice (Wien, 1939) and rats (Weed and Ecker, 1933, Wien, 1939, and Fitzhugh et al., 1950). Wien, 1939, observed gastrointestinal lesions after parenteral administration in rats and mice. Morikawa, 1961a, found no neuropathological changes in heavily exposed cats.

5.2.3 Alkoxyalkyl Mercury Compounds

Of the alkoxyalkyl mercury compounds, only *methoxyethyl mercury* has been experimentally investigated and associated with clinical intoxication. No cases of prenatal poisoning have been published. As in the case of aryl mercury compounds, the relative chemical instability of alkoxyalkyl mercury compounds indicates a possible mixed exposure with inorganic mercury in experimental and clinical cases.

Substituted alkoxyalkyl mercury compounds (mercurial diuretics) will be discussed in Section 5.2.4.

5.2.3.1 In Human Beings

Methoxyethyl mercury silicate has a local irritating effect on the skin (Wilkening and Litzner, 1952). A few cases of systemic poisoning after inhalation of methoxyethyl mercury compounds have been published (Wilkening and Litzner, 1952, Zeyer, 1952, Dérobert and Marcus, 1956, and Strunge, 1970). Most cases have shown symptoms from the gastrointestinal tract (gingivitis, abdominal pain, vomiting, diarrhea or constipation) and/or kidneys (albuminuria and red cells and casts in the urine), and/or unspecific symptoms, such as erethism, fatigue, headache, anorexia. One patient had a nephrotic syndrome (Strunge, 1970). Two persons showed pulmonary symptoms after inhaling methoxyethyl mercury oxalate or silicate (Zeyer, 1952, and Derobert and Marcus, 1956). Only 1 patient had objective neurological signs with tremor, dysgraphia, and motor and sensory disturbances in the legs, but no definite ataxia (Zeyer, 1952).

No report is available on the morphological changes in alkoxyalkyl mercury poisoning.

5.2.3.2 In Animals

Methoxyethyl mercury acetate is a vesicant when applied in concentrated solutions onto the skin (Lecomte and Bacq, 1949). Inhalation exposure of mice and rats to methoxyethyl mercury silicate dust produced severe pulmonary involvement (Hagen, 1955).

Lehotzky and Bordas, 1968, exposed rats to methoxyethyl mercury chloride intraperitoneally. The animals showed evidence of impaired weight gain, renal damage, and nervous symptoms, (ataxia, tremor, and palsy). At repeated administration of methoxyethyl mercury chloride to mice, Gage (in press) did not observe paralysis of the type seen at methyl mercury administration.

No report is available on the morphology of alkoxyalkyl mercury poisoning.

5.2.4 Other Organic Mercury Compounds

Intoxication by organic mercury compounds, which were not clearly specified, has been described in pigs (McEntee, 1950, and Loosmore, Harding, and Lewis, 1967), in cattle (Boley, Morrill, and Graham, 1941, Herberg, 1954, and Fujimoto et al., 1956), and in a horse (Edwards, C. M., 1942). Generally there were symptoms from the nervous system, the kidneys, and the gastrointestinal tract.

The toxicity of mercurial diuretics is generally low but cardiac toxicity, nephrotoxicity, and hepatic toxicity were reported (e.g., Hutcheon, 1965), as well as allergic reactions (e.g., Brown, 1957).

5.2.5 Summary

Alkyl, aryl, and alkoxyalkyl mercury compounds may cause local effects on the skin and mucous membranes.

Prenatal intoxication with *methyl* and *ethyl mercury* may give rise to severe mental and motor symptoms in man.

Short chain *alkyl mercury* compounds such as methyl and ethyl mercury give rise to *postnatal poisonings* dominated by neurological symptoms such as sensory disturbances, ataxia, concentric constriction of visual fields, and hearing loss. In some cases gastrointestinal and pulmonary symptoms and albuminuria have been reported. In ethyl mercury poisoning cardiac symptoms have been observed. In clinically manifested poisonings severe morphological damage to the cerebral and cerebellar cortex and peripheral nerves has been reported.

The number of poisonings due to *aryl* and *alkoxyalkyl mercury* is very limited and the description of the symptomatology is often contradictory. Both phenyl mercury and methoxyethyl mercury poisonings have been reported to produce various forms of kidney damage and neurological symptoms. Gastrointestinal and pulmonary symptoms have also been reported.

Chapter 6

"NORMAL" CONCENTRATIONS OF MERCURY IN HUMAN TISSUE AND URINE

Staffan Skerfving

6.1 Introduction

All human beings are exposed to small amounts of mercury through food, water, and air. Many are also exposed in other ways, e.g., by odontological and medical treatment or occupationally. Exposure also might occur through cigarette smoking (Maruyama, Komiya, and Manri, 1970). It is not possible to make a clearcut distinction between "normal" or "nonexposed" people and exposed ones. In this chapter, only data concerning persons reported not to have been subjected to any special kind of exposure will be presented. Levels in exposed persons without evidence of intoxication are dealt with in Chapters 7 and 8 in connection with levels in cases of intoxication. The intake of fish is a main source for the general population. Only a few studies have taken this into account.

From what has been said in Chapter 4 about the possible indices of mercury exposure and of levels in critical organs, it is obvious that levels in blood (cells and plasma), hair, and urine are of the greatest interest. Data will also be presented on the levels of mercury in some other tissues and organs.

6.2 Blood

6.2.1 Data on Fish Consumption Not Available

In an international study (WHO, 1966, partly reported by Goldwater, Ladd, and Jacobs, 1964), 812 samples of whole blood were collected by 18 laboratories in 15 different countries and analyzed in 1 laboratory by atomic absorption according to the method of Jacobs et al., 1960. The specimens obtained did not represent the population of any country. Information about the subjects who contributed the samples was limited to age, sex, and residence, rural or urban. The working definition of "normal" meant persons with no evident occupational, medical, or other unusual exposure to mercury. Some of the results are shown in Table 6:1. Seventy-seven percent of all samples had concentrations below 5 ng/ml, which was the analytical zero, 85% below 10 ng/ml, and 95% below 30 ng/ml. In 1.5% of all samples the level was 100 ng/ml or more. There were certain variations among countries but no difference in relation to age, sex, or residence, urban or rural. It was proposed in the report that the 95th percentile (about 30 ng/ml) should be regarded as the upper limit for "normal" concentrations.

6.2.2 Data on Fish Consumption Available

Data are given in Table 6:2 on mercury in blood cells and plasma in persons in Sweden with no occupational exposure, with none or low to moderate consumption of fish, and with predominantly low mercury levels (1 meal per week of salt water fish or less). As seen, people who do not usually eat fish had mercury levels in blood cells in the range of 2 to 5 ng/g. Persons having 1 meal a week of salt water fish had somewhat higher levels ranging to about 20 ng/g with a mean

TABLE 6:1

"Normal" Mercury Levels (atomic absorption spectrometry) in Whole Blood in Samples from Different Countries. (Data from WHO, 1966.)

Country	Total number of samples	Percent of samples <5 ng Hg/ml	Highest level ng Hg/ml
Argentina	49	80	30
Chile	35	69	30
Czechoslovakia	20	60	21
Finland	46	70	75
Israel	67	90	39
Italy	27	78	30
Japan	40	80	30
Netherlands	60	93	21
Peru	58	29	200
Poland	95	72	370
Sweden	30	90	90
Yugoslavia	67	72	270
U.A.R. (Egypt)	28	93	10
U.K.	30	93	75
U.S.	160	82	240
California	33	79	51
New York	87	83	45
Ohio	40	85	240
Total	812	77	370

TABLE 6:2

Total Mercury Levels in Blood in Subjects in Sweden without Occupational Exposure and with None (above line) or Low to Moderate Fish Intake (below line)

Number of samples	Analytical method[x]	Mercury level ng/g				Ratio C/P	Reference
		Blood cells (C)		Plasma (P)			
		Mean (±SE)	Range	Mean (±SE)	Range		
24	At	3.8 (± 0.16)	1.9-4.8	1.3 (± 0.15)	0.3-4.0	3.9	Tejning, 1970a
13	Ac	9.6 (± 1.3)	4.7-21	3.2 (± 0.5)	1.8-8.7	3.2	Birke et al (to be published)
83	Ac	10 (± 0.3)	5.4-17	2.3 (± 0.1)	1.1-7.5	4.6	Tejning, 1967a
23	At	6.9	1.9-14	1.6	0.8-3.4		Tejning, 1969c
14[xx]	Ac + At	8.6 (± 0.8)	4.8-15	2.5 (± 0.25)	1.3-4.5		Tejning, 1968a, 1970b
14[xxx]	Ac + At	12 (± 1.2)	6.0-21	1.9 (± 0.17)	1.1-3.0		Tejning, 1968a, 1970b

[x]Ac = Neutron activation analysis; At = Atomic absorption spectrometry.
[xx]Pregnant women. Sample taken immediately before delivery.
[xxx]Umbilical cord blood from newborn children to the women under[xx].

of about 10 ng/g. The levels in plasma are generally 3 to 5 times lower than those in blood cells.

Birke et al. (to be published) studied total mercury levels as well as methyl mercury concentrations in blood cells of 10 Swedes. The methyl mercury corresponded to ≤ 10 to 50% of the total mercury.

Tejning, 1968a and 1970b, has investigated blood mercury levels in pregnant Swedish women immediately before delivery and in umbilical cord blood. The cord blood cell levels were statistically significantly higher than those in the mothers, whereas the plasma levels were lower. The studies show that mercury passes through the placenta, though the biological significance in the difference of levels in the mothers and the newborns is not clear. The results of Tejning have recently been confirmed in Japan by Suzuki, Miyama, and Katsunuma, 1971b. They found in an examination of 9 subjects mean values for maternal blood cells of 22.9 ng/ml (S.D.: 11.9) and for umbilical cord blood cells, 30.8 ng/ml (S.D.: 21.6). Corresponding values for plasma were 12.4 (S.D.: 7.3) and 11.2 (S.D.: 7.2) ng Hg/ml.

6.3 Hair

In Table 6:3 some reported levels of mercury in the hair are presented. These subjects have not been exposed occupationally to mercury. Data on their fish consumption habits are not available. In the 2 studies with the largest samples, Perkons and Jervis, 1965, found in Canadians a mean concentration of about 1.8 μg/g, whereas Coleman et al., 1967, found in Englishmen an average level of 5.1 μg/g in men and 6.9 μg/g in women. In both studies there was a skew distribution toward low values. The Japanese studies indicate an average level of about 5 μg/g.

Two Japanese studies on *methyl mercury* in the hair have been published. Sumino, 1968b, reported that 37 men had an average level of 2.8 μg Hg/g and 26 women, 1.7 μg Hg/g. Ueda and Aoki, 1969 (table in Ueda, 1969), found an average level of 2.4 μg in 21 persons (all levels below 5 μg/g). In 6 subjects who ate only unpolished rice, the average total mercury level was 7.0 μg/g, of which a mean of 44% was methyl mercury.

6.4 Brain, Liver, and Kidneys

The levels that have been found (see Table 6:4) vary considerably among different materials. Besides differences in exposure, methodological differences must be considered. Since the levels are near the analytical zero of the method employed in many cases, great caution must be taken in estimating "normal" levels from data in this table.

In a recent study Glomski, Brody, and Pillay,

TABLE 6:3

Total Mercury Levels in Hair from the Scalps of Subjects without Reported Occupational Exposure (no data on fish intake available)

Country	Number of samples	Analytical method[x]	Mercury level µg/g Mean	Standard deviation	Range	Reference
Canada	776	Ac	~1.8		0-19	Perkons and Jervis, 1965
England	840	Ac	5.1[xx]	0.98		Coleman et al., 1967
			6.9[xxx]	0.37		
Japan	94	D	4.2		<0.99-<12	Yamaguchi and Matsumoto, 1966
	73	D	6	2.9	0.98-23	Hoshino et al., 1966
New Zealand	33	Ac	2.2	1.3	0.3-34	Bate and Dyer, 1965
	33	Ac	1.8	0.88	0.5-5.3	
Scotland	26	Ac	8.8			Nixon and Smith, 1965
	70	Ac	5.5		0.03-24	Howie and Smith, 1967
U.S.	33	Ac	7.6	11	0.1-33	Bate and Dyer, 1965

[x]Ac = Neutron activation analysis; D = Dithizone method
[xx] Males
[xxx]Females

TABLE 6:4

Total Mercury Level (wet weight; mean and/or range) in Different Organs from "Nonexposed" Subjects

Country	Number of subjects	Analytical method[x]	Mercury level ng/g Brain	Liver	Kidney	Reference
Germany	11	M	10-1,200	60-460	30-5,100	Stock, 1940
Japan	15	D	0-500	0-3,000	0-2,000	Takeuchi et al., 1962, Takeuchi 1961, 1968a
	10			400	600	Fujimura, 1964
	17	D	100-400	600-1,000	700-3,000	Matsumoto, Koya, and Takeuchi, 1965
Scotland	20-22	Ac	590[xx]	730[xx]	1,800[xx]	Howie and Smith, 1967
			24-3,000	3-4,000	2-16,000	
Sweden	5	Ac		20		Samsahl, Brune and Wester, 1965
	7	Ac		33		
U.S.	69	D		60	750	Butt and Simonsen, 1950
	15	D		740[xx]	4,200[xx]	Griffith, Butt, and Walker, 1954
	39	At	100	300	2,800	Joselow, Goldwater, and Weinberg 1967
			0-600	0-900	0-26,000	
	7	Ac	350[xxx]			Glomski, Brody, and Pillay, 1971
			40-1,300			

[x]M = Micrometric method; D = Dithizone method; Ac = Neutron activation analysis; At = Atomic absorption spectrometry
[xx]Values converted from dry weight to approximate wet weight by division by a factor of 5.
[xxx]Calculated from mean of all analyzed areas in each brain.

1971, analyzed different regions of brain in 7 specimens of autopsy material from the eastern part of the U.S. for mercury by neutron activation analysis. Mercury concentrations varied between 20 and 2,000 ng/g tissue, with the majority of the samples below 300 ng/g tissue. Trace concentrations of mercury were present in all examined regions of brain; the highest concentrations were generally found in cerebellar cortex and the lowest concentrations, in cerebral white matter.

6.5 Urine

As stated in Chapter 4, mercury concentration in urine is an unreliable index of an individual's exposure to mercury, especially at exposure to the alkyl mercury compounds. However, urinary levels have often been used to control exposure and to evaluate risks of intoxication in exposed subjects. In view of the wide use, surprisingly few investigations of "normal" concentrations of mercury in urine have been reported.

Because only small amounts of mercury are excreted in the urine after methyl mercury exposure, there is little need to take fish intake habits into account. Table 6:5 presents data from the international study mentioned in Section 6.2.1 (WHO, 1966, partly reported by Jacobs, Ladd, and Goldwater, 1964, atomic absorption method). Of the total of 1,107 samples, 79% had levels below 0.5 µg/l., which was the analytical zero of the method used, 86% below 5 µg/l., 89% below 10 µg/l., and 95% below 20 µg/l. 1.6 percent of the samples had concentrations higher than 50 µg/l. No systematic influence from age, sex, or residence (urban or rural) could be detected. In accordance with the general proposal, the 95th percentile, i.e., 20 µg/l., was regarded as the upper limit of "normal" concentrations.

6.6 Summary

All human beings are exposed to small amounts of mercury through food, water, and air. Fish intake habits seem to be an important factor for this "background" exposure. Some people are also exposed occupationally, by medical or odontological treatment or through consumption of contaminated food other than fish.

Many of the published investigations on "normal" mercury levels in "nonexposed" subjects do not give adequate definitions in regard to representation, possible sources of exposure,

TABLE 6:5

"Normal" Mercury Levels (atomic absorption spectrometry) in Urine in Samples from Different Countries. (Data from WHO, 1966.)

Country	Total number of samples	Percent of samples <0.5 µg Hg/l.	Highest value µg Hg/l.
Argentina	49	84	21
Chile	35	69	21
Czechoslovakia	20	85	11
Finland	46	67	30
Israel	83	87	95
Italy	25	76	37
Japan	40	85	45
Netherlands	60	87	15
Peru	64	50	107
Poland	98	71	158
Sweden	30	80	74
Yugoslavia	65	83	69
U.A.R. (Egypt)	14	64	12
U.K.	30	87	38
U.S.	434	82	221
California	31	87	15
New York	363	80	97
Ohio	40	93	221
Total	1,107	79	221

sampling procedure, and analytical methods, and thus do not allow for definite evaluation and comparison.

The general reported average level of mercury is below or about 5 ng/g in whole blood in "normal" subjects without any known special exposure. Considerably higher levels have been reported, however.

Data from Sweden indicate a ratio of 3 to 5 between concentrations in blood cells and plasma.

There is a considerable variation among reported levels of mercury in hair in different investigations and in different parts of the world. One study from Canada indicates a mean level of about 1.8 µg/g while another from the U.K. and some from Japan indicate levels 2 to 3 times as high.

The reported investigations of "normal" mercury levels in other organs are quite contradictory and do not allow for definite conclusions.

The general level of mercury in urine in "nonexposed" subjects seems to be below or about 0.5 µg/l. Here again, however, much higher values have been reported.

Chapter 7

INORGANIC MERCURY — RELATION BETWEEN EXPOSURE AND EFFECTS

Lars Friberg and Gunnar F. Nordberg

As criteria for the evaluation of effects the classic symptoms and signs of mercurialism have been employed. More subtle changes, called micromercurialism in the studies from the U.S.S.R. have also been taken into the account.

The emphasis in this review will be on dose-response relationships found in chronic exposure, but a few data will be given referring to acute exposure. Acrodynia has not been treated because no dose-response relationships seem to exist.

7.1 In Human Beings

A considerable number of studies relating exposure and effects have been published. The exposure usually has been evaluated through air measurements but often also through analysis of mercury in urine or blood. Most data are from exposure to mercury vapor but often it is not possible to decide to what extent exposure to aerosols of other forms of inorganic mercury also is involved. It has not been deemed possible to treat the different exposure forms separately.

7.1.1 Acute Effects

The main acute manifestations of inorganic mercury poisoning, as have been discussed in Chapter 5, are pulmonary irritation after exposure to mercury vapor and kidney injury after exposure to mercuric salts.

The concentration needed to give rise to acute pulmonary manifestations in human beings is not known. They have appeared after accidental exposure to very high concentrations. In a recent report by Milne, Christophers, and deSilva, 1970, it was estimated that only a few hours' exposure to between 1 and 3 mg Hg/m^3 had caused 4 cases of acute mercurial pneumonitis.

A detailed dose-response curve for acute poisoning with mercuric salts also is not known. Most reported cases are brought about by ingestion of bichloride of mercury with suicidal intent. Often several grams of mercury have been taken but severe poisoning has been reported after ingestion of less than 1 g of mercuric chloride (Sollmann, and Schreiber, 1936, Troen, Kaufman, and Katz, 1951, and Sanchez-Sicilia et al., 1963).

There is some information in the literature about mercury content in organs in human beings fatally intoxicated by mercuric mercury. Sollmann and Schreiber, 1936, reported 7 fatal cases in which the median concentrations in kidneys were 38 (range: 16 to 70) ppm wet weight and in liver, 20 (range: 3 to 32) ppm wet weight. In 3 fatal cases described by Sanchez-Sicilia et al., 1963, kidney values ranged from 9 to 19 ppm and liver values from 10 to 63 ppm. The last mentioned cases had been treated with BAL.

7.1.2 Chronic Effects
7.1.2.1 Relation Between Mercury in Air and Effects
7.1.2.1.1 Studies in General

Neal et al., 1937 and 1941, made early but comprehensive efforts to study relationships between exposure and symptoms in the felt-hat industry. The data have served as basis for establishing the industrial MAC-value for mercury in the U.S.

Although these studies have provided valuable information in regard to many aspects of mercury poisoning, several drawbacks and inconsistencies lessen their suitability as a basis for establishing MAC-values. For example, the air analyses of mercury were spot samples which cannot be used for scientific evaluation of concentrations below which symptoms of mercury poisoning will not occur. Apart from this, the authors' conclusion in the 1941 report that 0.1 mg Hg/m^3 "probably represents the upper limit of safe exposure" is not warranted, even by the data presented in their studies, as cases of mercurialism occurred also at just that exposure. This is confirmed further by the 1937 report in which they reported mercury poisoning among 6% of workers exposed to less than 0.09 mg Hg/m^3 of air.

Vouk, Fugas, and Topolnik, 1950, and Kesic and Haeusler, 1951, reported on mercury poisoning in a survey of 130 workers in a mercury mine, 59 workers in a smelting works, and 70 female workers in a felt-hat factory. Mercury concentrations in air (spot samples, dithizone method) varied between 1.2 to 5.9 mg/m^3 in the mine, 0.25 to 0.85 mg/m^3 in the smelting works, and 0.25 to 1.0 mg/m^3 in the hat factory. About one third of the workers in the mine and smelting works and about two thirds of the female workers in the felt-hat factory showed pronounced symptoms of chronic mercury intoxication. The authors compared all 3 groups of workers with a control group of 466 persons and did not find significant blood changes.

Bidstrup et al., 1951, found 27 cases of mercury poisoning among 161 workers in workshops for repair of direct current meters. The atmospheric mercury concentrations (spot samples, dithizone method) were summertime low, usually less than 0.05 mg/m^3. Wintertime values of between 0.1 to 0.3 were common. Values of up to 1.6 mg/m^3 could be found over the work desks. Friberg, 1951, reported 7 cases of pronounced tremor among 91 workers in a chlorine plant. The mercury exposure was usually below 0.1 mg/m^3 (spot samples, dithizone method), but could be as high as about 1 mg/m^3. Even these last mentioned studies do not provide suitable data for establishing a "safe" exposure to mercury. They tend to show that values around 0.1 to 0.2 mg Hg/m^3 of air can give rise to a considerable risk of chronic mercury poisoning. Turrian, Grandjean, and Turrian, 1956, examined 58 workers in a rectifier factory, a thermometer factory, and a chemical works in Switzerland. In 15 workers they observed tremor and mental disturbances. In 2 of those cases the average exposure (spot samples, dithizone method) was only between 0.01 to 0.06 mg Hg/m^3. The exposure for the rest of the workers varied between 0.1 to 0.6 mg Hg/m^3. A tendency toward hyperchromic anemia was seen in several workers. The value of the studies is limited, chiefly because only spot samples were taken for analysis and no controls were clinically examined.

Rentos and Seligman, 1968, reported 6 cases of suspect or definite mercury poisoning among 13 workers with an average daily exposure of between 0.08 to 0.68 mg Hg/m^3 (mean: about 0.5) but observed no symptoms among 9 workers with average daily exposures of 0.02 mg/m^3. Exposure was evaluated partly as 8-hour average values and partly as spot samples. Mercury vapor meters and dithizone methods were used. The authors concluded that a TLV value of 0.1 mg Hg/m^3 was supported, even if the data show that this level contains a safety factor of no more than 2. However, it seems more justified to conclude from the study that mercury poisoning occurred after exposure to concentrations above 0.2 to 0.3 mg Hg/m^3 and that mercury poisoning was not seen in a small number of the examined individuals after exposure to about 0.02 mg Hg/m^3. No conclusions at all can be drawn in regard to exposure to concentrations between 0.02 and 0.2.

A comprehensive study has been reported by Smith et al., 1970. They examined 567 workers exposed to mercury (in general, more than 90% as mercury vapor) in the manufacture of chlorine and a control group of 382 persons. More than half of the study group, all males, had worked between 6 and 14 years in the industry and they came from 21 different plants. Every worker was examined once during a one-year period (not necessarily at the same time) by plant physicians according to a predetermined procedure. At least 4 times a year blood and urine samples were examined for mercury (the methods used were those by Campbell and Head, 1955 and Jacobs, Goldwater, and Gilbert, 1961, respectively). The mercury concentrations in air were measured at different sampling places at least 6 times a year by means of ultraviolet meters. Time-weighted averages were calculated for each worker.

The methods used made it possible to correlate symptoms with air, blood, and urine concentrations of mercury, as well as to correlate air data with blood and urinary values. In this section, only correlations between air mercury concentrations and symptoms will be dealt with.

In Table 7:1 the mercury-exposed workers have been grouped according to their time-weighted average exposure levels. The prevalence of certain medical findings in relation to mercury exposure is illustrated in Figure 7:1. As can be seen from it, several findings reveal a clear dose-related response to mercury exposure, including signs and symptoms from the nervous system expected in mercury poisoning. For diastolic blood pressure there was a negative correlation with mercury exposure. For several findings there is no indication that even the lowest exposure (time-weighted average: < 0.01 to 0.05 mg/m^3) took place without effect. This demonstrates potential effects of even minimal exposures.

The authors reported several other results in which a significant correlation with mercury exposure was not found. Such findings included oropharyngeal signs, i.e., abnormalities of teeth and gums. The authors' general conclusions are,

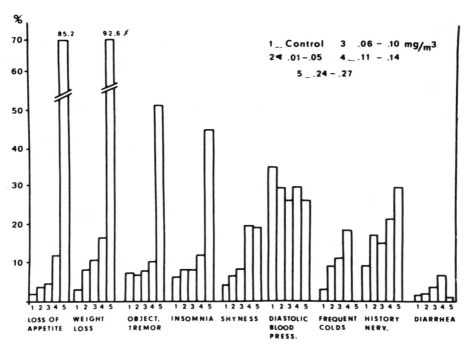

FIGURE 7:1. Percentage prevalence of certain signs and symptoms among workers exposed to mercury in relation to degree of exposure. (From Smith et al., 1970.) Data on diastolic blood pressure probably mean % below a certain level. This level is not given in the article.

TABLE 7:1

Mercury-Exposed Workers Grouped by Time-Weighted Average Exposure Levels. (From Smith et al., 1970.)

Exposure levels (mg/m³)	Number of workers	Percentage of exposed workers
<0.01	58	10.2
0.01-.05	276	48.7
0.06-.10	145	25.6
0.11-.14	61	10.7
0.15-.23	—	—
0.24-.27	27	4.8

"The data presented here show no significant signs or symptoms in persons exposed to mercury vapor at or below a level of 0.1 mg/m³. However, the data do raise a question regarding the adequacy of the safety factor provided by a TLV of this magnitude."

What should be considered of importance from the point of view of industrial TLV values is, of course, always a matter of judgment. The published data, however, no doubt point to the conclusion that a no-effect level for mercury exposure was not found and that time-weighted average exposures to below 0.05 mg Hg/m³ may produce medical effects. In interpreting the data, a problem arises regarding the comparability of the 2 groups. The authors were of the opinion that the study group and the control group were comparable. There is no reason to doubt their opinion, age distribution included, when the control group is compared with the study group as a whole. Unfortunately, there is no information concerning the comparability when the study group is broken down into subgroups. Furthermore, a possible bias, the "interviewer effect," in interpreting minor signs and symptoms should be mentioned. In all the studies the medical examinations were made by the factory physicians, who might have had some knowledge of the exposure situations for the different categories of patients. On the other hand, one has to appreciate the well-standarized questionnaire employed.

7.1.2.1.2 Russian Studies — Including Studies on Micromercurialism

The data in this section are taken partly from translations of Russian publications and partly from information obtained by personal contacts (by GFN) with scientists in the U.S.S.R. It should be emphasized that data from the Soviet Union are often presented in an abbreviated way compared to the usual format of the Western countries. Materials and methods are often so standardized that they are not presented in detail at the publication of the results. Much work is described in complete form in unpublished doctoral theses and only summarized in published articles. Such factors make a correct evaluation of the data difficult.

Extensive clinical studies performed by Trachtenberg and his collaborators in Kiev were published in a monograph, Trachtenberg, 1969. One study covered 574 people from Kiev, from 20 to over 50 years of age and 50% of both sexes. Approximately 500 of the studied persons had been exposed to low concentrations of mercury in their professions (in research institutes, industrial plants, hospitals, etc.) for more than 1 year (group A). The control group consisted of 68 persons, mostly clerks and service personnel working at similar places but without any direct contact with mercury (group B). The medical examinations were carried out by plant physicians.

Mercury concentrations in air of the subjects' working places were determined by a colorimetric method described by Poleshajev, 1956 (see also Alekseeva, 1957). The accuracy and precision of this method are not known but might be influenced by subjective factors (Chapter 2). In each workroom a number of spot samples (usually more than 60, 10 to 20-minute values) had been made. In this way, a number of minimum, maximum, and average values was obtained for the different rooms. The exposure conditions are given in Table 7:2.

TABLE 7:2

Mercury in Air of Workshops. (From Trachtenberg, 1969.)[x]

Group A: 506 persons

Work place	Mercury concentration in air (mg/m³)		
	Minimum	Maximum	Average
Production of measuring instruments	0.004 -0.008	0.015-0.12	0.01-0.04
Research institutes	≤ 0.01	0.055-0.08	0.02-0.05
Higher education	0.007 -0.015	0.01 -0.1	0.02-0.035
Production of rectifiers	0.007 -0.01	0.01 -0.065	0.02-0.03
Hospitals	≤ 0.025	0.015-0.17	0.01-0.04
Unspecified industries and institutes	0.0085-0.012	0.03 -0.15	0.015-0.05

Group B: 68 persons

Clerks and service personnel		0.01 or less	

[x]Four of the values differ from those given in Trachtenberg's monograph. They are corrections of printing errors, according to personal discussions with Trachtenberg.

Trachtenberg found an asthenic-vegetative syndrome in 51.2% (259 persons) in group A. Of these syndromes, he considered that 13.6% had an unspecific etiology while 37.6% could be traced to the mercury exposure. The asthenic-vegetative syndrome is not clearly defined but includes several neurasthenic symptoms. Trachtenberg is of the opinion that there is a difference between the asthenic-vegetative syndrome caused by mercury and that caused by some other etiology. The latter kind is generally not accompanied by the emotional lability predominant in patients displaying the syndrome with a mercury etiology. The emotional lability included, as a rule, increased excitability and susceptibility, mental instability, apathy, and a tendency to weep.

For the diagnosis of an asthenic-vegetative syndrome as a nosological unit of mercury etiology, other clinical findings such as tremor, enlargement of the thyroid, uptake of radioactive iodine, hematological changes, and excretion of mercury in urine were also used as supporting evidence. There is no mention in the monograph if, and to what extent, such findings were obligatory for the diagnosis of the asthenic-vegetative syndrome as mercury induced. By personal discussions with Trachtenberg, it was established that the following criteria were applied: a mercury value in the urine exceeding the normal limit (0.01 mg Hg/l.) or at least 8-fold increases in urinary concentrations after medication with unithiol. If these criteria were not fulfilled, the finding of 3 or more of the following objective symptoms was enough for a classification of "mercury etiology": tremor, thyroid enlargement, increased uptake of radioiodine in the thyroid, hematological changes, hypotension, labile pulse, tachycardia, dermographism, and gingivitis.

The prevalence of medical findings in groups A and B, respectively, is shown in Table 7:3. According to Trachtenberg, the findings come early, often within the first years of exposure. No differences between the groups which could be related to differences in exposure (Table 7:2) are seen. That Trachtenberg reported that he observed mercury-induced asthenic-vegetative symptoms in 40% of the controls (exposed to less than 0.01 mg Hg/m^3) is very surprising. He also found nearly the same prevalence of asthenic-vegetative symptoms caused by mercury in workers exposed for less than 4 years as in workers exposed for longer periods (34.6 vs. 40.6%). The comparability of groups A and B cannot be evaluated; it is known, though, that the workers in group B generally were somewhat older.

Another study referred to in the monograph is that reported earlier by Trachtenberg, Savitskij, and Sternhartz, 1965, covering workers involved in the production of vacuum tubes in Moscow. Apart from mercury, the workers were exposed to high

TABLE 7:3

Prevalence of Medical Findings in Workers Exposed to Mercury.[x] (From Trachtenberg, 1969.)

Medical findings	Group A (506 persons) %	Group B (68 persons) %
Asthenic-vegetative syndrome with unspecific etiology	14	7
Asthenic-vegetative syndrome due to mercury	38	40
Chest pains or palpitations	31	28
Enlargement of thyroid	14	4
Hypotension	32	28
Stomatitis	13	16
Liver disorders	19	20

[x]See also Table 7:2.

TABLE 7:4

Prevalence of Medical Findings in Exposed Groups and a Control Group. (From Trachtenberg, Savitskij, and Sternhartz, 1965, and Trachtenberg, 1969.)

Medical findings	Groups 1 and 2 %	Control group %
Insomnia, sweating, emotional lability	28-50	13
Tremor of hands and eyelids and enlargement of thyroid	28-37	8-12
Extensor strength of right hand dominant over that of left hand (Teleky symptom)	51	76

temperatures. By consulting the original publication and by discussing the study personally with Drs. Trachtenberg and Savitskij, the following details were obtained.

The data were taken from the yearly examinations of the workers in the industry. Three groups of workers were selected. Group 1 was exposed to average mercury concentrations between 0.03 to 0.04 mg/m^3 and normal temperatures (26 to 31° C in the summer and 16 to 24° C in the winter). Group 2 was exposed to yearly averages between 0.006 and 0.01 mg/m^3 and temperatures of 40 to 42° C in the summer and 28 to 38° C in the winter. A control group was not exposed to mercury, but to high temperatures, 38 to 42° C.

The study covered the period of 1955 to 1962. In group 1, 42 to 93 subjects were included and in group 2, 49 to 208. The control group consisted of 60 to 80 subjects.[x] In all 3 groups, about 70% of the workers were women. The age distribution within the groups varied somewhat from year to year, but was considered approximately the same for the different groups. About 45% of the subjects were between 30 to 40 years old and about 30% between 40 and 50 years old.

The prevalence of medical findings is seen in Table 7:4. In contrast to the earlier mentioned study by Trachtenberg, differences were found between the exposed groups and the control group. Concerning the Teleky symptom, some researchers in the U.S.S.R. believe that the relative strength of the extensors of the right hand compared with that of the left hand will decrease under the influence of toxic substances.

Trachtenberg and his collaborators studied the *function of the thyroid* by means of radioactive iodine. The results from an exposed group and a control group are given in Table 7:5. The workers in the exposed group were part of those workers in the production of measuring instruments mentioned in Table 7:2. As can be seen, the differences in uptake of radioactive iodine between exposed workers and controls are substantial, both for men and women.

Different blood examinations were made. No controls were examined but in the exposed workers, an anemia, increasing with time after exposure, was indicated (Figures 7:2 and 7:3). No age distribution was given but the workers who had been employed for the longest time probably were older than those employed only a short time. This might have had an influence on the results.

Certain studies were made concerning the odor perception of exposed workers and controls. Differences were found for odor thresholds, adaptation times, and recovery times between the groups. No data concerning methodology have been available to us and knowing the methodological difficulties with odor studies, we shall not comment further upon the results.

7.1.2.2 Relation Between Mercury in Urine and Effects or Exposure

7.1.2.2.1 Mercury in Urine and Effects

Several studies have related mercury excretion via urine with symptoms of mercury poisoning. In some early data reported by Neal et al., 1937, and 1941, some 30% of persons with mercurialism did not have mercury in urine at all. The validity of

---

[x]The number of members in the groups depended upon the presence of the worker in a certain area. If he changed to another area or to another job, he was no longer included. Likewise, all new arrivals to the area were included in the study. Hence, the fluctuation in the number of participants was great; here, summarily, is the range for the entire 7-year span.

TABLE 7:5

Relative Number of Workers (%) with Uptake of Less and More Than 25% of Radioactive Iodine (after 24 hours) in the Thyroid. (From Trachtenberg, 1969.)

Uptake of radioactive iodine	Groups exposed to mercury			Control groups		
	Men	Women	Total	Men	Women	Total
	(36)	(31)	(67)	(26)	(19)	(45)
<25%	39	29	34	84	68	78
>25%	61	71	66	16	32	22

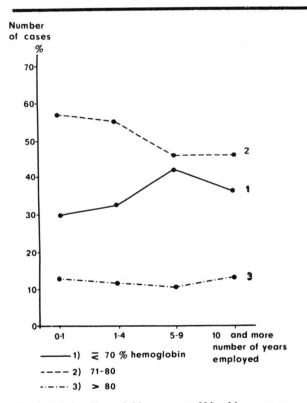

FIGURE 7:2. Hemoglobin content of blood in mercury exposed workers in relation to time of employment. (From Trachtenberg, 1969.)

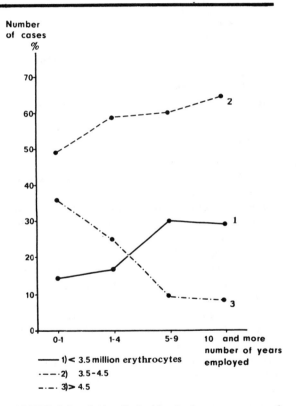

FIGURE 7:3. Red cells in blood of mercury exposed workers in relation to time of employment. (From Trachtenberg, 1969.)

these data must be questioned, partly in view of the fact that mercury could be detected in only less than half of all subjects examined (all exposed to well above 0.1 mg Hg/m³).

Friberg, 1951, in the above mentioned study of 91 workers in a chlorine plant, reported 7 cases of pronounced tremor. Four workers with pronounced tremor had been exposed to mercury vapor for about 25 years but had mercury levels in urine (dithizone method) of only 0.2 to 0.3 mg Hg/l. The other 3 had between 0.7 to 1.3 mg Hg/l. urine. Moderate tremor (10 cases) occurred without any clearcut association with urinary mercury levels (Figure 7:4). Several workers had a high mercury excretion without symptoms.

The study by Bidstrup et al., 1951, of 27

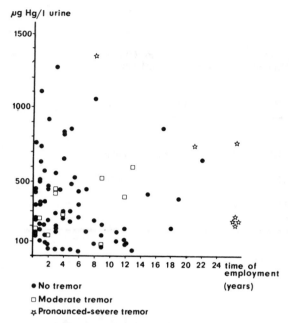

FIGURE 7:4. Prevalence of tremor in workers exposed to mercury in a chlorine plant. (From Friberg, 1951.)

persons with mercury poisoning showed that, as a rule, those with clinical evidence of mercury poisoning had a high excretion of mercury (dithizone method), often more than 1 mg of mercury in 24 hours. A low excretion was also seen, however. Three out of the 27 workers excreted less than 0.1 mg/24 hours. Sixteen out of 101 workers without symptoms excreted more than 300 µg of mercury/24 hours against 21 out of the 27 cases with signs of mercurialism.

Among 120 exposed workers in a thermometer workshop, Seifert and Neudert, 1954, reported 8 suspect and 1 definite mercury poisoning at very low urinary concentrations of mercury (0.04 to 0.06 mg Hg/l., dithizone method). The validity of the diagnosis of mercury poisoning in the study seems questionable, though. In several cases no diagnosis of mercury poisoning was made, despite symptoms, while in other cases the opposite was true. For example, one suspect case was diagnosed based only on a history of stomatitis, without any objective finding at the examination. No controls were examined. Also Turrian, Grandjean, and Turrian, 1956, did not find a correlation between urinary mercury levels (dithizone method) and symptoms. Neither did they find a correlation between urinary values and exposure. Ladd et al., 1966, made investigations on miners. Their findings led them to the conclusion that symptoms of poisoning can occur at low urinary mercury levels but will not necessarily occur, even when concentrations of mercury in urine are high.

Rentos and Seligman, 1968, reported high mercury concentrations (probably dithizone method) in the urine of 6 workers with suspected or definite symptoms of poisoning (0.34 to 4.3 mg Hg/l.). In 7 workers with a high exposure to mercury but without symptoms, the mercury concentration in urine was 0.2 to 2 mg/l. No mercurialism was reported among 9 controls only slightly exposed and with a mercury excretion averaging about 0.05 mg/l. urine.

Positive correlation between severity of poisoning and urinary concentrations (dithizone method) was observed by West and Lim, 1968, in 13 mill workers exposed to mercury vapor concentrations exceeding 1.2 mg Hg/m^3. The exposure, however, must have been extremely high. The median urinary level in this group of 13 cases of mercury poisoning was 1.2 mg Hg/l. and the highest concentration, 7.1 mg Hg/l. On the other hand, the authors observed low urinary levels (0.1 mg Hg/l.) in a worker with typical symptomatology of mercury intoxication, and levels between 0.2 to 1.1 mg Hg/l. urine in workers without symptoms. West and Lim concluded that urinary concentrations below 0.8 mg Hg/l. do not correlate well with presence of clinical symptoms of mercurialism; at levels above 0.8 mg Hg/l., however, the severity of manifestations correlates well with urinary levels.

Trachtenberg, 1969, expressed the opinion that mercury concentrations in the urine are of limited value in the individual case. Despite this, as was mentioned in Section 7.1.2.1.2, mercury concentrations above the normal value, 0.01 mg/l. urine, were considered by him as a supporting criterion for mercury etiology in clinical diagnosis of the asthenic-vegetative syndrome.

El-Sadik and El-Dakhakhny, 1970, reported on symptoms (mercury neurasthenia) in workers employed in a sodium hydroxide producing plant for periods from less than 6 months to more than 3 years. They did not find any correlation between symptoms and urinary mercury levels. One worker with a mercury concentration in urine of only 4 µg/l. was reported to have manifestations of mercurialism. Mercury levels (dithizone method) in urine were higher among those exposed for less than 6 months (48 to 132 µg/l.) than among workers exposed for more than 3 years (39 to 66 µg/l.). Air concentrations of mercury (dithizone method) ranged in 36 samples between 0.072 to 0.88 mg/m³, with an average of 0.3 mg/m³. Symptoms were also found in a control group of 10 people but to a lesser degree. Urinary mercury levels among the controls varied between 32 to 40 µg/l. The report does not give sufficient information for an evaluation of the medical findings. The relation between exposure and urinary excretion of mercury differed considerably from what has been reported in the study by Smith et al., 1970 (see Figure 7:5).

In their extensive study, Smith et al., 1970 (see Section 7.1.2.1.1), looked into associations between urinary mercury levels and medical findings. They did not give prevalence data for medical findings for different urinary mercury levels, but did mention that in spite of the strong correlations between time-weighted averages for the exposure and urine levels (see below), the correlations between urine levels and medical findings were in general much weaker, and usually were clear only in specific findings which most strongly correlated with air levels.

In summary, it can be stated that, although on a group basis high mercury levels lead to higher probabilities of mercury poisoning, in the individual case, high values of mercury can occur without symptoms, while symptoms can occur also in association with low levels of mercury in the urine.

As has been mentioned earlier (Chapter 5), chronic exposure to inorganic mercury can cause proteinuria, including a nephrotic syndrome. A clear dose-response relationship which would show that workers with proteinuria have had a higher exposure to mercury than workers without proteinuria has not been demonstrated. In reported cases with the nephrotic syndrome the urinary excretion of mercury has been high, as a rule 0.5 to 1 mg Hg/l. or even more. On the other hand, several workers without proteinuria excreted similar or higher amounts of mercury (Ledergerber, 1949, Friberg, Hammarström, and Nyström, 1953, Goldwater, 1953, and Kazantzis et al., 1962). Proteinuria also has been reported to have occurred after the use of mercury-containing ointments (see e.g., Young, 1960, and Silverberg, McCall, and Hunt, 1967). The mercury excretion in urine was high but no evidence of a dose-related effect has been reported. It has been suggested that the nephrotic syndrome may arise because of an idiosyncracy to mercury (see e.g., Kazantzis et al., 1962), but this question is by no means settled.

Joselow and Goldwater, 1967, found that a group of workers exposed to vapors and dust of

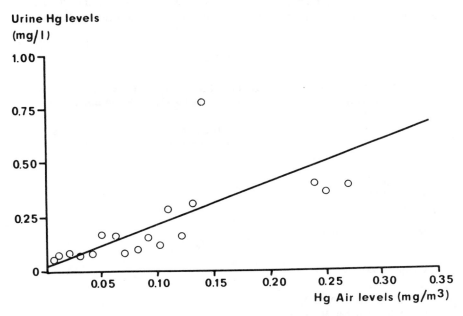

FIGURE 7:5. Concentrations of mercury in urine (uncorrected for specific gravity) in relation to time-weighted average exposure levels. (From Smith et al., 1970.)

phenyl and/or inorganic mercury excreted more protein on an average (9 mg protein/100 ml) than a control group (5.3 mg/100 ml). They also found a statistically significant correlation between excretion of protein and mercury. A wide individual scatter was evident, however.

Goldwater and Joselow, 1967, reported an association between excretion of mercury and coproporphyrin. Wada et al., 1969, found a correlation with coproporphyrin excretion and a negative correlation between urinary levels of mercury and levels of δ-aminolevulinic acid (ALA) dehydratase in erythrocytes and cholinesterase (ChE) in serum. Particularly the correlation with ChE activity may serve as an early sign of a biological effect of mercury, even if the data presented thus far do not allow any conclusions of a critical value.

Kosmider, Wocka-Marek, and Kujawska, 1969, reported on the usefulness of biochemical tests in the early detection of intoxications with metallic mercury. They examined 100 patients exposed to metallic mercury from 1 to 26 years and 100 controls in similar age groups not exposed to mercury. The patients were divided into 1 group with a mercury excretion in urine (dithizone method) of 40 to 120 µg/l. (group A) and another group with mercury excretion less than 40 µg/l. (group B). Several biochemical tests were carried out, e.g., lactic acid dehydrogenase (LDH), alkaline phosphatase, pseudocholinesterase, alanine-aminotransferase, electrophoretic protein studies in urine, lipoproteins in serum, cholesterol, and liver function studies (thymol and bromsulfalein tests).

The results of the clinical studies are given in Table 7:6. As can be seen, the prevalence of several symptoms is higher in the group with the highest urinary levels of mercury. There is no information in the report about the criteria used for the clinical damage beyond the statement that "liver damage" was the diagnosis when at least 2 liver function tests were positive. The group comparability is also not clear.

Several biochemical findings were observed which were associated with the exposure to mercury. The lactic acid dehydrogenase in 50 controls was on an average 295 units (range: 240 to 350) compared with 232 units (range: 122 to 288) in group A. Alkaline phosphatase was an average of 1.6 (range: 1.0 to 2.5) compared with 1.3 units (range 0.6 to 1.8) in group A. The alanine-aminotransferase in the controls was an average of 19 units (range: 6 to 40) compared with 42 (range: 12 to 94) in group A. All these findings were reported to be statistically significant ($p < 0.01$).

7.1.2.2.2 Mercury in Urine and Exposure

Urinary mercury measurements are used not only for diagnosing mercury poisoning but also for evaluating mercury exposure. There exists an abundance of data in the literature pointing to a positive association between occupational

TABLE 7:6

Prevalence (%) of Medical Findings in 2 Groups[x] of Workers with Different Urinary Mercury Levels. (From Kosmider, Wocka-Marek, and Kujawaka, 1969.)

Findings	Group A (40 workers) %	Group B (60 workers) %
Neurological disorders	58	23
Kidney damage	22	5
Liver damage	20	15
Disorders of the cardio-vascular system	48	20
Complex organ disorders	62	22

[x]Group A: 40 to 120 µg Hg/l
Group B: Less than 40 µg Hg/l

exposure and urinary mercury levels on a group basis (see, e.g., Goldwater, 1964).

The data from the study by Smith et al., 1970, are the most comprehensive and elucidative and are given in Table 7:7 and Figure 7:5 (for methods, see Section 7.1.2.1.1). As can be seen, there is a correlation on a group basis but with a wide individual dispersion. The average ratio between urinary (mg/l.) and atmospheric (mg/m^3) mercury as seen in Figure 7:5 is of the same order of magnitude (about 2) as reported in early studies by Storlazzi and Elkins, 1941. They found an average ratio between urinary mercury and atmospheric mercury of 2.6. For urinary mercury analyses, a modification of Stock's method (Stock and Lux, 1931) was used and could show a good recovery of added known amounts of mercury. The figures given correspond fairly well with results from the study by Tejning and Öhman, 1966 (Section 4.4.1.3.2) if due attention is paid to the fact that Tejning and Öhman expressed their values on a 24-hour basis, and not on a volume basis.

Armeli and Cavagna, 1966, showed a positive relationship between exposure and urinary excretion, but only for the first period of the workers' employment. They reported mercury levels in 94% of the workers exposed to air concentrations below 0.1 mg/m^3 to be less than 0.15 mg/l. urine. Air concentrations were not determined as time-weighted averages.

Trachtenberg and Korshun (personal communications) have provided some data on associations between mercury in air (the Poleshajev method) and mercury in urine (the Ginzburg method). The data, given in Table 7:8, are from 195 subjects randomly chosen from exposure group A in Table 7:2 and from 50 workers in a chlorine producing plant. If the values are compared with those reported by Smith et al., 1970, it can be seen that the urinary values given by Trachtenberg and Korshun for the same exposure are lower. A detailed comparison, however, is impossible, as no similar breakdown of the urinary values as shown in the Russian studies was attempted in the American studies. Furthermore, the data from the U.S.S.R. are not based on time-weighted averages. In view of these barriers, the values agree reasonably well.

A problem in studying the associations between exposure and urinary mercury levels is the degree to which urinary excretions of mercury may fluctuate, independently of exposure. Data by Friberg, 1961 (Figure 7:6), show that such fluctuations can be considerable. Wide diurnal and day-to-day variations have also been reported by Jacobs, Ladd, and Goldwater, 1964. Adjustment of urinary concentrations for specific gravity or creatinine excretion may help, but only to a very limited degree (Elkins and Pagnotto, 1965, Molyneux, 1966, and Smith et al., 1970).

In summary, available data show an association

TABLE 7:7

Relationship of Mercury Exposure to Mercury Levels in Urine, Uncorrected for Specific Gravity.[x] (From Smith et al., 1970.)

Time-weighted averages; exposure level groups (mg/m^3)	Number of workers	Percentage of group within urine level range					
		<0.01	.01-.10	(mg/l.) .11-.30	.31-.60	.61-1.0	>1.00
Controls 0.00	142	35.2	62.7	2.1	0	0	0
<0.01	29	6.9	86.2	6.9	0	0	0
0.01-0.05	188	6.9	66.0	24.5	2.7	0	0
0.06-0.10	91	0	62.6	30.8	6.6	0	0
0.11-0.14	60	3.3	18.3	31.7	16.7	23.3	6.7
0.24-0.27	27	0	14.8	29.6	44.5	7.4	3.7

[x]Expressed as percentage of each exposure level group within designated ranges of urine mercury levels

TABLE 7:8

Relationship of Mercury Exposure to Mercury Levels in Urine, Uncorrected for Specific Gravity. (Trachtenberg and Korshun, personal communication.)

Exposure level (average of at least 60 spot samples, mg/m³)	Number of workers	Percentage of group within urine level range (mg/l.)			
		≤0.01	0.011-0.03	0.031-0.05	>0.05
0.01-0.05	195	46	39	9	6
0.03-0.04	50	48	36	12	4

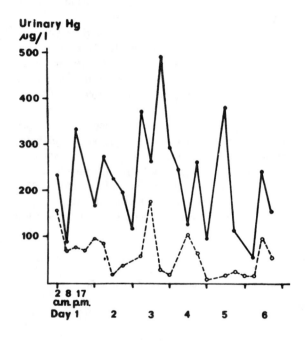

FIGURE 7:6. Variations within the 24-hour excretion of mercury in 2 workmen with mercury poisoning. (From Friberg, 1961.) Exposure to mercury had ceased 1 to 2 months previously.

between mercury exposure and mercury concentrations in urine on a group basis. A concentration of about 0.1 mg/m³ in air with a weekly exposure of 40 hours should correspond to about 0.2 mg Hg/l. urine. On the other hand, it is obvious that a urinary mercury level can not be predicted on an individual basis, even if exposure is measured as time-weighted averages.

7.1.2.3 Relation Between Mercury in Blood and Effects or Exposure

7.1.2.3.1 Mercury in Blood and Effects

There are few convincing studies relating mercury levels in blood with symptoms. Published data tend to point in the same direction as for urinary mercury levels, meaning that blood is not a good indicator for a quantitative evaluation of risks in the individual case. Some of the information at hand, both published and unpublished, could be commented upon.

Joselow and Goldwater, 1967, found a possible association on a group basis, but not for the individual subject, between mercury in blood and slight proteinuria. In a report by Benning, 1958, no association between blood levels and symptoms was reported in workers exposed to about 0.2 to 0.4 mg Hg/m³ (mostly vapors of metallic mercury). Similar lack of evidence of an association comes from a report on an investigation of miners by Ladd et al., 1966. The study by Smith et al., 1970, was reported to have shown a correlation on a group basis between blood values and symptoms. As for urinary levels (Section 7.1.2.2), the correlations were weaker than the correlation between air values and symptoms.

Vostal and Clarkson (unpublished data) observed a group of 6 women working in glass pipette calibration by metallic mercury. The working conditions allowed the transfer of mercury into their homes and, consequently, 24-hour continuous exposure. All of them showed typical symptoms of mercury poisoning. Their levels of mercury were 15.8, 13.9, 10.3, 4.8, and 4.1 μg Hg/100 ml red cells and correlated with the severity of the symptoms. Comparative levels of unexposed persons from the same localities were lower than 1 μg Hg/100 ml red cells.

7.1.2.3.2 Mercury in Blood and Exposure

Reports by Beani, 1955, and Goldwater, 1964, have indicated a positive group correlation between exposure to mercury and blood levels. This association was found also in the extensive study of Smith et al., 1970 (Table 7:9), but there seems to be a considerable dispersion.

On a group basis an association between mercury levels in blood and in urine has been shown by Ladd et al., 1966, Joselow, Ruiz, and Goldwater, 1968, and Smith et al., 1970. The last mentioned data (see Figure 7:7) point to a ratio of about 0.3 between blood mercury (μg Hg/l.) and urinary mercury (μg Hg/l.) (see Figure 7:7). This is in good agreement with data by Benning, 1958 (dithizone method) from which a median quotient of 0.31 can be calculated between blood and urinary levels from 28 subjects from whom blood and urinary samples were taken at the same time for analysis. The individual variation was great, with the range varying between 0.01 to 10.7 and the semiquartile range between 0.11 to 0.66.

It should be mentioned that Joselow, Ruiz, and Goldwater, 1968, showed a positive correlation between mercury in blood and mercury in parotid saliva.

7.1.2.4 Relation Between Mercury in Organs and Effects or Exposure

There are no data that give dose-response relationships and it is not possible to relate a certain exposure or effect to certain concentrations in organs. A recent article by Takahata et al., 1970, can be mentioned, however. They examined the mercury content (neutron acti-

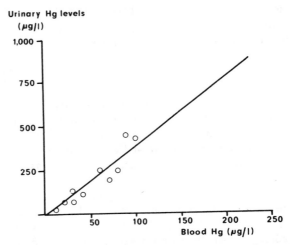

FIGURE 7:7. Relationship of concentrations of mercury in blood and in urine (uncorrected for specific gravity). (From Smith et al., 1970.)

TABLE 7:9

Relationship of Mercury Exposure to Blood Mercury Levels.[x] (From Smith et al., 1970.)

Time-weighted averages; exposure Level groups (mg/m³)	Number of workers	Percentage of group within blood level range			
		(μg/100 ml)			
		<1	1-5	6-10	>10
Controls 0.00	117	69.3	30.7	0.0	0.0
<0.01	27	33.3	63.0	3.7	0.0
0.01-0.05	175	20.6	74.9	4.0	0.6
0.06-0.10	77	10.4	81.8	6.5	1.3
0.11-0.14	53	3.8	22.6	26.4	47.2
0.24-0.27	26	0.0	19.2	26.9	53.9

[x]Expressed as percentage of each exposure level group with designated ranges of blood mercury levels.

vation) in brain of 2 deceased persons with mercurialism who had been exposed for several years to high concentrations of mercury in a mercury mine (in one case, continuously to 0.9 to 2.7 mg Hg/m^3). Before they died they had been away from mercury exposure for some years, the number of which was not specified. In the one case, the mercury content in different parts of the brain varied between 4 to 34 ppm and in the other, between 3 to 18 ppm wet weight. Even if the data do not give information about relations between dose and response, they are of value in showing that the biological half-life in the brain with all probability is long and that the distribution in this organ is uneven (see also Section 4.3.1.2).

7.1.3 Conclusions

Though a large number of studies have been published on the relation between exposure and effects in human beings, data giving valid information on both exposure and effects are unfortunately not at all so numerous. It seems reasonable to conclude, however, that prolonged exposure to mercury vapor at around 0.1 mg/m^3 can give rise to mercury intoxication. There is also evidence from studies both in the U.S. and in Eastern countries that concentrations below this value may not be without effect. In fact, medical findings have been reported at considerably lower concentrations, but it is difficult to know the significance of such findings on the basis of published data. New, extensive epidemiological studies using better epidemiological techniques and more unconventional methods are strongly needed. Of particular importance would be to try to study effects at very low exposures as seen in the U.S.S.R., using the same or improved methods. It might well be that concentrations considered without medical significance today will have to be re-evaluated considerably.

Data concerning urinary and blood levels of mercury do not lend themselves to a quantitative evaluation of exposure or effects on an individual basis. On a group basis, however, there is a quantitative correlation between exposure (probably recent exposure) and urinary and blood levels. An evaluation of exposure and also of risks can be achieved through repeated urinary or blood analysis. The average ratio between urinary (mg Hg/l.) and atmospheric mercury (mg Hg/m^3) during industrial exposure seems to be 2 to 2.5.

7.2 In Animals
7.2.1 Acute Effects

Many reports on the acute toxicity of mercury are available (see Chapter 5) but only a few give a careful description of the relation between dose and toxic manifestations. Some investigations compare a group of animals given mercuric salt only with another group given an additional drug or treatment which influences the acute toxicity. Work aiming at the selection of the most effective drug for the treatment of mercury poisoning is beyond the scope of the present report (see reviews by Swensson and Ulfvarson, 1967, and Winter et al., 1968) and it will not be discussed here.

7.2.1.1 Injection

The classic way to evaluate the acute toxicity of a compound is to find the LD_{50} value. For water soluble salts of mercuric mercury this value is 5 to 6 mg Hg/kg if the substance is injected as a solution by the intravenous or the intraperitoneal route to mice (Wien, 1939, Swensson, 1952, and Hagen, 1955), and about 12 mg Hg/kg by the subcutaneous route (Eberle, 1951, and Reber, 1953). The toxicity for rats is probably similar, as indicated by the results of, for example, Swensson and Ulfvarson, 1967, and Parizek and Ostadalova, 1967. Somewhat lower values have been reported by Lapp and Schafé, 1960, who considered 1.5 mg $HgCl_2$ i.p. (about 1.1 mg Hg/kg) as the minimal lethal dose and by Surtshin, 1957, who found 3 mg $HgCl_2$/kg to be a lethal dose. For rabbits, about 3 to 10 mg Hg/kg has been reported to be a lethal dose (Menten, 1922, and Hesse, 1926). For a comparison with LD_{50} for other mercury compounds, see Table 8:3.

Changes in the kidneys and other organs have been observed after injection of both lethal and sub-lethal doses of mercuric mercury. Alterations in the proximal convoluted tubule of the kidneys have been reported after intravenous injection of 0.1 to 0.2 mg $HgCl_2$/kg (Menten, 1922). As has been mentioned already in Chapter 5, the effect of intravenously injected $HgCl_2$ is very much dependent upon factors such as the rate of injection. It is, therefore, difficult to give a clearcut dose-response relationship. As further examples, however, it can be mentioned that Mudge and Weiner, 1958, have reported a diuretic action in dogs after the injection of 1 mg Hg/kg and that Simonds and Hepler, 1945, found an i.v.

injection of 2 mg Hg/kg to be necrotizing to the renal tubule in dogs. Haber and Jennings, 1964, showed a sex difference in the sensitivity of the kidney of rats injected intravenously with $HgCl_2$. Male rats injected with 0.4 mg Hg/kg had histological changes in the proximal kidney tubules to a greater extent than female rats given the same dose. Lapp and Schafé, 1960, studied 3 groups of rats given (I) 0.5 mg $HgCl_2$/kg, (II) 1.0 mg $HgCl_2$/kg, and (III) 1.5 mg $HgCl_2$/kg, respectively, as a single intraperitoneal injection. In groups I to II there was an increase in urinary volume after the injection. No animal died in these groups. In group III the animals developed anuria a few days after the injection and died, if not killed. Histological examination of the kidneys 2 days and longer after the injection disclosed changes in all groups, the severity of which was dose-related. The changes were reversible in groups I and II and were not seen at survival times exceeding 7 days. Changes in the uptake of trypan-blue in the kidney tubule were also observed in all groups. As mentioned in Chapter 5, functional impairment and concomitant histological changes in the proximal convoluted tubules have been detected at dose levels of 1.25 mg $HgCl_2$/kg and higher (Mustakallio and Telkkä, 1955, Rodin and Crowson, 1962, and Taylor, 1965).

Davies and Kennedy, 1967, detected an increased number of cells in the urine concomitantly with mild histological lesions in the kidney tubules in rats given a s.c. injection of 0.75 mg $HgCl_2$/kg and more pronounced changes at 0.9 and 2.4 mg $HgCl_2$/kg. Similar studies with repeated doses were performed by Prescott and Ansari, 1969 (see Section 7.2.3.1).

Kosmider, Kossmann, and Zajaczkowski, 1963, detected enzymatic changes in the blood of rabbits poisoned by i.v. injection of 3 mg/kg of mercuric chloride.

The above mentioned data concern mercuric mercury. There are not many reliable data on the toxicity of mercurous mercury. Injections of suspensions of calomel (HgCl) in water to animals and man have been described by Lomholt, 1928, and Rosenthal, 1928. Macroscopical and microscopical tissue changes were observed in kidneys, liver, and colon (Kolmer and Lucke, 1921, Almkvist, 1928, and Lomholt, 1928). It is evident from these studies that by such injections higher doses of mercury can be tolerated than is the case with injections of mercuric mercury. This difference probably is due to the slow resorption of the relatively insoluble compound from the injection site. Injections of finely dispersed metallic mercury also have been made by Lomholt, 1928, under which circumstances much higher amounts of mercury could be tolerated. However, in this case, a still more prominent deposition of the mercury at the injection site was observed. Injections of mercury in the form of mercury vapor directly by the intravenous route have been performed by Magos, 1968 (see Section 4.1.1.1.1.1). Toxic effects of these low dose injections were not reported.

7.2.1.2 Oral and Percutaneous Exposure

The LD_{50} for oral ingestion of mercuric mercury has not been well established. Lehman, 1951, found the LD_{50} by the oral route to be 37 mg $HgCl_2$/kg in the rat. For mercurous mercury Lehman reported symptoms of mercurialism in rats given 210 mg of calomel/kg body weight, but no animals died. This finding is in accord with the low oral absorption of mercurous mercury (see Section 4.1.1.2.2). Ingestion of large doses of metallic mercury (several g/kg) by rats (Bornmann et al., 1970) did not give rise to any toxic effects. This is probably a result of the very poor absorption of metallic mercury from the gastrointestinal tract (see Section 4.1.1.1.2).

Skin absorption of mercuric, mercurous, and metallic mercury can cause lethal poisoning in animals (Schamberg et al., 1918, and Wahlberg, 1965a). Wahlberg, 1965a, performing a well controlled study, found that both the percutaneous penetration and toxicity of potassium iodomercurate (K_2HgI_4) were somewhat higher than for mercuric chloride ($HgCl_2$). A dose corresponding to 250 mg Hg/kg was applied to the skin in both cases.

7.2.1.3 Inhalation

The toxicity of mercuric and mercurous mercury when inhaled has not been studied much and the existing data pertain only to the toxicity of Hg^o vapors. Ricker and Hesse, 1914, exposed mice, guinea pigs, rats, and rabbits to almost saturated mercury vapor at room temperature. The mice died after 36 to 50 hours of continuous inhalation, the guinea pigs after 3 1/2 to 4 1/2 days, the rats after 6 to 9 days, and the rabbits after 2 to 6 1/2 days of continuous inhalation. It is not known whether the mercury vapor concentration was the same in all experiments, as it was not measured. Fraser, Melville, and Stehle, 1934, exposed dogs 8 hours daily to 1.9 to 20 mg Hg/m^3. Death occurred after 2 to 16 days (mean 8 days) in 6 dogs exposed to 12.5 mg/m^3. Ashe et al., 1953, exposed 14 rabbits for 1 to 30 hours to 29 mg Hg/m^3. After 5 exposures of 6 hours each (i.e., totally 30 hours) 1 rabbit died. The others survived and were killed 6 days after the experiment had been initiated. At histological examination prominent changes were observed in the lungs, the liver, the colon, and the heart. Still more severe changes took place in the kidneys and brain.

7.2.2 Chronic Effects
7.2.2.1 Injection

Kolmer and Lucke, 1921, reported some perivascular infiltration in the brain and tubular damage in the kidneys but no damage in the nerve cells of rabbits given 6 or more repeated intramuscular injections of mercuric chloride or mercuric benzoate 0.4 to 0.5 mg Hg/kg 3 times a week.

Prescott and Ansari, 1969, observed no changes in renal tubular cell counts in urine when they administered s.c. 0.1 mg HgCl$_2$/kg daily to rats for 7 days. When they gave rats 0.5 mg HgCl$_2$/kg daily for 4 to 14 days they observed an abnormally large amount of renal tubular cells in the urine and also elevated levels of urine glutamic oxaloacetic transaminase (GOT) activity. The changes were also seen in groups of animals given greater amounts of mercury. In a group given 2 mg HgCl$_2$/kg elevation of serum GOT-activity was also seen. Histological changes appeared in the animals given repeated doses of 0.5 mg HgCl$_2$/kg but not in animals given 0.1 mg HgCl$_2$/kg. The histological changes, the urine GOT-activity, and the increased number of renal tubule cells in urine were most marked during the first days of treatment and later diminished in spite of continued exposure.

7.2.2.2 Oral and Percutaneous Exposure

Enders and Noetzel, 1955, reported microscopically evident calcification foci in the brain and histological kidney damage in rats given daily oral doses of 100 to 200 mg HgCl$_2$/kg. The rats were kept undernourished at a body weight of only 50 g during the experiment (up to 10 months' exposure). Several animals were reported to have died from the treatment. It is indeed strange that any of the animals could survive such enormous daily doses, exceeding considerably the dose which has been reported by others to be the LD$_{50}$ (see Section 7.2.1.2)! Fitzhugh et al., 1950, reported on rats given 40 ppm of mercuric acetate (about 33 ppm Hg^{2+}) in the diet for 1 year. Slight light microscopical changes were observed in the kidneys, which contained 16 µg Hg/g wet weight. In another group, given 160 ppm (about 130 ppm Hg^{2+}) for 1 year, moderate changes occurred in the kidneys. The mercury concentration in the kidneys of this group was 49 µg/g. Weight changes of males were seen after 12 weeks and onward in relation to the control group. Studies on percutaneous exposure up to 4 weeks have been reported by Wahlberg, 1965a (see Section 7.2.1.2). Further reports covering work on chronic percutaneous toxicity from which the possibility of simultaneous inhalation of mercury has been excluded are not available.

7.2.2.3 Inhalation
7.2.2.3.1 Studies in General

Fraser, Melville, and Stehle, 1934, exposed dogs to mercury 8 hours a day and observed that deaths resulted from about 40 days' exposure to concentrations of 6 mg Hg/m^3. Symptoms of mercurialism such as gingivitis, diarrhea, and loss of weight developed after about 15 days of exposure to 3 mg Hg/m^3. One dog was exposed much longer to that concentration and died after a period of 20 weeks. Preceding his death, the dog suffered gum ulcerations, ataxia, tremor, weight loss, and diarrhea.

Ashe et al., 1953, studied rats and rabbits exposed to different concentrations of mercury vapor for differing lengths of time up to 83 weeks. They used a colorimetric method (Cholak and Hubbard, 1946) to control the exposure and to analyze the mercury concentration in tissues and urine. As discussed in Chapter 2, the precision and accuracy of such methods vary with the concentration in the tissues. For the lower tissue concentrations reported by Ashe et al., 1953, a considerable error cannot be excluded. The animals were exposed 7 hours/day, 5 days/week. The results of histological examination and determination of mercury concentrations in tissues of rabbits are seen in Table 7:10. It is evident from the Table that the severest tissue damage was found in the kidney and brain. Less severe damage was observed in the lung, the liver, and the heart.

TABLE 7:10

Concentrations of Mercury (mg/100 g [x]) and Extent of Tissue Damage in Organs of Rabbits Exposed to Hg° -Vapor. (Data from Ashe et al., 1953.)

Exposure time [xx] weeks	Air concentration mg/m³	n	Kidney Conc.	Kidney Damage	Liver Conc.	Liver Damage	Brain Conc.	Brain Damage	Lung Conc.	Lung Damage	Blood Conc.	Urine mg/24 hr	
1	6.0	1	7.000	++	0.200	+	0.005	++	0.760	++	0.011	0.282	Damage to the heart, ++, was seen in most animals
2-3	6.0	2	15.115	++(+)	0.280	+(+)	0.286	++	0.402	+	0.021	0.376	
4-5	6.0	3	13.417	++	0.457	+	0.848	++	0.641	+(+)	0.052	0.463	
6-8	6.0	4	13.450	+++	0.495	++	1.390	++	0.380	+	0.202	0.139	
10-11	6.0	2	16.000	+++	0.845	++(+)	1.700	++(+)	1.330	++	0.103	0.037	
2-3	0.9	4	1.850	(+)	0.085	—	0.055	(+)	0.112	—	0.014	0.020	Damage to the heart, +, was seen from the 5th week
4-5	0.9	5	2.820	+	0.156	—	0.079	+(+)	0.405	(+)	0.021	0.024	
6-8	0.9	11	3.135	++	0.271	(+)	0.121	+	0.107	(+)	0.037	0.027	
10-12	0.9	4	3.750	++	0.480	—	0.136	++	0.146	(+)	0.017	0.032	
1	0.1	1	0.067	—	0.014	—	—	—	—	—	—	0.003	
4	0.1	1	0.620	—	0.012	—	—	—	0.051	—	0.003	0.003	
8	0.1	1	0.330	—	0.021	—	—	—	0.023	—	0.005	0.003	
9	0.1	1	0.477	—	0.056	—	—	—	0.075	—	0.004	0.003	
15-17	0.1	2	0.412	—	0.029	—	0.013	—	0.029	—	0.003	0.002	
26-28	0.1	2	0.760	—	0.115	—	0.005	—	0.051	—	0.009	0.004	
30-37	0.1	2	0.516	—	0.044	—	0.007	—	0.027	—	0.009	0.002	
46	0.1	4	0.360	—	0.065	—	0.012	—	0.020	—	0.003	0.003	
56-68	0.1	2	0.358	—	0.112	—	0.033	—	0.053	—	0.002	0.002	
82-83	0.1	2	0.318	—	0.125	—	0.045	—	0.039	—	0.012	0.002	

[x] probably wet weight
[xx] exposure was for 7 hrs/day 5 days/week
— No pathological changes
+ Definite but mild pathological changes
++ Moderate pathological changes
+++ Marked cellular degeneration with some necrosis

At concentrations of 0.9 mg Hg/m^3 and higher, damage was observed in the kidney and brain already after a few weeks of exposure. However, even after the most extended exposures to 0.1 mg/m^3, there were no microscopically detectable injuries. At the last mentioned exposure level and time, the mercury concentration in the kidneys was about 4 ppm and in the brain, about 0.3 ppm wet weight. A large individual variation was evident in the original values, but appears less prominent in Table 7:10, where only mean values are given for groups of animals representing certain time intervals. In the rabbits exposed to 0.9 mg/m^3 and in which microscopical evidence of intoxication was present, values of about 20 to 50 ppm were found in the kidneys, and 1 to 2 ppm in the brain. It appears from Table 7:10 that there is a reasonably good correlation between exposure and blood as well as urine values. In addition, the correlation is good between blood and urine values and the extent of tissue damage. These aspects of the data as well as complementary data reported by Ashe et al. have been discussed in Section 4.5.1. Ashe et al. also used rats in studies similar to those mentioned above for rabbits. At exposure to 0.1 mg Hg/m^3 for 67 to 72 weeks, the kidney concentration was about 10 ppm. The brain concentration was not given. No pathological changes were observed. In 2 dogs exposed according to the above mentioned weekly schedule for 61 and 83 weeks to 0.1 mg Hg/m^3, the kidney concentration was also about 10 ppm and no pathological changes were seen. Neither the behavior of the animals nor the renal function was studied in any of the experiments by Ashe et al., 1953.

A number of enzymatic changes in the blood, heart, liver, and kidneys of rabbits have been described by Jonek, Kosmider, and their associates (Jonek, 1964, Jonek and Grzybek, 1964, Jonek and Kosmider, 1964, Jonek, Pacholek, and Jez, 1964, Kosmider, 1964, 1965, and 1968). Exposure was carried out for 30 days, 1.5 hours/day, 11.6 mg/m^3. The authors did not describe how the mercury vapor concentration was measured during exposure, but did state that the urinary mercury level was measured by a dithizone method (Rolfe, Russell, and Wilkinson, 1955). The 24-hour mercury excretion in urine was 117 to 125 µg during the last part of the exposure. Three out of 12 animals died during the exposure and all animals showed salivation and apathy. Some weight loss also was noticed during the exposure.

Behavioral effects have been observed by Armstrong et al., 1963, on pigeons and by Beliles, Clark, and Yuile, 1968, on rats. The latter authors exposed rats to 17 mg Hg/m^3 for a total of 22 exposures of 2 hours each during 30 days. They recorded an increase in escape response latency and a decrease in avoidance response. Forty-five days after termination of exposure the rats resumed a normal performance of the test. Histological changes in the CNS with perivascular infiltration of lymphocytes in the medulla oblongata were observed in the exposed group. No changes "which could be attributable to the experimental procedure" were observed in lungs, kidney, or liver.

7.2.2.3.2 Russian Studies — Including Studies on Micromercurialism

In the Russian literature, a number of effects on various organs and functions have been reported for different animal species. Many experiments have included exposure to very low concentrations of mercury vapor for considerable periods of time. Since such work is urgent, an attempt will be made below to give an account of it, but the same difficulties in evaluating the data as were mentioned in Section 7.1.2.1.2 on human data from the U.S.S.R. are valid here (i.e., concerning how the data were obtained, etc.).

Trachtenberg, 1969, reported on extensive animal experiments. He exposed different groups of animals to different concentrations of mercury vapor for different periods of time. Generally, the exposure was for 6 hours a day, 6 days a week, and the exposure levels were checked by the Poleshajev method. In addition to gross observations for evident symptoms, more detailed studies were performed, such as tests for liver and thyroid functions, changes in the higher nervous activity, and morphological changes.

In guinea pigs (number of animals not stated), exposure to mercury vapor (1 mg Hg/m^3) caused a steady decrease of body weight already 5 days after the start of the exposure. A less prominent effect on body weight of white mice was reported at 1 month's exposure to 0.04 mg Hg/m^3. In exceptionally sensitive mice, symptoms such as tremor and paresis of hind limbs were reported after 3.5 months of exposure to mercury vapor of 0.02 mg Hg/m^3. Similar signs were reported in several of the rabbits exposed for 1 year to 0.01 to

0.04 mg Hg/m³. However, these data are difficult to evaluate because the number of animals is not known and the findings are reported only for individual animals. Moreover, no comparative study was made in relation to control groups.

In another series, 30 mice were exposed to mercury vapor, 0.45 mg Hg/m³. Nine of the 30 mice showed paresis of the hind limbs after 55 days of exposure. In these mice some studies on hemoglobin levels and blood corpuscles were also performed. Observations included anisocytosis and Jolly's bodies of the erythrocytes but the author stated that the changes were unimportant.

A number of investigations into the action of mercury on different reactive groups of tissue proteins have been made in the U.S.S.R. (Salimov, 1956, and Galojan, 1959). Trachtenberg, 1969, also made such investigations. In one of his series, 56 white rats (120 to 150 g) were exposed to mercury concentrations varying from 0.01 to 0.03 mg Hg/m³ (average: 0.014 mg Hg/m³). Another 54 rats served as controls and were not exposed.

The incorporation of amino acids into the plasma proteins in the exposed rats was found to be decreased by measuring the incorporated activity in aliquots of plasma proteins (precipitated with trichloracetic acid) at different hourly intervals after the injection of ^{35}S labeled methionine. In 16 animals killed after 143 days of exposure, an average value of 4.2% of administered activity per g body weight was found in 10 mg of precipitated plasma protein 18 hours after injection of ^{35}S methionine. In control animals the average value was 9.3%. Similar results were reported also for soluble liver proteins. The remaining rats were used for further studies on the protein synthesis by determining the rate of incorporation of radioactivity into the plasma proteins after 166 days of exposure. The results presented in Figure 7:8 show not only a decreased level of ^{35}S counts but also that the maximum incorporation appeared later, reflecting a slower rate of incorporation in the exposed group. The findings were interpreted as a disturbance in the liver function in synthesizing plasma proteins.

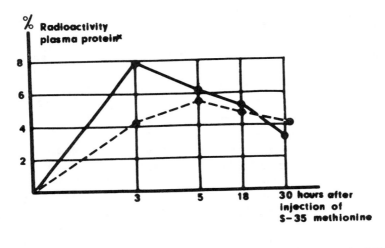

FIGURE 7:8. Incorporation of S-35 into plasma proteins of mercury exposed rats and controls at different times after injection of radioactive methionine S-35. (From Trachtenberg, 1969.)

Another investigation by Trachtenberg, 1969, which is related to the function of the liver, concerned the increased frequency of positive thymol tests in guinea pigs exposed for 104 days to 0.01 to 0.03 mg Hg/m^3 of mercury vapor (mean: 0.014 mg/m^3). In the mercury exposed group, 12 out of 14 animals were positive with a mean value of 12 units (range: 10 to 16 units). In the control group, 2 out of 14 animals were positive, with a mean value for the whole group of 5 units. An increased frequency of positive thymol reactions has also been reported in human beings by Kosmider, Wocka-Marek, and Kujawska, 1969.

The same animals were subjects for an investigation of the ability of the liver to convert dehydroascorbic acid to ascorbic acid. A statistically significant reduction in this process was seen, as the mercury exposed animals had only about 25% ± 5% of the reduction ability of the control animals. The concentration of ascorbic acid in the liver was also decreased.

In another study the sulfhydryl (SH) group content in soluble liver proteins was investigated in rats exposed to low concentrations (0.01 to 0.03 mg Hg/m^3) of mercury vapor for 150 to 180 days. A decrease in relation to a control group was seen both for "total" SH group content of denaturated (urea treated) proteins and in so-called "free" or "reactive" SH group content of liver proteins.

Investigations concerning mercury induced changes in the central nervous system and the higher nervous activity have been performed (Ivanov-Smolenskij, 1939, 1949, Ochnjanskaja, 1954, Sadcikova, 1955, Gimadejev, 1958, Drogitjina, 1959, 1962, and Kournossov, 1962). Trachtenberg's 1969 studies also included the higher nervous activity of mammals under the influence of long-term exposure to mercury vapor. In one series of experiments, cats were exposed to 0.085 to 0.2 mg Hg/m^3 (first series), 0.01 to 0.02 mg Hg/m^3 (second series), and 0.006 to 0.01 mg Hg/m^3 (third series). The number of cats in each series was not given in the monograph, but a minimum number of animals for the first series is 4 cats, for the second series 4 cats, and for the third series, 2 cats.

The results varied according to the individual cat's response to the test situation. For 3 cats in the first series, a clear effect on several of the parameters measured was observed already during the second week of exposure. For example, the latent period for response to light was at least doubled. The exposure was continued up to 8 weeks, whereby the effects increased. During the 8th week, the latency period for response to light was more than 5 times as long as the original period. However, after termination of exposure, a normalization was seen. Even at that time, only 5 to 8 nonreinforced signals were necessary for the cat to give up the conditioned reflex, whereas before the experiment 20 to 40 such signals had been necessary.

In the second series, 2 cats showed similar but less pronounced changes than those of the first series. During the first 8 weeks of exposure, no changes were observed in the parameters measured. In one animal the changes appeared after 10 weeks (see Figure 7:9) and in the other one after 22 weeks. In the third series some less prominent differences were reported. Whether these were significant in comparison to original values is not clear from the data given.

Trachtenberg, 1969, stated that his material concerning changes in the conditioned reflexes of cats was consistent with observations by Gimadejev, 1958, on rabbits, and by Kournossov, 1962, on rats. In the last mentioned investigation, disturbance in the higher nervous function was seen at concentrations as low as 0.002 to 0.005 mg Hg/m^3. As this is probably the lowest concentration of mercury which has been reported to have an effect on mammals, it seems reasonable to look for more details in the work by Kournossov. The following data are taken partly from Kournossov, 1962, and partly from personal discussions with Kournossov. He exposed rats in 4 groups (5 rats in each group) to different concentrations of mercury vapor. I: 0.02 to 0.03 mg Hg/m^3; II: 0.008 to 0.01 mg Hg/m^3; III: 0.002 to 0.005 mg Hg/m^3; IV: 0.0000 to 0.0003 mg Hg/m^3. Exposure lasted 6.5 hours daily, 6 days a week. Mercury concentrations in chambers were checked by Poleshajev's method. Tests for conditioned reflexes were performed for 3 months without mercury exposure. The studies on changes in conditioned reflexes were performed according to the technique described by Kotlyarevskij, 1954, in a book on methods generally used in the U.S.S.R., and in Ryazanov's review, 1957. Further details on experimental conditions and procedures were obtained from personal contacts with Kournossov.

The temperature in the exposure chambers varied between 20.5 to 28° C and the relative

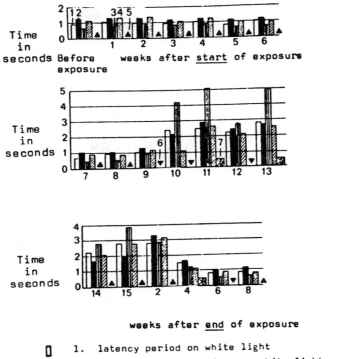

1. latency period on white light
2. time of running to food on white light
3. latency period on buzzer
4. time of running to food on buzzer
5. absence of reaction on blue light (normal)
6. unability to differentiate between blue light and white (pathological)
7. refusal to perform the test in some of the trials

FIGURE 7:9. Changes in conditioned reflexes of a cat before, during, and after exposure to mercury vapor 0.01 to 0.02 mg Hg/m³ for 6 days a week, 6 hours daily. (From Trachtenberg, 1969.)

moisture was 80 to 90%. The motor-nutritional reflexes were developed in Kotlyarevskij's chamber provided with acoustical and light signals. A Plexiglas® door, attached to one of the walls, had to be raised by the animal to gain access to the feeding box. As the lower end of the door was forced forward by the rat's motor activity, a lever attached to the door bore against a pneumatic system recording the force of the rat's motor activity. A similar system connected to the floor of the box permitted the recording of all of the movements of the rat inside the box. The animal was trained to discriminate according to a predetermined pattern of consecutive bell signals (food reinforcement), light signals (food reinforcement), and buzzers (no reinforcement).

Changes of several parameters observed when testing conditioned reflexes are seen in Figure 7:10. During the first month of exposure, animals from the first group exhibited an increased pushing strength in the test, but no substantial changes in the latency periods were observed. During the second month of exposure, the pushing strength returned to the original value, whereas the latency period for one of the stimuli increased. During the third and fourth months of exposure, a diminution of the activity as measured by the pushing strength was seen and the failures to respond to stimuli increased considerably. The rats eventually refused to perform the test. Similar but less pronounced changes were observed in group II. Even in group III there were deviations from original values but not until 2 1/2 months after the beginning of exposure (see Figure 7:10 III). In group IV no significant deviations from the original values occurred. At the end of the experiments, morpho-

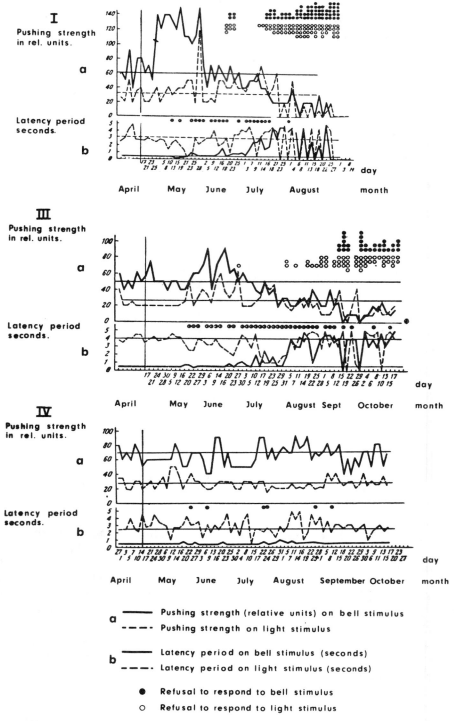

FIGURE 7:10. Registration of conditioned reflexes of 3 rats from groups exposed to: I: 0.02 to 0.03 mg Hg/m^3; III: 0.002 to 0.005 mg Hg/m^3; and IV: 0.0000 to 0.0003 mg Hg/m^3. (From Kournossov, 1962, and personal communication.)

logical examinations and mercury determinations in organs were performed. Analysis of mercury in tissues of 2 to 3 animals from each group according to a modification of Poleshajev's method showed in the kidneys: series I, about 1 ppm wet weight; series II, 1-2 ppm; series III, about 0.6 ppm; series IV, about 0.08 ppm; and in an entirely unexposed control (series V), about 0.04 ppm. In the brains the following concentrations were found: series I, 0.1 to 0.2; series II, 0.1-0.2; series III, 0.06 to 0.08; series IV, 0.00; and series V, 0.00 to 0.01 ppm wet weight. For a comparison of these values with the results of other studies on the accumulation and retention of mercury, see Chapter 4.

The data on conditioned reflexes agree with earlier observations of changes in conditioned reflexes at exposure to mercury vapor at 0.035 mg Hg/m^3 (Gimadejev, 1958, 1962). Gimadejev mentioned 2 phases in the higher nervous activities: "In the beginning an increase of the stimulation process, followed by the development of a spreading cerebral inhibition" (quoted in Medved, Spynu and Kagan, 1964).

Trachtenberg, 1969, reported on the decline of the concentrations of ascorbic acid in the adrenal glands of rats exposed to mercury vapor in concentrations of 0.007 to 0.02 mg Hg/m^3. A statistically significant decrease in the concentrations was observed 8 to 20 weeks after the beginning of exposure in young rats and 15 to 20 weeks after exposure in older rats. An increase in the weight of the adrenal glands was also noted and was statistically significant ($p < 0.01$) in comparison with a control group in young rats after 15 weeks of exposure and longer. In the old rats the comparative increase in weight of the adrenal glands was only statistically significant at the longest survival time (20 weeks). See Table 7:11.

Trachtenberg, 1969, reported on studies on the uptake of radioactive iodine in the thyroid of rats chronically exposed to mercury vapor at different concentrations. Several experiments demonstrated an increased intake of radioiodine in relation to pre-exposure values. Measurements of the uptake of radioactive iodine in the thyroid were performed according to a method described by Gabelova, 1953. 0.15 microcurie of I-131 was administered subcutaneously to a rat; then external measurements of radioactivity over a window in a lead shield by means of a G.M. tube were performed. A series of 15 rats exposed to 0.01 to 0.03 mg Hg/m^3 for 105 days showed a significant and clearcut difference in the iodine uptake compared to a control group and compared to pre-exposure values (see Table 7:12).

Increased uptake of radioactive iodine in the thyroid is usually considered an indication of hyperfunction of the organ. This commonly gives rise to an increased metabolic rate and an increased oxygen consumption. However, this was not the case in Trachtenberg's experiments. He reported that there was almost no change in the oxygen consumption of the animals during the experiment. Before the experiment, it was 1.7 ml/hour/kg body weight, whereas after 3 months of exposure, it was 1.8 ml/hour/kg. The author proposed the hypothesis that mercury inhibits the thyroxine activity of the blood. Even if the uptake of iodine and production of thyroxine in the thyroid are high, there will be no effect on metabolism.

All data by Trachtenberg, 1969, on radioiodine in the thyroid speak in favor of an increased uptake. In contrast to this observation, there are unpublished results by Dr. Avetzkaja, Donezk, U.S.S.R. During his visit in Kiev GFN had discussions with Drs. Avetzkaja and Trachtenberg. Dr. Avetzkaja spoke of her unpublished observations on 3 series of rats (10 rats in each series) exposed to mercury vapor for 3.5 to 5 months. Series I: no exposure; series II: 0.02 mg Hg/m^3; series III: 0.2 mg Hg/m^3. The animals were given a subcutaneous injection of ^{131}I and killed 24 hours later. The thyroid was dissected and the radioactivity was measured in a well-type scintillation detector. The following values were obtained (percent of injected dose ± S.D.): series I: 26 ± 2.6%; series II: 21 ± 2.1%; and series III: 10 ± 1%. These data demonstrate a dose-related *decrease* in the uptake of ^{131}I in the thyroid, i.e., the opposite from what was illustrated by several of Trachtenberg's investigations. It is difficult to account for this difference. During the discussion in Kiev, Trachtenberg explained the difference between his results and those of Dr. Avetzkaja by the differences in exposure time. In some of his series he did note a tendency to lower values at longer and more pronounced exposure. As a further example of the peculiarities observed with regard to the action of mercury on the thyroid, the reverse relation between Hg^o exposure and

TABLE 7:11

Weight of Adrenal Glands in Rats Exposed to Hg°-Vapor 0.007 to 0.02 mg/m.³ (Trachtenberg, personal communication.)

	Weight of adrenal glands					
Weeks of exposure	n	Exposed mg/100 g body wt.	S.D.	n	Controls mg/100 g body wt.	S.D.
1	8[x]	24.2	3.8	7[x]	25.6	5.4
	6[xx]	18.7	1.7	7[xx]	19.7	2.2
2	7	26.8	4.6	8	24.6	4.7
	6	18.2	3.2	7	19.1	2.6
4	7	25.6	5.0	6	25.6	2.5
	7	17.2	5.3	7	17.9	4.4
6	5	30.1	4.4	5	26.9	4.6
	7	21.2	2.2	5	19.8	5.3
8	6	34.7	10.5	7	26.4	6.1
	5	23.3	7.6	5	21.0	4.3
10	5	38.0	7.5	8	27.6	6.2
	6	24.0	7.4	6	18.7	5.9
15	6	40.4	9.9	6	26.9	4.4
	6	24.1	5.9	7	20.0	4.9
20	7	47.3	5.5	6	25.6	11.4
	7	24.9	4.2	7	18.1	4.3

[x]The first row of values for every weeks' measurements listed refers to the younger rats, 5 to 7 months old at the beginning of the experiment.
[xx]The second row of values refers to the older rats 18 to 20 months old at the beginning of the experiment.

thyroid diseases reported by Baldi, 1949, may be mentioned.

Changes in the ECG of rabbits exposed to low concentrations (probably 0.01 to 0.03 mg Hg/m³) of mercury vapor have been reported by Trachtenberg, 1969. During the first month, a tendency to tachycardia was noted, a reflection of increased sympathetic tonus according to Trachtenberg. After 1 or 2 months a change in beat frequency was noted, and after 3 months of exposure, all animals had bradycardia (220 to 250 beats/min — normal pre-exposure values about 390 beats/min). The author interpreted the bradycardia as due to an increased vagal tonus. Diminution of the potentials of the different ECG waves was also observed (P-wave from 0.12 V to 0.05 V and R-wave from 0.36 to 0.24 V after 70 days of exposure). The mercury exposed rabbits also showed a different reaction from nonexposed animals when pituitrin was injected. ST-T changes were seen then in the ECG of mercury exposed animals.

Trachtenberg, 1969, reported on the following experiments related to the immunological defense mechanism of the body. Groups of white rats exposed for periods up to 246 days to average concentrations of 0.01 to 0.02 mg Hg/m³ showed a lower rise in agglutination titer after immunization than control rats receiving the same immunization but no mercury exposure. In one case a titer of 1:6880 was found in control rats whereas the titer was only 1:524 in mercury exposed rats. These data were considered to indicate that the immune defense properties in the

TABLE 7:12

Uptake of Radioactive Iodine in the Thyroid Glands of Rats at Different Time Intervals After Injection of Radioactive Iodine (Exposure: 0.01 to 0.03 mg Hg/m^3, 6 hours daily, 6 days per week). (From Trachtenberg, 1969.)

Group of animals	Number of animals	Time of measurement	Uptake of ^{131}I in percent of injection dose (hours after injection of ^{131}I)						
			2 h	6 h	12 h	24 h	48 h	72 h	96 h
Mercury exposed	15	Preexposure values	12.6±0.5	14.9±0.8	18.9±0.7	29.8±1.1	27.8±1.1	25.4±1.4	20.8±0.9
		After 105 days exp.	37.1±3.3	89.4±8.1	83.2±7.5	68.1±3.2	39.2±3.6	20.4±1.7	17.4±1.0
Controls	15	Preexposure values	10.5±0.9	13.3±1.4	16.7±0.3	25.1±0.9	22.6±2.1	19.0±1.2	18.3±0.9
		After 105 days	12.1±1.3	14.5±1.1	18.3±1.1	29.8±1.0	23.9±0.9	19.0±0.6	16.7±0.8

blood of mercury exposed animals might be different from those in the blood of unexposed animals.

Morphological alterations were reported by Trachtenberg, 1969, for a number of organs in animals exposed to low concentrations of metallic mercury vapor and vapors from organic mercury compounds, especially ethyl mercury phosphate. In his 1969 monograph Trachtenberg expressed the opinion that the changes were similar regardless of the chemical form of the mercury and did not group his material with regard to the mercury compound. He usually does not report the frequency of findings in different exposure groups. Most of the animals studied were chronically exposed to low concentrations of mercury vapor, i.e., 0.01 to 0.05 mg Hg/m^3. Because of the above mentioned difficulties, it is impossible to draw definite conclusions with regard to dose-response relationships from this material on morphological alterations.

The action of mercury on the testicles has been probed by Sanotskij et al., 1967, and Phomenko (unpublished data). They observed changes in the reproductive function of male rats after comparatively brief exposure to mercury vapor of high concentrations.

Kournossov, 1962, studied morphological alterations in different organs of rats chronically exposed (6.5 months) to mercury vapor in concentrations (Poleshajev's method) of 0.02 to 0.03 mg Hg/m^3 (group I), 0.008 to 0.010 mg Hg/m^3 (group II), 0.002 to 0.005 mg Hg/m^3 (group III), and 0.0000 to 0.0003 mg Hg/m^3 (group IV). He reported mild changes in the brain of the animals from groups I, II, and, to a lesser degree, also in group III. The changes consisted of perivascular and pericellular edema and vacuolization of some cells in the cortex. By Nissl staining, swelling and vacuolization of the cytoplasm of nerve cells in the pyramidal and granular layers of the cerebral cortex were shown. Similar changes were observed in subcortical nuclei and in the brain stem. In group IV and group V, a control group, no changes were observed.

7.2.3. Conclusions

The LD_{50} for injected mercuric mercury is about 5 mg Hg/kg, and much higher for oral exposure. Percutaneous exposure can also give rise to poisoning. For mercurous mercury compounds the LD_{50} is higher irrespective of mode of administration. Acute effects of inorganic mercury are primarily on the kidneys, where acute intravenous doses lower than 0.5 mg Hg/kg give rise to histological changes and excretion of renal tubular cells in rats. Effects of injected doses of mercuric or mercurous mercury on other organs such as liver and colon also have been reported, as well as enzymatic changes in plasma. With ingestion of mercury salts, higher doses are required to cause poisoning. Similar changes to those mentioned above occur but effects on the gastrointestinal tract are more prominent. When exposure is by inhalation of mercury vapor, acute effects occur in the lung, brain, liver, kidney, and colon. Concentrations of about 10 mg/m^3 may be fatal or give rise to evident symptoms within 1 or a few days' exposure.

A number of effects on various organs have been recorded as resulting from long-term exposure to inorganic mercury. After long-term oral exposure to mercury salts damage to the kidneys has been observed at dose levels in the diet exceeding 30 ppm Hg^{2+}. Still much higher doses of mercuric mercury are necessary to cause death at long-term exposure by the oral route.

By inhalation of mercury vapor a lethal effect on experimental animals has been obtained after a few months of daily 8-hour exposure to concentrations of a few mg of mercury/m^3 of air. Pathological changes in kidneys and brain of animals have been evoked by similar exposure to concentrations of about 1 mg/m^3 and even lower. Enzymatic changes in the blood, heart, liver, and kidneys have also been reported, but these experiments have only been performed with high concentrations of Hg-vapor.

In the Russian literature weight loss and toxic signs in several animal species have been reported at exposure levels comparable to those mentioned above. The Russian scientists have also described similar but less frequent changes at much lower concentrations. In addition, changes in the functions of several organs of rats or rabbits such as CNS (conditioned reflexes), thyroid (increased uptake of ^{131}I), heart (changes in ECG), liver (changes in the thymol test, protein synthesis, SH-group, and ascorbic acid content), adrenal glands (diminished ascorbic acid content and a slight weight increase), and in the immunological response of the body have been stated to have occurred after exposure for several months to concentrations of 0.01 to 0.03 mg Hg/m^3. Changes in conditioned reflexes have been reported even at concentrations in the air of 0.002 to 0.005 mg Hg/m^3 when rats were exposed for several months.

The significance of the reported changes is difficult to evaluate for several reasons discussed earlier. Just as suggested with regard to the human data, it would be likewise of importance to try to study effects in animals at exposure levels as low as those studied in the U.S.S.R., using the same or improved methods.

Chapter 8

ORGANIC MERCURY COMPOUNDS – RELATION BETWEEN EXPOSURE AND EFFECTS

Staffan Skerfving

8.1 Alkyl Mercury Compounds
8.1.1 Prenatal Exposure

Cases caused by intra-uterine exposure to alkyl mercury compounds have mainly shown damage to the nervous system. It is not known at what stage of pregnancy the lesions were induced.

8.1.1.1 In Human Beings
8.1.1.1.1 Methyl Mercury

The cases of prenatal poisoning with methyl mercury from Minamata occurred in families with heavy consumption of fish (Harada, 1968b). Of 22 victims, 17 were born into families who fished regularly in the contaminated area. In 14 of the families postnatal cases also occurred. One child was fed with commercially produced baby food, 3 had mixed feedings, and the rest were breast-fed.

The frequency of cerebral palsy in the area around the Minamata Bay was high, 5 to 6% of the total number of births. In 1 village, 12% of the children had cerebral palsy. The expected frequency of cerebral palsy was 0.1 to 0.6% (Harada, 1968b).

There are data on hair total mercury levels of children with cerebral palsy and of their mothers from the area around the Minamata Bay (Harada, 1968b). The samples were taken at the time of the first examination, when the children were 1 to 6 years old, and 2 to 3 years later. The levels in the children at the first examination were 5 to 100 $\mu g/g$ of hair and in mothers, 2 to 190 $\mu g/g$. The method of analysis was not stated. There was no correlation between the ages of the children and the levels in the hair or between the levels in the children and in their mothers. No data on the exposure between birth and sampling are available. It is not possible to draw any conclusions about mercury levels in hair of the poisoned children at the time of birth. The children might well have been exposed considerably postnatally.

In a study made several years after the epidemic (1962 to 1963), 15 mothers had neurological signs such as paresthesia and positive Romberg sign (Harada, 1964). Harada, 1968b, stated that numbness in extremities and neurological symptoms had been observed during pregnancy in only 5 of the mothers. The symptoms disappeared soon, except in 1 case. No data are available on the frequency of similar symptoms in control groups in Japan. In the Minamata Report, 1968, none of the mothers were clinically evaluated as having a typical case of the Minamata disease. Recently, Murakami, 1971, reported that 1 of the mothers had been recognized as a victim of the disease.

An investigation of mercury levels in hair samples from clinically healthy children was performed in connection with the Minamata epidemic (Harada, 1968b). The method of analysis was not stated in the publication. In 2 out of 12 clinically healthy infants (2 to 6 months of age) the levels were 89 and 160 $\mu g/g$ of hair, respectively, and in the others, 22 $\mu g/g$ or below. In the group of 13 children between 1 to 6 years of age, 2 had 43 and 48 $\mu g/g$, respectively, and the rest had 25 $\mu g/g$ or below. In 18 mothers and their clinically healthy children, concentrations of 0.5 to 63 and 0 to 43 $\mu g/g$ hair, respectively, were found. The levels in breast milk from 17 of those mothers were below 0.2 $\mu g/g$, which was stated to have corresponded to levels found in samples from another area. In 6 healthy children and in 6 children and 10 adults with cerebral palsy from other parts of Japan, levels below 7 $\mu g/g$ were found (in 1 subject, 12 $\mu g/g$).

In Niigata, no definite case of prenatal poisoning occurred (Tsubaki, 1971). One case of cerebral palsy was reported. The mother had consumed fish from the Agano River during 7 to 9 months of the pregnancy (Tsubaki et al., 1967a). The father had symptoms of poisoning (Matsuda et al., 1967). The infant had 77 μg Hg/g hair at 5 months of age. The mother had 290 $\mu g/g$ hair at 2 1/2 months after the delivery (Matsuda et al., 1967, and Tsubaki, 1971). Nothing was stated about exposure between delivery and sampling.

Also in the Niigata area, pregnant women and newborn infants and their mothers were studied (dithizone analyses). None of 57 pregnant women had levels over 50 $\mu g/g$ (Matsuda et al., 1967). Nine mothers of newborn babies had levels above

50 µg/g, 4 above 100 µg/g and 1, 200 µg/g or more. One infant out of 14 had a level in the interval of 100 to 150 µg/g, while all of the others had below 50 µg/g. Tsubaki et al., 1967a, reported that 81 pregnant women had been studied, of whom 4 had levels in the range 51 to 110 µg/g. Nothing abnormal was observed in any of the children. It is likely that some of the levels mentioned above were reported twice or thrice. The babies were studied less than 2 1/2 years after the start of the epidemic.

Engleson and Herner, 1952, described a case of mental retardation in a child whose mother had eaten porridge made from methyl mercury dicyandiamide dressed seed during pregnancy. While the mother had no symptoms of poisoning, the father and a brother had neurological symptoms.

Snyder, 1971, reported a case of prenatal intoxication in an infant whose mother had consumed regularly during the third to sixth months of pregnancy meat from hogs fed with seed grain treated with methyl mercury. The mother did not show any neurological signs or symptoms and had normal visual fields. Postnatal exposure was excluded, as the child was never breast-fed and received only commercially prepared baby food. Some analytical data on the congenital case reported by Snyder have been provided by Curley et al., 1971, and the Center for Disease Control, 1971. Analyses were made by an atomic absorption method. The pork contained 28 mg Hg/kg. Amniotic fluid obtained during the last third of the pregnancy contained less than 0.02 µg Hg/g (detection limit of the method employed). A hair sample from the mother had a mercury level of 310 µg/g. Two samples of serum, taken 3 weeks apart, were reported to have contained 2.9 and 0.47 µg/g, respectively. When compared to the hair level, the serum levels are unexpectedly high.

8.1.1.1.2 Ethyl Mercury

Ten cases of prenatal poisoning by ethyl mercury have been reported from the U.S.S.R. by Bakulina, 1968. The mothers had shown symptoms of poisoning by ethyl mercury chloride during pregnancy or up to 3 years prior to delivery, and the children showed various degrees of physical and mental retardation. No detailed medical histories are available. It is not known to what extent postnatal exposure to mercury was important for the development of symptoms. In 2 mothers, however, levels of mercury in breast milk of 0.3 and 0.75 mg/l. were reported, meaning a possible exposure for the babies of 0.05 to 0.1 mg Hg/kg body weight/day. Mercury was determined by a colorimetric precipitation method. It is not known in what form mercury was present in the breast milk.

8.1.1.2 In Animals

Only a few animal experiments on prenatal alkyl mercury poisoning have been reported.

8.1.1.2.1 Methyl Mercury

Moriyama, 1968, exposed rats to methyl mercury chloride and methyl mercury methyl sulfide in varying doses before and during their pregnancies. The methodology is described so superficially that conclusions cannot be drawn. He also exposed pregnant cats to methyl mercury chloride and methyl mercury methyl sulfide in doses of 0.5 to 1 mg/kg body weight/day for 3 to 57 days. No controls were included in the experiment. In spite of weaknesses in methodology, it is apparent that the highest dose induced fetal damage if administered late in pregnancy. Poisoned mothers did not give birth to healthy offspring.

Murakami, 1969, mentioned a study by Tatetsu et al., 1968, on pregnant rats given methyl mercury methyl sulfide. The mothers were said to have had clinical symptoms but no morphological changes while the offspring were healthy at birth but had morphological changes in their central nervous systems. Matsumoto et al., 1967, and Nakamura and Suzuki, 1967, reported on pregnant rats given methyl mercury, but the study does not permit any conclusions.

Nonaka, 1969, reported an electron microscopical study of full term rat fetuses and 100-day-old litters of mothers who had received 2 mg Hg/kg body weight/day as methyl mercury orally during their pregnancies. Both mothers and litters were free from clinical symptoms. Although subcellular changes were reported, it is questionable whether the methods used can allow such conclusions (Berglund et al., 1971).

Spyker and Sparber, 1971, reported that, with a dose of 2 mg/kg body weight of methyl mercury dicyandiamide injected on day 7 or 9 of gestation into mice, 490 of 498 surviving fetuses appeared morphologically normal on day 18 of pregnancy. When 4 or 8 mg/kg were injected, an increased number of apparently normal neonates were killed by the mother. Surviving offspring were tested by behavioral techniques on day 30. An open field test revealed significant effects in many but not in all of the surviving neonates. Neurological symptoms developed 2 1/2 months later. The reason that the differences in behavior and symptomatology were found only in part of the exposed offspring is not clear.

Sobotka, Cook, and Brodie, 1971, analyzed eye opening, righting reflex, general activity, and body weight in neonatal rats from mothers injected with single doses of 0.1, 0.5, and 2.5 mg Hg/kg as methyl mercury chloride on days 6 to 15 of gestation. No major neurotoxic symptoms occurred in mothers or litters, and only subtle developmental neurochemical changes were observed. The exposed groups showed "maturation acceleration" (i.e., earlier eye opening and enhanced development of clinging ability). Small regional changes in nonspecific cholinesterase activity, serotonin, and norepinephrine levels in brain were found at 28 days of age.

Frölén and Ramel (to be published) administered about 3 mg Hg/kg body weight as methyl mercury dicyandiamide intraperitoneally to mice on day 10 of their pregnancies. The number of dead fetuses and resorbed litters was significantly higher in the experimental group than in a control group. It must be emphasized that methyl mercury injected intraperitoneally induces peritonitis.

Khera (quoted by Clegg, 1971) gave mice methyl mercury chloride orally in doses of 0.1, 1, 2.5, and 5 mg Hg/kg from day 6 through day 17 of pregnancy; 5 mg/kg resulted in reduced litter size. At 2.5 mg/kg the litter size was normal but all in the litter died within 24 hours postpartum. At 1 mg/kg the newborns appeared normal but the development of the cerebellum was retarded morphologically on days 7 to 14 of life. The effect was not observed later in postnatal development. No effects were observed at 0.1 mg/kg/day.

Oral administration of methyl mercury chloride to pregnant rats on days 7 to 20 was reported to have resulted in decreased weight of offspring when 6 mg/kg/day was given. Marked reduction of litter size took place when a dose of 8 mg/kg per day was administered (Courtney, quoted by Clegg, 1971).

8.1.1.2.2 Ethyl Mercury

Morikawa, 1961b, and Takeuchi, 1968b, described 3 cats given orally 2 to 3 mg/kg body weight/day of *bis*-ethyl mercury sulfide (it is not clear whether the dose means mercury or the compound) during the latter part of their pregnancies. Two of the mothers had clinical symptoms of alkyl mercury poisoning and all of them had morphological changes in the central nervous system. One out of 8 kittens was clinically intoxicated and all of them had morphological damage in the central nervous system.

Okada and Oharazawa, 1967, administered subcutaneously ethyl mercury phosphate in doses of 5 to 40 mg Hg/kg to pregnant mice on day 10. There was reduced litter weight on day 19. Oharazawa, 1968 (quoted by Clegg, 1971), gave 40 mg/kg of the same substance on the same day to the same species. Litter size was unaffected but the offspring were undersized and 32% had cleft palates.

8.1.1.3 Conclusions

Poisoning has been observed in children of mothers exposed to methyl and ethyl mercury compounds. Besides the obvious transplacental exposure of the fetus, the possibility of a postnatal exposure through breast milk has been indicated (see also Section 4.4.2.1.1.2.2).

The children were born to mothers heavily exposed to alkyl mercury. No further information is available regarding the exposure of the mothers during pregnancy. Postnatal cases occurred in the families of about half of the children poisoned prenatally by *methyl mercury*. None of the mothers of affected children was classified as "methyl mercury poisoned" according to the criteria used at the epidemic in Minamata. It seems that some neurological symptoms and signs were present in several of the mothers, but the relevance of these cannot be evaluated. Recently it has been stated that 1 mother was recognized as having a case of the Minamata disease. Because the neurological damage was definitely much more severe in the children than in the mothers, it seems reasonable to assume that the fetus is more susceptible than the pregnant woman.

No information is available on the levels of mercury in blood or hair of mothers of poisoned children at the time of delivery. At least 4 pregnant women, or women who had just given birth to healthy children (when observed during up to 2 1/2 years after the onset of the epidemic), had levels above 100 μg Hg/g hair and at least 9 such subjects had levels above 50 μg/g.

In one study, clinical symptomatology of infants born to mothers poisoned by *ethyl mercury* was reported to occur up to 3 years after onset of symptoms in the mothers. The information about the clinical picture is scanty. Experimental studies on prenatal alkyl mercury poisoning are limited and even more limited in conclusive value. When pregnant animals have been exposed, reduced litter size and/or weight, fetal death, resorption, neonatal death, morphological lesions in the CNS, and neurological symptoms have been reported in mice, reduced litter weight and morphological lesions in rats, and morphological and CNS lesions and neurological symptoms in cats. It is not possible to draw definite conclusions regarding toxic exposures. There are several indications that the fetus is more susceptible than the pregnant animal.

8.1.2 Postnatal Exposure
8.1.2.1 In Human Beings

Since organ levels and effects, exposure and organ levels, as well as exposure and effects have been documented very seldom on an intra-individual basis in poisoned persons, they will be considered separately.

8.1.2.1.1 Relation Between Organ Levels and Effects

Whole blood, or blood cell level, is considered to be the best available index of exposure to and retention of alkyl mercury (Section 4.5.2.1). If external contamination can be excluded, hair levels can also be used, especially at constant exposure. Besides the levels in these index tissues, the levels in the critical organ system, i.e., the nervous system, and in those organs which particularly accumulate mercury, i.e., kidney and liver, will be considered.

8.1.2.1.1.1 Blood
8.1.2.1.1.1.1 Methyl Mercury Exposure

Symptoms reported — Lundgren and Swensson, 1948 and 1949, have reported a whole blood total mercury level of about 4 μg/g (dithizone method) in a worker fatally poisoned by methyl mercury through inhalation.

Tsuda, Anzai, and Sakai, 1963, Ukita, Hoshino, and Tanzawa, 1963, and Okinaka et al., 1964, have reported 1 case of methyl mercury poisoning after a treatment for mycosis with methyl mercury thioacetamide. Mercury levels in whole blood were 1.0 to 1.8 μg/g after 5 months (dithizone method).

Another case of poisoning after repeated application of methyl mercury thioacetamide solution for 2 months in the treatment of mycosis was reported by Suzuki and Yoshino, 1969. Mercury content in whole blood was still 0.12 μg/ml 9 months after the onset of symptoms and cessation of exposure.

Blood mercury levels have been reported for a total of 17 adults with manifest neurological symptoms from the Niigata epidemic. Based on data given by Tsubaki et al., 1967a, Kawasaka et al., 1967, Matsuda et al., 1967, and Tsubaki (personal communication), Berglund et al., 1971, have calculated the relationship between time elapsed since the onset of symptoms and whole blood mercury levels (Figure 8:1). It was estimated by extrapolation from the diagram that the level at onset of symptoms should have been at or above 0.2 μg/g.

There are several uncertainties in the estima-

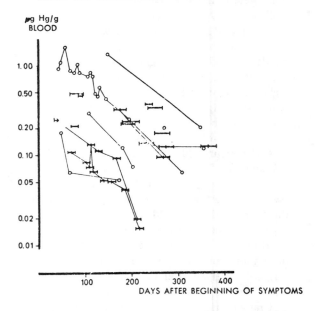

FIGURE 8:1. Relation between total mercury level in whole blood and the time elapsed after onset in cases of methyl mercury poisoning from Niigata. (From Berglund et al., 1971.) In cases where there was uncertainty regarding date of onset or of sampling or both, the total uncertainty concerning the time that elapsed after the onset of symptoms has been indicated as an interval. When repeated analyses apply to the same patient, the figures (where appropriate the middle points of an interval) have been joined together. (Data according to Tsubaki personal communications, Matsuda et al., 1967, Kawasaka et al., 1967, and Tsubaki et al., 1967a.)

tion. The analyses were made by a dithizone method, and no data are available on the reliability of the analytical procedure for blood samples. However, in those cases in which analyses were made on several samples taken at different times from the same patient, the reliability of the data might be indicated by the relatively rectilinear course of the blood clearance when plotted in a semilogarithmic diagram. Even so, a systematic error cannot be excluded. Another problem is the lack of information on the time at which exposure stopped. It is known that in some cases there was an exposure continuing after onset of symptoms, but it seems that this was not true in the patients with the lowest blood mercury levels. The decrease in blood mercury levels in patients with repeated sampling indicates that, for them, probably no significant exposure occurred during the sampling period. In several patients hair samples were taken closer than blood samples to the onset of symptoms (see Figure 8:3). The decline in hair mercury levels with time makes it reasonable to assume that in those cases no significant exposure occurred after the onset of symptoms.

Similarly, it cannot be excluded that the exposure had stopped before the onset of symptoms. As there is no information available on this possibility, this assumption is considered unjustified.

Curley et al., 1971, and the Center for Disease Control, 1971, have reported some serum mercury levels in a family in New Mexico, U.S., exposed for 3 1/2 months by ingestion of meat of swine fed methyl mercury dicyandiamide treated seed. In 3 severely poisoned persons, 8 to 20 years of age, serum samples obtained about 1 month after onset of symptoms contained 1.9 to 2.9 μg Hg/g (an atomic absorption method was used). Samples of cerebrospinal fluid (CSF) from 1 of the persons ranged 3.3 to 3.5 $\mu g/g$. Because the blood cell level is expected to be about 10 times the plasma level in persons heavily exposed to methyl mercury (Section 4.2.2.1.2), the reported serum levels are extremely high. The same is true of the CSF level, though the number of CSF analyses published is limited.

Herdman, 1971, reported a case of suspected methyl mercury poisoning. A women had consumed 0.35 kg of swordfish (about 1 mg Hg/kg) per day for 21 months when she began to experience dizziness, tremor of the hands and the tongue, mispronunciation of words, and loss of memory and of reading comprehension. At a neurological examination a wide-based gait was noted. The diagnosis was psychoneurosis. The exposure was calculated at about 0.35 mg Hg/day. The swordfish diet was repeated 3 to 6 weeks at 2 to 3 times a year for 5 years. Whole blood mercury level (method not stated) in a sample obtained 4 months after the last dietary period of 4 weeks was 0.060 $\mu g/ml$. It is difficult to know whether the symptoms and signs in this case were caused by the methyl mercury exposure. The clinical picture is not in accordance with those seen in poisoned Japanese people from Minamata and Niigata or in cases of occupational poisoning from other parts of the world.

Symptoms not reported – Lundgren, Swensson, and Ulfvarson, 1967, examined 9 workers without symptoms in a factory producing a methyl mercury compound. The average mercury level in whole blood was 0.1 (range: 0.07 to 0.180) $\mu g/g$. Tejning, 1967b, found 0.013 to 0.170 $\mu g/g$ in blood cells in a similar group of 66 workers.

Berglund et al., 1971, have compiled mercury levels in blood cells in material from Sweden (Birke et al., 1967, and to be published; Tejning, 1967c and 1968b, and Skerfving, 1971) and Finland (Sumari et al., 1969), describing subjects exposed to methyl mercury through consumption of contaminated fish. None of the persons investigated had any symptom of methyl mercury poisoning. The compiled material is presented in Figure 8:2. It can be seen that out of a total of 227 subjects, 60 persons had levels above 0.1 $\mu g/g$ in blood cells, 19 above 0.2 $\mu g/g$, 6 above 0.3 $\mu g/g$, 4 above 0.4 $\mu g/g$, and 3 above 0.5 $\mu g/g$. The 2 highest figures were in the range 1.1 to 1.2 $\mu g/g$ (corresponding to 0.60 to 0.65 $\mu g/g$ whole blood).

Blood levels (neutron activation analysis) were reported in 20 persons who had eaten contaminated fish but showed no signs of poisoning (Mastromatteo and Sutherland, 1970). Persons who had stopped fish consumption 5 months prior to the investigation had levels of 20 to 85 ng/g whole blood. A level of 155 ng/g was found in 1 individual with a recent intake of fish.

McDuffie, 1971, analyzed (atomic absorption method) whole blood samples from 42 subjects in the U.S. who had been on a high fish diet including moderate to high amounts of tuna and swordfish (average mercury levels about 0.25 and 1 mg/kg, respectively). The calculated average

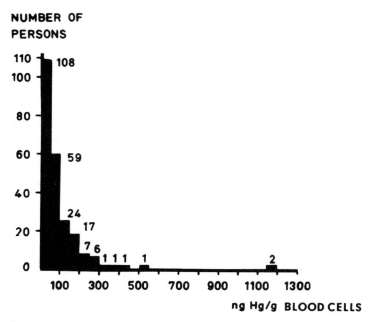

FIGURE 8:2. Total mercury levels in blood cells of 227 persons in Sweden and Finland who consumed large amounts of fish or who ate fish with a high methyl mercury level, or both. (Figure from Berglund et al., 1971, based on data from Birke et al., 1967, Tejning, 1967c, 1968b, Sumari et al., 1969, and Skerfving, 1971.)

exposure was 29 µg/day/person (150 lbs., about 70 kg). The exposure range was 7 to 74 µg/day. The average whole blood level was 0.01 µg/g (3 values over 0.03 µg/g). A control group of 18 subjects had levels of 0.002 (range 0 to 0.009) µg/g. The author remarked that the blood levels found were probably too low because of analytical problems (McDuffie, 1971, and in press).

8.1.2.1.1.1.2 Ethyl Mercury Exposure

Symptoms reported — Katsunuma et al., 1963, have reported 4 suspected cases of intoxication due to exposure to dust of ethyl mercury chloride and phenyl mercury acetate or a mixture of these compounds. One case caused by exposure to di-ethyl mercury had also occurred. The main symptoms were gingivitis, tremor, and neurasthenic symptoms. No ataxia or sensory disturbances were reported. The whole blood total mercury level at onset of symptoms was 0.65-1.7 µg/g. Repeated analyses indicated a biological half-life of 3 to 4 months after the end of exposure. The levels in urine were 81 to 220 µg Hg/l. The mixed exposure makes an evaluation of the etiology of the symptoms impossible.

Suzuki et al., in press, have given data on organ levels in a 13 year-old boy who was said to have suspected symptoms of ethyl mercury poisoning. The patient had a protein-loosing enteropathy. He was given intravenously massive doses (Section 8.1.2.1.2.2) of human plasma containing 0.01% sodium ethyl mercury thiosalicylate as a preservative. Neither the symptomatology nor the reason for his subsequent death was stated. Probably the last transfusion was given 5 days prior to his death. The blood cell total mercury level (atomic absorption) was 12.5 µg/g and the plasma level, 1.7 µg/g. Inorganic mercury as determined by the method of Magos and Cernik, 1969, made up 12 and 20% of the total mercury in blood cells and plasma, respectively. The total mercury level in urine was 5.3 mg/l., of which 44% was inorganic. It is difficult to evaluate to what extent the fact that the mercury was administered intravenously influenced the blood levels.

Symptoms not reported — Suzuki et al., in press, reported on 4 persons without clinical symptoms or signs of alkyl mercury poisoning after they had received transfusions of human plasma containing sodium ethyl mercury thiosalicylate (Section 8.1.2.1.2.2) because of surgery for cancer in the pancreas, gas gangrene, or ileus. The levels in blood cells were about 0.1 to 0.7 µg/g and in plasma, 0.05 to 0.4 µg/g, 11 to 22 days after the last transfusion. In blood cells only a few percent of the mercury was inorganic while in the plasma the corresponding fraction was above 50%. The total mercury excretion in the urine was 0.05 to 0.6 mg/24 hours, almost completely as inorganic mercury. In 2 of the patients the biological half-life in blood cells was calculated at about 1 week while the plasma level was steady when studied for about 1 month.

8.1.2.1.1.2 Hair
8.1.2.1.1.2.1 Methyl Mercury Exposure

Symptoms reported — Hair *total mercury* levels have been reported for a total of 36 patients from the Niigata epidemic. Just as for blood levels, Berglund et al., 1971, have calculated the relationship between time elapsing between onset of symptoms and sampling, and hair mercury levels (Figure 8:3), based upon data of Kawasaka et al., 1967, Matsuda et al., 1967, Tsubaki et al., 1967b, and Tsubaki, personal communication. By extrapolation from the diagram, the level at onset of symptoms in these cases was estimated to have been at or above 200 µg/g, and in 1 case, as low as about 50 µg/g. The uncertainties discussed above in connection with the blood mercury levels are also valid for the estimation of a toxic hair level. Tsubaki, 1971, reported that 7 additional cases of methyl mercury poisoning had been diagnosed in the Niigata area. From a diagram in the paper it is clear that the lowest hair level in a poisoned person, even after addition of the new cases, was 50 µg/g.

Suzuki and Yoshino, 1969, reported a case of poisoning from local treatment of mycosis by a methyl mercury preparation. The hair mercury level was 100 µg/g 9 months after the time of cessation of exposure and onset of symptoms.

Jervis et al., 1970, Eyl, 1971, and the Center for Disease Control, 1971, reported hair mercury levels (neutron activation analysis) in the family in New Mexico, U.S., the members of which had been exposed by ingestion of meat of swine which had eaten methyl mercury dicyandiamide treated seed (Section 8.1.2.1.1.1.1). Two of the victims, aged 8 and 20 years, were reported to have been comatose. They had hair levels of 1,400 and 2,400 µg/g, respectively.

Herdman, 1971, reported hair levels for the

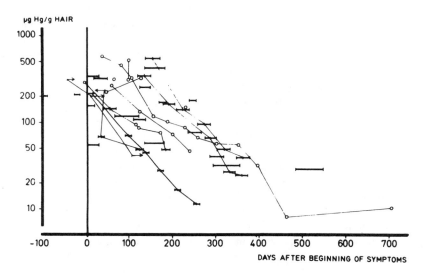

FIGURE 8:3. Relation between total mercury level in the hair and time that elapsed after onset of the disease in patients with methyl mercury poisoning in Niigata. (From Berglund et al., 1971.) In cases where there was uncertainty regarding date of onset or of sampling, or both, the total uncertainty has been indicated by an interval. Where repeated analyses were made for the same patient, the values (where appropriate the middle points of the interval) were joined. In cases in which sampling was done before onset, this has been shown as a negative number of days. (Data according to Tsubaki personal communications, Matsuda et al., 1967, Kawasaki et al., 1967, and Tsubaki et al., 1967b.)

woman who ate such extreme quantities of swordfish and suffered symptoms suspected to have been brought about by methyl mercury poisoning (Section 8.1.2.1.1.1.1). Four months after a 4-week period of the swordfish diet, she had 42 µg/g of hair (method not stated). A hair sample divided into 3 segments showed 30 to 39 µg/g in the different parts.

There are a few data specifying *methyl mercury* levels in hair in poisoned individuals. Levels given in the Niigata Report, 1967 (Matsuda et al., 1967, and Tsubaki et al., 1967b), might indicate that at least half of the total mercury consisted of methyl mercury. Sumino, 1968b, reported levels in 7 patients. Total mercury ranged from 59 to 420 and methyl mercury from 23 to 150 µg/g, which, calculated for each single case, corresponded to 13 to 67% methyl mercury out of total mercury. Data have also been given by Takizawa and Kosaka, 1966, and Takizawa, 1970. In 1 patient the total mercury level was 340 µg/g, and the methyl mercury level was 93 µg/g.

Symptoms not reported — In connection with the Japanese epidemics, hair *total mercury* levels were investigated in a great number of persons. Matsushima and Mitzoguchi, 1961 (quoted in Berlin, Ramel, and Swensson, 1969), analyzed hair samples from 967 fishermen at the Minamata Bay. Among 85% of them, the levels were above 10 µg/g, in 20% above 50 µg/g, and in 2 persons, above 300 µg/g (the highest value was 920 µg/g). In a control group from Minamata City, 30% had over 10 µg/g but no one had over 50 µg/g. The method of analysis is not clear. Though most of the subjects investigated were free from symptoms, some poisoned subjects might have been included.

Hair mercury levels of a total of 1,458 persons have been reported from Niigata (Matsuda et al., 1967, dithizone method). The principles for the sampling are not fully clear, but it seems that relatives of poisoned subjects, heavy fish consumers, and individuals with certain symptoms were included (Berglund et al., 1971). Berglund et al., 1971, concluded that at least 127 persons had levels above 50 µg/g, at least 36 above 100 µg/g, at least 6 above 200 µg/g, and at least 3 above 300 µg/g. Recently, Tsubaki, 1971, reported that 7 additional cases of poisoning had been diagnosed in Niigata among exposed persons who had earlier been considered nonpoisoned. The author stated that most persons having hair levels above 200 µg/g were diagnosed as poisoned. In a diagram 1 person having about 200 and one having 300 µg/g were indicated as asymptomatic. About 20 healthy persons had levels above 100 µg/g and about 60 above 50 µg/g.

Berglund et al., 1971, have compiled hair mercury levels in persons without symptoms from Sweden (Birke et al., 1967, and to be published, and Tejning, 1967c) and Finland (Sumari et al., 1969). The compilation is shown in Figure 8:4. All persons were or had been exposed to methyl mercury through consumption of contaminated fish. Eight subjects had levels above 30 µg/g and four above 50 µg/g. The highest level found was 180 µg/g.

Jervis et al., 1970, and Mastromatteo and Sutherland, 1970, reported hair mercury levels (neutron activation analyses) from 33 persons in Ontario, Canada. In 24 subjects who had eaten fish from contaminated waters at some time during the last year prior to the analyses (up to 5 times a week), levels up to 96 µg/g were found. Four individuals showed levels of 50 µg/g or more. As in many cases no contaminated fish had been consumed for 5 months; the levels probably had been higher earlier. In 9 subjects who had not eaten fish from contaminated waters, levels of less than 2 µg/g up to 14 µg/g were found.

Hair samples from 42 persons who had been on high tuna and swordfish diets in New York, U.S., were analyzed by McDuffie, 1971 (Section 8.1.2.1.1.1.1). The levels (atomic absorption method) averaged 8.9 µg/g (range: 0.8 to 41 µg/g). The Center for Disease Control, 1971, has reported mercury levels in hair from 2 exposed but healthy members of the family in New Mexico, U.S., who ate meat from hogs fed methyl mercury dicyandiamide treated grain. The levels (neutron activation analysis) were 190 and 330 µg/g.

Ui and Kitamura, 1971, have reported hair mercury levels in 31 fishermen from Italy and France. The average total mercury levels in groups from different areas ranged from 1.9 to 5.8 µg/g (atomic absorption analysis), of which 28 to 79% consisted of *methyl mercury*.

Ueda and Aoki (quoted by Ueda, 1969) and Ueda, Aoki, and Nishimura, 1971, reported on methyl mercury levels in hair of 37 subjects who had consumed fish from a contaminated river but who had no symptoms of poisoning. The average methyl mercury level was 6.2 µg Hg/g. Eight samples were above 10 µg/g, the highest, 25 µg/g.

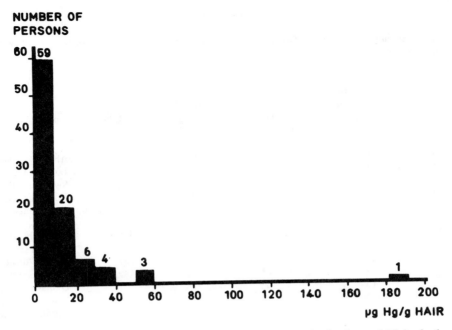

FIGURE 8:4. Total mercury levels in hair in 93 persons in Sweden and Finland who consumed large amounts of fish or who ate fish with a high methyl mercury level, or both. (Figure from Berglund et al., 1971, based on data from Birke et al., 1967, Tejning, 1967c, and Sumari et al., 1969.)

In 7 of those subjects the total mercury was also determined (neutron activation analysis). An average of 84 (range: 53 to 120)% of the total mercury consisted of methyl mercury.

8.1.2.1.1.2.2 Ethyl Mercury Exposure

Kawasaka et al., 1967, mentioned that a person poisoned by sodium salt of ethyl mercury thiosalicylate had 170 µg Hg/g hair. Ethyl mercury was demonstrated by thin layer chromatography. No further information was given about the case.

Suzuki et al., in press, found a hair total mercury level of 187 µg/g, of which 5% was inorganic, in the victim of suspected poisoning by intravenous administration of human plasma containing sodium ethyl mercury thiosalicylate (Section 8.1.2.1.2.2).

8.1.2.1.1.3 Brain, Liver, and Kidney
8.1.2.1.1.3.1 Methyl Mercury Exposure

Table 8:1 lists *total mercury* levels found in brain, liver, and kidney from persons poisoned by methyl mercury. The relation between the mercury concentration in the brain and the time between onset of symptoms and death is plotted in Figure 8:5 (from cases reported by Takeuchi,

1968a). If a biological half-life of 85 days in brain is assumed (Åberg et al., 1969), the lowest level at onset of symptoms might be estimated at about 6 µg/g.

In Table 8:2, total and *methyl mercury* levels in brain in cases of poisoning have been compiled. From 60 to 100% of the mercury found was in the form of methyl mercury.

8.1.2.1.1.3.2 Ethyl Mercury Exposure

Welter, 1949 (quoted by Tornow, 1953), found total mercury levels of about 9 µg/g (method not stated) in brain tissue from 2 persons who had died of occupational alkyl (probably ethyl) mercury poisoning.

In 22 subjects who died after ingestion of bread baked with seed treated with ethyl mercury p-toluene sulfonanilide, the mercury concentration (method not stated) in liver was 66 ± 19 µg/g (Jalili and Abbasi, 1961).

Hay et al., 1963, described a case of occupational poisoning by ethyl mercury chloride. The patient died 18 weeks after the onset of symptoms and 25 weeks after the end of exposure. The mercury concentration (method not stated) in the liver was 17 and in the kidney, 82 µg/g. In the cerebellum the level was 0.97 µg/g, in cerebral

TABLE 8:1

Total Mercury in Organs in Cases of Methyl Mercury Intoxication (all analyses made with dithizone methods)

No. of cases	Time after onset of symptoms, days	Brain μg Hg/g[1]	Liver μg Hg/g	Kidney μg Hg/g	References
1	30	5	20	30	Ahlmark, 1948
1	30	5 (4-10)	14	3	Lundgren and Swensson, 1948
1	21	12	39	27	Höök, Lundgren, and Swensson, 1954
12	19-100	2.6-24	22-71	21-140	Takeuchi, 1961, and 1968a
1	21	30 (20-48)	21	51	Tsuda, Anzai, and Sakai, 1963
1	13	13~79	88~140		Okinaka et al., 1964
1		15	20	18	Ordonez et al., 1966
1	40	13 (7-20)		13	Hiroshi et al., 1967, Tsubaki, (personal com.)
1	126	11 (8-14)	36	47	Hiroshi et al., 1967

[1] When several brain samples were analyzed the mean is stated, with the range in parentheses.

TABLE 8:2

Total Mercury and Methyl Mercury (MeHg) Levels in Brain in Cases of Methyl Mercury Intoxication. Each Value Refers to One Case

Time after onset of symptoms, days	Total Hg[1] μg/g	MeHg[1] μg Hg/g	References
70	7.8	9 (7-13)	Sumino, 1968b
45	25	16 (13-19)	Sumino, 1968b
40	13 (7-20)	10 (9-13)	Tsubaki (personal communication)
97	11 (8-14)	8 (6-9)	Tsubaki (personal communication)
30	16[2]	16[2]	Grant, Moberg, and Westöö, to be published

[1] When several brain samples were analyzed the mean is stated, with range in parentheses.
[2] Dehydrated and xylol-extracted tissue from basal ganglions.

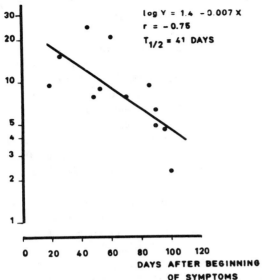

FIGURE 8:5. Relation between total mercury in the brain and the time that elapsed after onset of symptoms in autopsy cases of methyl mercury poisoning from Minamata. (Figure from Berglund et al., 1971, based on data from Takeuchi, 1961, and 1968a.)

cortex and white matter 6.9 to 7.2 µg/g and in corpus callosum 62 µg/g.

Suzuki et al., in press, analyzed several organs for total and inorganic mercury in the case of the boy suspected to have been poisoned by an ethyl mercury preservative in human plasma (see Section 8.1.2.1.2.2). The total mercury levels were 13 to 24 µg/g in different parts of cerebrum, cerebellum, mesencephalon, and spinal cord, without any clearcut difference among different regions. In a spinal ganglion and in N. ischiadicus, 9 µg/g were found. In the cerebrum about 35% of the total mercury consisted of inorganic mercury. The liver contained 69 µg total mercury/g (31% inorganic), the renal cortex, 35 µg/g (69% inorganic) and the renal medulla, 43 µg/g (51% inorganic).

8.1.2.1.1.4 Conclusions

In the adult cases of *methyl mercury* poisoning, the extrapolated total mercury levels in whole blood at the onset of symptoms seem to have been 0.2 µg/g, or higher. Levels as high as 4 µg/g have been reported. In the estimation of the level 0.2 µg/g, some uncertainties are involved. The accuracy of the analytical method used is not known. Most of the blood samples were taken after the onset of symptoms and the time for the cessation of the exposure was not well established.

High mercury levels in blood have been reported in subjects without symptoms of poisoning. In Sweden and Finland about 70 persons had mercury in blood cells corresponding to over 0.05 µg/g whole blood, 24 over 0.1 µg/g, 4 over 0.2 µg/g, and 2 in the range of 0.5 to 0.6 µg/g whole blood. In Canada, levels up to about 0.16 µg/ml whole blood have been found in persons who consumed large amounts of fish.

The mercury levels in hair extrapolated to the time of the onset of symptoms seem to be about 200 µg/g or above. In 1 case the level might have been lower, about 50 µg/g. In most of the cases, the levels seem to have been well above 200 µg/g, and levels as high as 2,400 µg/g have been reported. Uncertainties similar to those for blood are applicable for the estimations of levels at the onset of symptoms.

High levels of mercury in hair have been reported in exposed but apparently healthy subjects. From Japan, levels above 50 µg/g were found in at least 60 persons, above 100 µg/g in at least 20, and 200 µg/g or above in 2 persons. In 4 persons who had eaten contaminated fish in Sweden and Finland, levels above 30 µg/g hair have been found. In 1 clinically healthy subject the level was 180 µg/g. From Canada, levels of 50 to 100 µg/g have been reported in 4 subjects, and from the U.S., 190 and 330 µg/g, respectively, in 2 subjects.

As concluded in Section 4.5.2.1, hair levels are about 300 times higher than whole blood levels. The lowest blood level (0.2 µg/g) assumed to have been present at the onset of symptoms is thus only one third of the blood concentration corresponding to the hair level 200 µg/g. Possible reasons for the different relation between levels in blood and hair have been discussed in Section 4.5.2.1. The value 0.2 µg/g corresponds well to the lowest hair level of 50 µg/g found in 1 poisoned subject.

The total mercury levels in brain reported in patients who died from methyl mercury intoxication indicate concentrations above 5 µg/g at onset of symptoms.

In the case of *ethyl mercury* the available information is much more limited. A blood cell total mercury level of 12 µg/g was reported in a suspected case in a boy who had received massive doses intravenously in a human plasma solution. It is difficult to evaluate the influence of the route of administration upon the level. The hair total mercury level in the same case was 190 μg/g and the central nervous system levels averaged 18 μg/g. In other, less well-studied cases, about half as high CNS levels have been reported.

8.1.2.1.2 Relation Between Exposure and Effects
8.1.2.1.2.1 Methyl Mercury Exposure

A review of the available exposure data from the epidemics in Minamata and Niigata has been made by Berglund et al., 1971. Fish was consumed daily or at least several times a week by most of the patients (Kitamura, 1966, quoted by Nomura, 1968, and Matsuda et al., 1967). The amount of fish ingested at each meal was up to 250 to 500 g. The level of mercury in fish and shellfish is difficult to establish. Berglund et al., 1971, concluded that some of the published data concerning Minamata (Kurland, Faro, and Siedler, 1960, Kitamura, 1968, Takeuchi, 1968a, and Irukayama, personal communication) indicated an average level of about 20 mg Hg/kg shellfish and fish. On the basis of the data given by Kawasaka et al., 1967, Matsuda et al., 1967, and Sato, 1968, it was concluded that the mercury levels in fish from Niigata on the average seemed to have been about 5 mg Hg/kg fish, with wide variations. Tsubaki, 1971, reported that 45% of a sample of 195 fish from Niigata had levels above 1 mg/kg.

Takeuchi, 1970, asserted that fishermen and their families used to eat 200 g a day of fish containing about 20 mg Hg/kg wet weight. The exposure would then have been 4 mg Hg/day.

Sato, 1968, calculated the exposure for 27 patients from Niigata. The median exposure was estimated at about 1.5 (range: 0.23 to 4.8) mg Hg/day. There was no correlation between estimated intake of mercury and levels in hair or severity of symptoms. This might be explained by the uncertain estimations of intake and levels in fish.

Wahlberg, Henriksson, and Karppanen, 1970, reported a history of visual disturbances and, at examination, sensory disturbances in a woman who had eaten goosander (*Mergus merganser*) eggs containing methyl mercury for many years. The low exposure (0.5 to 6 mg of mercury as methyl mercury each spring) and the low mercury level in hair (2.8 μg/g) make an association between the methyl mercury and the symptoms and signs improbable (Berglund et al., 1971).

Birke et al., 1967, and to be published, observed no symptoms or signs of poisoning in a few persons exposed to methyl mercury up to 0.8 mg of mercury/day for 5 years through consumption of contaminated fish.

8.1.2.1.2.2 Ethyl Mercury Exposure

Powell and Jamieson, 1931, administered ethyl mercury thiosalicylate (Merthiolate®) intravenously to 21 subjects in doses up to about 250 mg mercury. One individual received 900 mg during 9 days. The subjects were observed for 1 to 62 days after the administration. In 1 person a "nephritis" was reported and in another, a thrombophlebitis. No other symptoms were observed.

In Iraq, 331 persons were poisoned through 2 to 3 months' ingestion of bread baked with seed treated with ethyl mercury *p*-toluene sulfonanilide (Jalili and Abbasi, 1961, and Dahhan and Orfaly, 1964). The mercury level in the seed was about 15 mg/kg. From the U.S.S.R., Drogitjina and Karimova, 1956, reported 6 cases of poisoning caused by ingestion of bread baked with ethyl mercury chloride treated seed. Four of the persons, of whom 2 died, had eaten contaminated bread for 2 weeks before the onset of symptoms. In a similar incidence in the U.S.S.R., 7 persons fell ill 5 to 10 days after a 10-day consumption of bread baked with seed treated with ethyl mercury chloride (Mnatsakonov et al., 1968). A 13 year-old-boy was poisoned (nausea, headache, visual hallucinosis, and later epileptiform seizures) 2 to 3 hours after ingestion of 200 to 250 g of peas containing 95 mg Hg/kg as ethyl mercury chloride (Slatov and Zimnikova, 1968). The dose corresponds to 25 mg mercury.

Dinman, Evans, and Linch, 1958, studied 20 workers (aged 25 to 33 years) exposed 8 hours a day, 5 days a week to dust of ethyl mercury chloride or phosphate as well as solvent solutions of ethyl and phenyl mercury acetate. The monthly average levels of mercury in air ranged from 0.03 to 0.1 mg/m^3. Mercury analyses were made by a modified dithizone method. There were no significant objective findings in repeated medical examinations and no statistically significant differences in subjective complaints of the exposed workers compared with those of unexposed controls.

Suzuki et al., in press, reported on a boy suffering from protein-loosing enteropathy, who, after transfusions of human plasma preserved with sodium ethyl mercury thiosalicylate showed symptoms suspected to be caused by alkyl mercury poisoning. He later died but the cause of death was not stated. During 3 months a total of 9,000 ml of plasma was administered daily or weekly. The level of total mercury in the plasma solution was 32 µg/ml, of which 0.2 µg/ml was inorganic. The total dose of mercury was calculated at about 280 mg. The distribution of the dose over the 3-month period was not stated, but if it is assumed that it was even, the exposure can be calculated at 0.14 mg/kg body weight per day. The time of onset of symptoms was not denoted in the report. Takeuchi, 1970, used this case for calculations of an acceptable exposure. However, the figures given for duration of exposure and for dose of mercury were quite different from those given by Suzuki et al., in press.

Suzuki et al., in press, also gave some data on 4 other patients treated with plasma because of surgery. The doses were 3 to 210 mg mercury as sodium ethyl mercury thiosalicylate. The patients were stated not to have symptoms of poisoning. Since the duration of administration was not given, the exposure intensity cannot be calculated.

Hill, 1943, reported 2 fatal cases of poisoning in clerks in a storehouse for *di-ethyl mercury*. The air mercury level at their working place was (spot samples) about 1 mg/m³ (method of analysis not stated). They hád worked there for about 3 and 5 months, respectively.

Substituted alkyl mercury compounds will be discussed in Section 8.4.

8.1.2.1.2.3 Conclusions

The data available for quantitative evaluation of exposure to alkyl mercury compounds in adults with neurological symptoms of poisoning are scanty and contradictory.

The exposure through consumption of *methyl mercury* contaminated food cannot be quantitatively evaluated with certainty.

For *mono-ethyl mercury* the data are uncertain and quite contradictory. While an exposure as low as 25 mg of mercury given orally in a single dose has been reported to have been poisonous, doses as large as 250 mg and even 900 mg administered intravenously have been said not to have resulted in damage to the nervous system.

In the only study of workers exposed to ethyl mercury mainly by inhalation, the levels (total mercury) in air were 0.03 to 0.1 mg Hg/m³. The exposed subjects were reported to have been free of symptoms of poisoning. However, the exposure was mixed, including phenyl mercury, so conclusions cannot be used for evaluating the risks connected with ethyl mercury.

Exposure to *di-ethyl mercury* vapor at a concentration of about 1 mg/m³ during the working day for a few months has been reported to have resulted in a fatal poisoning.

8.1.2.1.3. Relation Between Exposure and Organ Levels

On the basis of data reported by Birke et al., 1967, and to be published, and Tejning, 1967a and c, 1969c, and 1970a, Berglund et al., 1971, made an estimation of the relationship between exposure to *methyl mercury* through consumption of contaminated fish and levels of mercury in blood cells (Figure 8:6). It is obvious that the mercury level in blood cells is dependent upon intake of methyl mercury via fish. Berglund et al., 1971, stated that the slope of the dotted regression line in the Figure is probably too small due to an over-estimation of the average mercury content in the fish used for the calculations. The steeper regression line is probably more correct. It should be stressed, however, that the information on what happens at higher exposure levels is scanty.

If, from the relationship shown in Figure 8:6, an estimation is made to find what long-term exposure to methyl mercury would give a whole blood mercury level of 0.2 µg/g, one arrives at about 0.3 mg/day, which would correspond to about 4 µg Hg/kg body weight/day in a 70 kg man.

McDuffie, 1971, and in press, made an estimation of the relation between exposure to methyl mercury and total mercury levels in whole blood and hair in 42 persons who had extreme intakes of tuna and swordfish. There was a significant correlation between calculated exposure and levels in blood (adjusted for exposure time) although the variation was considerable. An exposure of 60 µg/day corresponded to a level of about 0.02 µg/ml whole blood. This is only about half the level obtained by the corresponding exposure according to the conclusions drawn by Berglund et al., 1971. However, McDuffie remarked that the analytical method used (atomic absorption) probably gave levels which were too low. The correlation between exposure and hair levels was not particularly good.

An estimation of the relationship between exposure and blood levels can also be made from

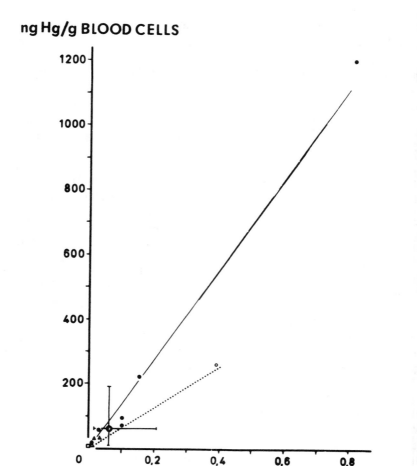

FIGURE 8:6. Relation between total mercury concentrations in blood cells and exposure to methyl mercury through fish. (From Berglund et al., 1971.)

the tracer dose experiments on man presented in Sections 4.3.2.1.2.1 and 4.4.2.1.1.2.1. Those studies indicated that about 1% of the total body burden was eliminated daily, about 10% of the total body burden was localized in the head, probably to a great extent in the brain, and about 5% could be found in the blood. From those data it can be calculated that the continuous long-term exposure to about 0.2 mg Hg/day as methyl mercury would give a whole blood level of 0.2 µg/g. This figure is in reasonable agreement with 0.3 mg/day, arrived at in calculations made from epidemiological data by Berglund et al., 1971. The level in brain corresponding to 0.2 µg/g whole blood would be about 1.3 µg/g. If the lowest level in brain reported for patients who died from methyl mercury poisoning, 5 µg/g, is used for similar calculations, the corresponding exposure would be 0.8 mg/day and the corresponding whole blood level, 0.8 µg/g (Berglund et al., 1971). Considering the uncertainties in the estimations and particularly the fact that the brain levels originated from subjects who died of methyl mercury poisoning while the whole blood levels were derived from surviving patients, including slightly intoxicated people, Berglund et al., 1971, concluded that the agreement among the different ways of calculating the toxic exposure was reasonable.

In conclusion, although uncertainties exist, it seems reasonable to assume that the continuous long-term methyl mercury exposure needed to reach a whole blood total mercury concentration of 0.2 µg/g is about 0.3 mg Hg as methyl mercury/day, corresponding to about 4 µg/kg/day in a 70 kg man.

8.1.2.2 In Animals
8.1.2.2.1 Single Administration

In Table 8:3 (from Berglund et al., 1971) the LD_{50}'s for different mercury compounds have been accumulated. In most of the experiments included, the latency period between a single exposure and the onset of symptoms observed in alkyl mercury poisoning has not been safely covered. It is evident that information from such experiments is of limited interest.

Steinwall and Olsson, 1969, induced damage to the blood-brain barrier within 24 hours by injection of 200 mg Hg/kg body weight intravenously into rats. Due to the high dose in an acute experiment, conclusions on the relevance of the damage to clinical methyl mercury poisoning cannot be drawn. Similar effects were induced by mercuric chloride.

8.1.2.2.2 Repeated Administration
8.1.2.2.2.1 Methyl Mercury Exposure

In Table 8:4 (from Berglund et al., 1971) available data have been compiled on the toxicity of methyl mercury compounds at repeated exposure of different species. It must be realized that the exposure time in some of the experiments was short and poisoning probably could occur at lower exposure. The lowest toxic exposures have been reported for monkeys (0.3 to 0.7 mg Hg/kg body weight/day) and cats (0.2 to 0.6 mg/kg/day).

Table 8:5 (from Berglund et al., 1971) shows available data on total mercury levels in the brain in animals poisoned by methyl mercury compounds. The exposures (both time and intensity) varied widely among the different experiments and various methods of analysis were employed. After scrutinizing the background of experimental conditions and results, it is reasonable to assume that poisoning may occur at a brain mercury level of approximately 10 µg/g.

8.1.2.2.2.2 Ethyl Mercury Exposure

In Table 8:6 data on ethyl mercury compounds are compiled. In comparison with Table 8:4, these data show that the toxicity of ethyl mercury does not differ practically from that of methyl mercury.

Levels of mercury in organs in ethyl mercury poisoned calves have been reported by Oliver and Platonow, 1960. Three heavily exposed animals with neurological symptoms and histological lesions in CNS had brain levels of 12 to 29 µg/g (dithizone method). Two animals with histological kidney lesions had levels of 29 and 60 µg/g kidney, respectively. Itsuno, 1968, found 17 to 29 µg Hg/g brain (dithizone method) in rats poisoned by ethyl mercury compounds.

8.1.2.2.2.3 Other Alkyl Mercury Exposure

Itsuno, 1968, made oral toxicity studies on rats with regard to several higher alkyl mercury compounds, including n-propyl, iso-propyl, n-butyl, tert-butyl, n-amyl, iso-amyl and n-hexyl mercury salts. The doses were 5 to 15 mg Hg/kg/day and the exposure times, 10 to 50 days. The number of animals in each group was probably 1 to 4. The exposure to methyl and ethyl mercury salts under these conditions caused neurological symptoms. Among the higher alkyl mercury compounds, only n-propyl mercury compounds caused intoxication. In rats poisoned by n-propyl mercury compounds, brain mercury levels ranging from 21 to 32 (in 1 case 2) µg/g (dithizone method) were measured.

8.2 Aryl Mercury Compounds
8.2.1 Prenatal Exposure

Murakami, Kameyama, and Kato, 1956, found an increased frequency of fetal death and malformation (on day 14) when pregnant mice were given phenyl mercury acetate intravaginally or subcutaneously (on days 7 and 8, respectively) in a single dose corresponding to about 3 mg Hg/kg. In the kidneys of the mothers, changes suggestive of "acute mercury poisoning" were said to have been present.

Piechocka, 1968, reported a reduction in litter size in rats given food containing 8 mg Hg/kg as phenyl mercury acetate for 6 months, compared to controls. In the exposed rats, no clinical symptoms of poisoning were apparent.

TABLE 8:3

LD_{50} for Different Mercury Compounds. (From Berglund et al., 1971, with some additions.)

Compound[1]	Animal species	Administration route[2]	LD_{50} mg Hg/kg body weight	Observation period, days	Reference
1. INORGANIC Hg COMPOUNDS					
1.1. $HgCl_2$	Mouse	i.p.	5	7	Swensson, 1952
	Mouse	i.p.	7	1	Swensson, 1952
	Mouse	i.v.	6	5	Wien, 1939
	Mouse	i.p.	6	14	Hagen, 1955
2. ALKYL Hg COMPOUNDS					
2.1. *MeHg*					
2.1.1. MeHgCl	Mouse	i.p.	14	7	Swensson, 1952
	Mouse	i.p.	17	14	Hagen, 1955
2.1.2. MeHgOH	Mouse	i.p.	17	7	Swensson and Ulfvarson, unpublished data
2.1.3. MeHg dicyandiamide	Mouse	i.p.	8	7	Swensson, 1952
	Rat	or.	12	10	Lundgren and Swensson, unpublished data
	Rat	or.	10	30	Lundgren and Swensson, unpublished data
2.1.4. MeHg toluenesulfonate	Mouse	i.p.	15	1	Lundgren and Swensson, 1950
2.1.5. MeHg propandiolmerkaptide	Mouse	i.p.	29	7	Swensson and Ulfvarson, unpublished data
2.2 *EtHg*					
2.2.1. EtHgCl	Mouse	i.p.	12	7	Swensson, 1952
	Mouse	i.p.	15	14	Hagen, 1955
	Rat	or.	23	10	Lundgren and Swensson, unpublished data
2.2.2. EtHg dicyandiamide	Mouse	i.p.	7	7	Swensson, 1952
2.2.3. EtHg toluenesulfonate	Mouse	i.p.	14	1	Lundgren and Swensson, 1950
2.2.4. EtHg phosphate	Mouse	or.	61	7	Sera, Murakami, and Sera, 1961
	Mouse	s.c.	63	7	Sera, Murakami, and Sera, 1961
2.3. *Other*					
2.3.1. Isopropyl HgOH	Mouse	i.p.	12	7	Swensson and Ulfvarson, unpublished data
3. ARYL Hg COMPOUNDS					
3.1. $PhHgNO3$	Mouse	i.v.	16	5	Wien, 1939
3.2. PhHg acetate	Mouse	i.p.	8	7	Swensson, 1952
	Mouse	or.	26	7	Sera, Murakami, and Sera, 1961
	Mouse	s.c.	37	7	Sera, Murakami, and Sera, 1961
	Rat	or.	22	30	Lundgren and Swensson, unpublished data

TABLE 8:3 (continued)

Compound[1]	Animal species	Administration route[2]	LD_{50} mg Hg/kg body weight	Observation period, days	Reference
3.3. *PhHg dinaphthyl methanedisulfonate*	Mouse	i.p.	8	14	Goldberg, Shapero, and Wilder, 1950
	Mouse	or.	21	14	Goldberg, Shapero, and Wilder, 1950
3.4. *PhHg catecholate*	Mouse	i.p.	18-36	14	Hagen, 1955
4. ALKOXYALKYL Hg COMPOUNDS					
4.1. *MeOEtHg acetate*	Rat	or.	16	10	Lundgren and Swensson, unpublished data
	Rat	or.	10	30	Lundgren and Swensson, unpublished data
4.2. *MeOEtHg silicate*	Mouse	i.p.	30	14	Hagen, 1955
4.3. *MeOEtHgCl*	Mouse	or.	47	7	Sera, Murakami, and Sera, 1961
	Mouse	s.c.	60	7	Sera, Murakami, and Sera, 1961

[1] MeHg = Methyl mercury, EtHg = Ethyl mercury, PhHg = Phenyl mercury; MeOEtHg = Methoxyethyl mercury
[2] i.p. = intraperitoneally; i.v. = intravenously; or. = orally; s.c. = subcutaneously

8.2.2 Postnatal Exposure
8.2.2.1 In Human Beings

Data available on dose-response relationships for aryl mercury compounds are limited. There are additional difficulties in evaluating the risks of exposure to aryl mercury compounds. The clinical picture is not as exactly defined as that for alkyl mercury poisoning (Section 5.2.2.1.2). Furthermore, the chemical instability of aryl mercury compounds results in exposure not only to the organomercury compound but also to elemental mercury vapor. Hypersensitivity or idiosyncracy (Section 5.2.2.1.3) has not been treated because no clear dose-response relationships seem to exist.

Phenyl mercury compounds have been widely used for local treatment of cutaneous infections (e.g., Levine, 1933, Greaves, 1936), infections in the vagina (e.g., Biskind, 1933, 1935, and Stuart, 1936) and for intravaginal contraceptives (Baker, Ranson, and Tynen, 1938, Jackson, 1938, and Eastman and Scott, 1944). The mercury concentrations in the solutions or jellies applied have ranged from 0.02 to 0.8 g Hg/l. The only negative effect reported has been chemical burns at concentrations of 0.6 g Hg/l. or above (Levine, 1933, and Biskind, 1935). Intravaginally applied phenyl mercury compounds are absorbed and mercury is eliminated in the urine (Biskind, 1933, and Eastman and Scott, 1944). However, no systemic symptoms have been reported.

Janson, 1929, reported a case of acute poisoning after a few hours' inhalation of dust of phenyl mercury nitrate. Weed and Ecker, 1931, gave 250 cc of a saturated 1:1,250 solution of phenyl mercury nitrate orally to 1 person. There were no signs of intoxication. The dose corresponds to about 120 mg of mercury. Birkhaug, 1933, took a total of about 100 mg of mercury as crystalline phenyl mercury nitrate in 4 oral doses within 24 hours. About 30 hours after the beginning of the administration, he felt slight abdominal pain and loose passages occurred. In other experiments "repeated series" of phenyl mercury nitrate were taken in doses corresponding to about 6 mg of mercury twice or thrice daily for periods of 1 week. No symptoms or signs of mercury intoxication were noted. Nothing was said about examination for kidney lesions, however.

Tokuomi, 1969, very briefly mentioned that a person had taken 50 cc of a 6.6% solution of a phenyl mercury compound corresponding to

TABLE 8:4

Toxicity for Different Methyl Mercury (MeHg) Compounds at Repeated Exposure. (From Berglund et al., 1971, with some additions.)

Animal species	No. of animals	Compound/ source	Administration route[1]	Exposure mg Hg/kg body weight/ day[2]	Total dose mg Hg/kg body weight	Duration of exposure, days	Time until onset of symptoms[3]	Reference
Mouse	10	MeHgCl	or.	15	105	7	8-12	Suzuki, 1969
	10	MeHgCl	or.	10	70	7	<30	Suzuki, 1969
	10	MeHgCl	or.	6.8	48	7	No symptoms until 30 days	Suzuki, 1969
Rat	3	MeHgI	or.	2-4	75	29	21-28	Hunter, Bomford, and Russell, 1940
	2	MeHgNO$_3$	or.	2-4	66	29	21-28	Hunter, Bomford, and Russell, 1940
	?	MeHgCl or MeHg di- cyandiamide	i.p.	1-3		28	21-35	Swensson, 1952
	?	MeHgCl or MeHg di- cyandiamide	i.p.	0.5		28	No symptoms until 42 days	Swensson, 1952
	12	MeHgCl	or.	10-20	55-135	9	<10-13	Kai, 1963
	12	MeHgSHgMe	or.	10-20	58-155	9	<10-27	Kai, 1963
	58	MeHgCl or MeHgSMe	or.	2-10	15-130	8-28	8-13 in animals that got totally 65-130 mg/kg. Of 9 animals that got totally 100 mg/kg, none ill	Moriyama, 1968
	10?	Shellfish	or.	5-10	120-250		20-48	Takeuchi, 1968b
	?	MeHgSMe	or.	5-10	100-200			Takeuchi, 1968b
	?	MeHgSHgMe	or.	5-10	100-200			Takeuchi, 1968b
	?	Shellfish	or.		94			Takeuchi, 1968b
	10	MeHgOH	i.p.	0.9-1.7	50-60	35-56	21	Berglund, 1969
	5	MeHgSMe	or.	2-3	100		50	Miyakawa et al., 1969
	5	MeHgSMe	or.	10	130-160	19-20	19-20	Miyakawa and Deshimaru, 1969
	8	MeHgOH	or.	~1	150-210	140-225	~130	Berglund et al., to be published

TABLE 8:4 (continued)

Animal species	No. of animals	Compound/source	Administration route[1]	Exposure mg Hg/kg body weight/day[2]	Total dose mg Hg/kg body weight	Duration of exposure, days	Time until onset of symptoms[3]	Reference
Rabbit	8	MeHgOH	or.	~0.5	110-140	170-225	No symptoms	Berglund et al., to be published
	2	MeHgCl	or.	0.8-1.6	24	19	<47-66	Kai, 1963
	2	MeHgSHgMe	or.	1.0-1.9	22-30	20	<46-63	Kai, 1963
Cat	18	Shellfish	or.	1-3 (-10)	20-60		14-84	Takeuchi, 1961
	?	MeHgSHgMe	or.	1.2-2.0	21-34			Takeuchi, 1968b
	?	MeHgSMe	or.	1.5-1.8	21-26			Takeuchi, 1968b
	9	MeHgCl	or.	0.8-1.6	8-56	5-35	<17-50	Kai, 1963
	3	MeHgI	or.	0.6-1.4	14-25	10-28	<17-55	Kai, 1963
	2	MeHgOH	or.	1.0-2.0	30-38	20-26	<120-124	Kai, 1963
	6	MeHgSHgMe	or.	0.5-1.7	13-25	12-25	20-40	Kai, 1963
	1	MeHgMe	or.	0.8-2.4	79	60	<93	Kai, 1963
	3	MeHg cysteine	or.	1.1-1.2	22	20	<57	Kai, 1963
	1	MeHg glutathione	or.	1.0-3.0	56	28	<63	Kai, 1963
	2	Shellfish	or.	1.0	43-44	51-55	52-58	Kai, 1963
	2	Shellfish	or.	0.3	20-24	68-97	No symptoms until 100 days	Kai, 1963
	7	Fish	or.	0.3-0.6	28-33	60-83	60-83	Albanus et al., to be published
Dog	5	MeHgOH	or.	0.4-0.5	32-43	69-75	69-75	Kai, 1963
	1	MeHgCl	or.	1.6	16	10	<23	Kai, 1963
	1	MeHgSHgMe	or.	1.7	22	13	<28	Kai, 1963

[1] or. = orally; i.p. = intraperitoneally.
[2] The exposure has been recalculated on an every day basis.
[3] < indicates spontaneous death or killing of animals. The symptoms must have appeared at earlier date.

TABLE 8:5

Mercury Level in the Brain at Methyl Mercury Intoxication with Neurological Symptoms in Different Species. (From Berglund et al., 1971, with some additions.)

Animal species	No. of animals	Hg level µg/g Mean value	Range	Reference
Mouse	8	28	11-61	Saito et al., 1961
	20	~30	20-40	Suzuki, 1969
	10	~40	25-55	Suzuki, 1969
Rat	12	49		Takeshita and Uchida, 1963
	8	16	11-19	Berglund et al., to be published
Ferret	4	27	7-39	Hanko et al., 1970
Cat	4	14		Takeuchi, 1961
	3	10	8-12	Kai, 1963
	5	11	2-19	Yamashita, 1964
	3	6	2-12	Yamashita, 1964
	7	21	3-60	Yamashita, 1964
	2	9	8-10	Kitamura, 1968
	5	13	8-19	Kitamura, 1968
	2		23-32	Rissanen, 1969
	2	9		Skerfving and Nordberg, unpublished data
Dog	5	19	4-32	Yoshino, Mozai, and Nakao, 1966a
Pig	2	23		Piper, Miller, and Dickinson, 1971
Monkey	4	15	12-19	Nordberg, Berlin, and Grant, 1971

1,250 mg mercury. Great amounts of mercury were said to have been eliminated in the urine. It is not clear whether any clinical symptoms were observed. Laboratory investigations as well as biopsy of the kidney were said to have shown normal conditions. Massmann, 1957, investigated a factory for production of phenyl mercury pyrocatecholate. The levels of total mercury in air in various locations of the factory ranged from 0.2 to 3.2 mg/m^3 as determined by a dithizone method in spot samples. In 21 workers (22 to 62 years of age) exposed for 1 month to 6 years to dust of phenyl mercury compounds, urine mercury levels were 0.4 to 6 mg/l. In 4 cases organic mercury in the urine was determined separately. During exposure, the total mercury levels in these cases ranged from 0.5 to 1.5 mg/l., of which 70 to 90% was organomercury. On the whole, 9 of the 21 investigated persons had subjective complaints (frequent voidings, 5; insomnia, 2; anorexia, 2; and frequent numbness in hands, 2). Ten workers had slight objective signs (gingivitis or paradentosis, 10; cardiac signs, 2; fine finger tremor, 3; albuminuria, 2, associated in 1 case with isosthenuria, azotemia, and slight hypertension). None of these complaints were considered to be clearly due to the phenyl mercury exposure.

Goldwater et al., 1964, reported on a person sprayed over his eyes, neck, arms, and clothes with a solution containing 12% phenyl mercury acetate by weight. The only symptoms observed were second degree chemical burns and albuminuria (maximum 30 mg/100 ml, traces for 2 weeks). Mercury analyses were performed by atomic absorption spectrophotometry. The level in gastric washing was 1 mg/l. The maximum urinary elimination of mercury was noted during the first

TABLE 8:6

Toxicity for Different Ethyl Mercury (EtHg) Compounds at Repeated Exposure

Animal species	Compound	No. of animals	Administration route[1]	Exposure[2]	Duration of exposure, days	Total dose, mg Hg/kg body weight	Time until onset of symptoms, days	Type of symptoms	Reference
Mouse	EtHg phosphate and EtHgCl	?	Inhal.	10-30 mg Hg/m³	3-5 hours		<3-5 hours	Death	Trachtenberg, 1969
	EtHgCl	<20	or.		Single dose	25-50	10-20	Kidney lesions seen by light and electron microscope	Meshkov, Glezer, and Panov, 1963
	EtHgSHgEt	?	or.	10-15 mg/kg/day[3]	≥10-≤25	150-250[3]	≥10-≤25	Neurological Histological lesions in CNS	Takeuchi, 1968b
Cat	EtHgCl	3	or.	2-3 mg/kg/day[3]	~30-46	70-110[3]	~16	Neurological Histological changes in CNS	Morikawa, 1961a
	EtHgSHgEt	3	or.	2-3 mg/kg/day[3]	~46-76	140-220[3]	~21	same as above	
	(EtHg)₂HPO₄	3	or.	2-3 mg/kg/day[3]	~37-44	80-150[3]	~18	same as above	
	EtHgSH(NH) NH₂HBr	3	or.	2-3 mg/kg/day[3]	~59-84	70-150[3]	~34	same as above	
	EtHgEt	3	or.	2-3 mg/kg/day[3]	~48-66	90-150[3]	~24	same as above	Yamashita, 1964
	EtHgI	3	or.	0.9-1.1 mg Hg/kg/day	24-29	24-30	26-30	Neurological	
	EtHg phosphate	5	or.	0.8-1.5 mg Hg/kg/day	13-43	19-33	18-38	Neurological	Takeuchi, 1968b
	EtHgSHgEt	?	or.	2.6-3.2[3]	≥13-≤22	43-56[3]	≥13-<22	Neurological Histological lesions in CNS	
Rabbit	EtHg compound	?	or.	2-4 mg/kg/day[3]	3.5-14 months	>40-≤120	20-30	Neurological ECG changes	Tokuomi, 1969 Trachtenberg, Goncharuk, and Balashov, 1966
	EtHgCl[4]	?	Inhal.	0.04 mg Hg/m³ for 6 hours daily					
	EtHg acetone	9	i.p.	~1-7 mg Hg/kg/day	6-53	~50-100		Neurological Histological kidney and heart damage	Schmidt and Harzmann, 1970

TABLE 8:6 (continued)

Animal species	Compound	No. of animals	Administration route[1]	Exposure[2]	Duration of exposure, days	Total dose, mg Hg/kg body weight	Time until onset of symptoms, days	Type of symptoms	Reference
Dog	EtHgSC$_6$H$_4$COONa	4	i.v.	1 mg Hg/kg x 13 (0.3 mg Hg/kg/day)	40	13		*No symptoms* until day 47. No definite histological changes	Powell and Jamieson, 1931
Sheep	EtHg p-toluene-sulfonanilide	3	or.	0.4-1.2 mg Hg/kg/day	12-33	12-17	6-31	Gastrointestinal and neurological	Palmer, 1963
Calf	EtHg p-toluene-sulfonanilide	1	or.	5 mg Hg/kg/day	18	38	36	Neurological, ECG Histological changes in CNS, kidneys and heart	Oliver and Platonow, 1960
		1	or.	23 mg Hg/kg/day	58	25	23	Neurological Histological changes in CNS	
		1	or.	47 mg Hg/kg/day	42	9	3	Gastrointestinal and neurological Histological changes in CNS	

[1] Inhal. = inhalation; or. = orally; i.p. = intraperitoneally; i.v. = intravenously.
[2] The exposure has been recalculated on an every day basis.
[3] Not clear whether this is mercury or compound.
[4] Organic mercury compounds, mainly EtHgCl.

24 hours (10 mg, or 8.5 mg/l.). The patient was treated with BAL for 1 week. The urinary mercury level was above 2 mg/l. for 1 week and ranged from 0.1 to 1.6 mg/l. during the additional 40 days studied. Initially almost all the mercury in blood was said to have been in the blood cells, while later the major fraction was said to have been in the plasma. The maximum whole blood mercury level was 0.25 μg/ml (day 4). The level was above 0.1 mg/ml for 10 days and then < 0.01 to 0.05 mg/ml for the 40 days studied.

In a study of workers exposed to phenyl mercury salts, Ladd, Goldwater, and Jacobs, 1964, reported on clinical examinations (including a test for albuminuria), air mercury measurements (elemental mercury vapor by vapor meter and total mercury by an iodine-iodide method), and blood and urine mercury determinations (atomic absorption spectrometry). In 1 factory with air mercury levels (spot samples) of less than 0.08 mg/m^3 in all locations but 1, which showed 40 mg/m^3, 23 workers exposed to phenyl mercury benzoate had blood levels of < 10 to 90 ng/ml (2 out of 8 values below 10 ng/ml) and urinary levels of < 1 to 790 μg/l. The only clinical abnormality noted was eosinophilia (4 to 13%) in 7 subjects. In 21 workers exposed to phenyl mercury benzoate in another plant, blood levels up to 66 ng/ml (18 values below 5 ng/ml) and urinary levels below 240 μg/l. (8 out of 20 values below 0.5 μg/l.) were found. The air mercury levels recorded in different localities at different times ranged from 0 to 0.5 mg/m^3. Comparisons of measurements of elemental mercury vapor and total mercury indicated that the air mercury was almost exclusively present as elemental mercury. Apart from a history of dermatitis in 12 employees, no signs of poisoning were observed. In a plant handling phenyl mercury acetate, air total mercury levels below 0.1 mg/m^3 (in all but 2 locations, nothing detected) were found. In 23 workers blood mercury concentrations ranged < 5 to 550 ng/ml (13 values below 5 ng/ml) and urine levels, < 0.5 to 220 μg/l. (14 below 0.5 μg/l.). For 9 of the workers, analyses were made twice in 2 months. In several cases, the samples varied more than 20 times. Another group of 24 workers without evidence of toxic effects from the exposure to phenyl mercury oleate had urine mercury levels ranging from 100 to 700 μg/l.

In the material presented by Ladd, Goldwater, and Jacobs, 1964, no obvious relation can be detected between air mercury levels and blood or urine levels or between levels in blood and urine.

The same team has reported observations on other workers exposed to phenyl mercury (Goldwater, Jacobs, and Ladd, 1962, Jacobs, Ladd, and Goldwater, 1963, Ladd, Goldwater, and Jabobs, 1963, and Jacobs, Ladd, and Goldwater, 1964). Since the exposure was mixed with inorganic mercury, no conclusions on the toxicity of phenyl mercury can be made out of these studies.

Jacobs and Goldwater, 1965, investigated blood and urine mercury levels in subjects exposed in a room painted with paint containing 0.02% Hg as phenyl mercury acetate. They concluded that little if any mercury was absorbed by the painters during the painting job and that the absorption by the occupants of the painted room was insignificant.

Cotter, 1947, and Brown, 1954, described symptoms in subjects exposed to phenyl mercury compounds. However, the relations between exposure and symptoms were questionable and/or the urinary samples analyzed for mercury were taken after, in most cases long after, the end of the exposure. The latter remark is true also for the case described by Bonnin, 1951.

Summary — It is difficult to summarize the dose-response relationships for aryl mercury compounds from the data published.

Ingestion of 100 mg of mercury as phenyl mercury has been reported to cause only slight gastrointestinal symptoms. In another case, laboratory investigation and renal biopsies showed normal conditions after the person had ingested as much as 1,250 mg of mercury. Obviously, the oral toxicity of phenyl mercury compounds is rather low.

There is evidence for absorption of mercury from phenyl mercury compounds applied on the surface of the skin or into the vagina. It is not known whether there is also inhalation of mercury along with the skin application. One heavily exposed subject showed a transient albuminuria.

In a few studies air mercury levels have been reported. Because only spot sample levels have been provided, no conclusion can be drawn.

In a few cases, urinary mercury levels in poisoned individuals have been reported. In all but 1 case either the symptoms were questionable or the samples (spot samples) were taken after the end of exposure. In 1 exceptional case of a massive single exposure to phenyl mercury in which the only symptoms of intoxication were albuminuria and chemical burns, the initial mercury level in urine was 8.5 mg/l. and in whole blood 0.25 μg/ml. There is no information concerning the dose necessary to produce these levels. On the other hand, levels of up to 6 mg Hg/l. in spot samples of urine and 0.6 μg/ml in whole blood have been published for phenyl mercury exposed workers considered to be free of symptoms of poisoning.

8.2.2.2 In Animals

In Table 8:3 the acute LD_{50}'s for phenyl mercury compounds have been summarized together with those for other organic mercury compounds.

In Table 8:7 other experimental toxicity studies of phenyl mercury compounds have been put together. Although numerous studies have been performed, few conclusions are possible. In the rat, subcutaneous exposure to 6 mg Hg/kg/day for 14 days may induce a decrease in weight gain as compared to controls (Wien, 1939). In the investigation made by Fitzhugh et al., 1950, a level of phenyl mercury acetate (corresponding to 0.1 mg Hg/kg) was given to rats in the food for 2 years and induced slight histological kidney lesions in some animals. No data in terms of exposure/kg body weight/day were given, but the exposure must have been very small. On the other hand, 10 mg Hg/kg of food produced severe kidney damage. In rabbits, a subcutaneous exposure to 1 to 2 mg Hg/kg body weight/day for 9 to 11 days induced no symptoms or histological lesions in the kidney (Weed and Ecker, 1933). The central nervous system of cats tolerates an oral exposure of 2 to 3 mg/kg body weight/day for 25 to 52 days (Morikawa, 1961a). Nothing was stated about the kidneys, however. Piglets which received orally 2.3 to 4.6 mg Hg/kg body weight/day for 14 to 63 days were clinically diseased and had histological lesions in the kidney, liver, and gastrointestinal tract (Tryphonas and Nielsen, 1970).

Hagen, 1955 (see also Section 4.1.2.2.1), exposed mice by inhalation to phenyl mercury acetate dust with different particle sizes. With the particle diameter of 0.6 to 1.2 micron, death occurred after about 1 hour. With larger diameters, 2 to 40 microns, no poisoning occurred in 30 hours. No data on air mercury or air dust levels were given. After exposure to phenyl mercury pyrocatecholate (particle size about 1μ), death occurred within 1.2 and 12 hours in 2 experiments. In the first one, the air mercury vapor level was less than 1 mg/m^3 (dust was not determined) and in the second, the dust level was 80 mg/m^3. A histological examination revealed pulmonary edema. Rats were generally half as susceptible as mice. In Hagen's experiments, death occurred earlier at exposure to phenyl mercury (and also to methoxyethyl mercury, see Section 8.3.2) than at methyl, and much earlier than at ethyl mercury exposure. At exposure to alkyl mercury compounds, the levels of vapor (80 and 17 mg Hg/m^3, respectively) were much higher than at phenyl mercury exposure (less than 1 mg Hg/m^3). In the case of phenyl mercury, however, the dust levels were high.

Information about organ levels at phenyl mercury poisoning is even more inconclusive than that about exposure. In groups of rats with occasional slight renal changes after 1 1/2 to 2 years of exposure to concentrations of 0.1 mg Hg/kg of food, the mean kidney levels were 2.3 μg Hg/g (Fitzhugh et al., 1950). In groups exposed to 10 mg Hg/kg of food for the same time and with severe forms of kidney damage, the average kidney level was 30 μg Hg/g. Tryphonas and Nielsen, 1970, found 160 to 370 μg Hg/g kidney in piglets showing histological evidence of renal damage. The exposure was 2.3 to 4.6 mg Hg/kg body weight/day orally for 14 to 63 days.

8.3 Alkoxyalkyl Mercury Compounds
8.3.1 In Human Beings

Dérobert and Marcus, 1956, described a person who, a few hours after a 2 to 3 hour inhalation of dust of methoxyethyl mercury silicate, displayed pulmonary and gastrointestinal symptoms, and later, evidence of renal damage and neurasthenic symptoms (the last 2 mentioned symptoms months and years after exposure). One week after the exposure the urine mercury level was 1 mg/l.

8.3.2 In Animals

For alkoxyalkyl mercury compounds, toxicity studies other than LD_{50} determinations (see Table 8:3) are practically nonexistent.

Hagen, 1955 (see also Sections 4.1.2.3.1 and 8.2.2.2), exposed mice to methoxyethyl mercury silicate dust. Death occurred after 1.2 to 14 hours in different experiments. The dust levels in air were 1,200 and 50 mg Hg/m^3 in 2 experiments in

TABLE 8:7

Toxicity for Different Phenyl Mercury (PhHg) Compounds at Repeated Exposure

Animal species	Compound	No. of animals	Administration route[1]	Exposure[2]	Duration of exposure, days	Total dose, mg Hg/kg body weight	Time until onset of symptoms, days	Type of symptoms	Reference
Mouse	PhHgNO$_3$[3]	6	or.	~500 mg Hg/l. of water	7			*No symptoms* until 14 days	Weed and Ecker, 1931
		10	or.	~300 mg Hg/l. of drinking water	70			*No symptoms*	Birkhaug, 1933
		6	or.	~500 mg Hg/l. of water	14			Two experimental and 1 control animals died; One experimental animal had diarrhea	Weed and Ecker, 1933
Rat	PhHgNO$_3$[3]	8	or.	~500 mg Hg/l. of water	14		8-9	Death without obvious reason in 1 animal, diarrhea in 3	Weed and Ecker, 1931
		8	or.	~500 mg Hg/l. of water			8-10	One animal died without obvious reason; 3 died with slight hemorrhages in the intestines	Weed and Ecker, 1933
		5	s.c.	6 mg Hg/kg/day	14	84		Depression of growth rate, No definite histological lesions	Wien, 1939
	PhHg acetate	6	i.p.	0.9 mg Hg/kg/day	14	12		*No symptoms*. No histological lesions in kidney, liver spleen and adrenals	Eastman and Scott, 1944
		6	i.p.	1.8 mg Hg/kg/day	14	24		*No symptoms*. No histological lesions in kidney, liver spleen and adrenals	
		12	or.	0.1 mg Hg/kg food	2 years		2 years	*No symptoms*. Slight, histological lesions in kidneys[4]	Fitzhugh et al., 1950
		12	or.	2.5 mg Hg/kg food	365		365	*No symptoms*. Slight, histological lesions in kidneys[5]	
		12	or.	10 mg Hg/kg food	365		365	Pronounced histological lesions in kidney[6]	
		?	i.p.	0.5 mg Hg/kg/day	28	14	14	Thinner, sluggish, and apathetic	Swensson, 1952

TABLE 8:7 (continued)

Animal species	Compound	No. of animals	Administration route[1]	Exposure[2]	Duration of exposure, days	Total dose, mg Hg/kg body weight	Time until onset of symptoms, days	Type of symptoms	Reference
Rat	PhHgCl	3	or.	3-10 mg Hg/kg/day	4-50	40-300		*No symptoms*	Itsuno, 1968
	PhHg acetate	3	or.	3-9 mg Hg/kg/day	4-50	40-300		*No symptoms*	
	PhHgBr	6	or.	5-7 mg Hg/kg/day	30-50	200-300		*No symptoms*	
	PCMB[7]	3	or.	3-7 mg Hg/kg/day	20-50	100-250		*No symptoms*	
	$(Ph)_2$Hg	4	or.	6-10 mg Hg/kg/day	50	~400		*No symptoms*	
	PhHg acetate	120	or.	1-8 mg Hg/kg food	180			*No symptoms.* Reduced number of litters in the 8 mg Hg/kg group	Piechocka, 1968
Guinea pig	$PhHgNO_3$[3]	?	or.?	~3-20 mg Hg/kg/day	180			*No symptoms*	Tokuomi, 1969
		4	or.	~500 mg Hg/l. of water	7			*No symptoms* until 21 days	Weed and Ecker, 1931
		?	or.	~500 mg Hg/l. of water	10			*No symptoms*	Weed and Ecker, 1931
	PhHg dinaphthyl-methane disulfonate	4	or.	0.02 mg Hg/kg/day	180	3.6		*No symptoms*	Goldberg and Shapero, 1957
	PhHg dinaphthyl-methane disulfonate	4	or.	0.2 mg Hg/kg/day	180	36		*No symptoms*	
Rabbit	$PhHgNO_3$[3]	4	or.?	~1 mg Hg/kg x 4	24	4		*No symptoms*	Weed and Ecker, 1931
		1	or.	~2 mg Hg/kg x 4	24	8		*No symptoms*	
		4	i.p.	~2 mg Hg/kg x 3	12	8		*No symptoms*	
		4	s.c.	1-2 mg Hg/kg x 9-11	48-52	40-50		*No symptoms.* No histological lesions in kidney	
		2	s.c.	~0.3 mg Hg/kg/day	28			*No symptoms.* No histological lesions in kidney	Wien, 1939

TABLE 8:7 (continued)

Animal species	Compound	No. of animals	Administration route[1]	Exposure[2]	Duration of exposure, days	Total dose, mg Hg/kg body weight	Time until onset of symptoms, days	Type of symptoms	Reference
	PhHg acetate	4	i.v.	0.5 mg Hg/kg/day	14	7		*No symptoms*	Eastman and Scott, 1944
		8	i.p.	0.04 mg Hg/kg/day	70	2.8		*No symptoms*	
		8	i.p.	0.08 mg Hg/kg/day	70	5.6		*No symptoms*	
Cat	PhHg acetate	3	or.	2-3 mg Hg/kg/day[8]	25-52	75-200[8]		*No symptoms*. No definite histological lesions in CNS	Morikawa, 1961a
Piglet	PhHg acetate	23	or.	0.2-4.6 mg Hg/kg/day	1-90	2.7-68		*No symptoms*.[9] No histological lesions	Tryphonas and Nielsen 1970
		10	or.	2.3-4.6 mg Hg/kg/day	14-63	32-230	10-31	Diarrhea. Weight loss. Histological lesions in gastrointestinal tract, kidney and liver	

[1] or. = orally; s.c. = subcutaneously; i.p. = intraperitoneally; i.v. = intravenously
[2] The exposure has been recalculated on an every day basis.
[3] According to Wien, 1939, the substance used was the basic salt $C_6H_6HgNO_3 \cdot C_6H_6HgOH$.
[4] Slight in females, very slight in males. No lesions after 1 yr of exposure.
[5] Slight lesions in females, no in males. Exposure for 2 yr induced moderate lesions in females, slight in males.
[6] Pronounced lesions in females, no in males. After exposure for 2 yr pronounced lesions in females, slight in males. The lowest exposure that induced lesions in 1 yr in both sexes was 160 mg Hg/kg of food!
[7] Sodium para-chloro mercury benzoate.
[8] Not clear whether this is mercury or compound.
[9] Depression of growth rate was observed in animals fed 0.8 mg Hg/kg/day.

which death occurred after 6 and 14 hours, respectively. Air mercury vapor (organic and inorganic) level in 1 experiment was about 1 mg Hg/m^3. Autopsy revealed pulmonary edema. Lehotzky and Bordas, 1968, studied the effect of methoxyethyl mercury chloride in rats. Approximately 1.2 mg Hg/kg/day was administered intraperitoneally for a total of 50 days. After 30 days a decreased growth rate was observed and neurological symptoms occurred in several animals. Histological changes "as those found after mercury bi-chloride intoxication" were noted in the kidneys. Similar changes were said to have been present in animals receiving 0.12 mg Hg/kg body weight/day for a total of 80 days.

Lehotzky and Bordas, 1968, studied further behavioral effects. For the group receiving about 1.2 mg Hg/kg body weight/day there was a significant effect as compared to controls on one of the performance tests already within 10 days. At the lower dose, 0.12 mg Hg/kg body weight/day, effects on another test occurred late in the experiment when at least some of the animals in each group already had neurological symptoms.

Hapke, 1970, very briefly stated that rats and mice given (period of exposure not stated) food containing 5.5 mg Hg/kg as methoxyethyl mercury silicate did not take care of their offspring in a normal way. Rats administered 3.5 mg Hg/kg/day (route and period of exposure not stated) were described as having a decreased learning ability when tested in a labyrinth.

8.4 Other Organic Mercury Compounds

As was said in Section 4.2.2.4, the compound 1-bromomercuri-2-hydroxypropane (MHP) labeled with ^{197}Hg or ^{203}Hg has been used for studies of the morphology and function of the spleen. The compound is a substituted alkyl mercury compound. The toxicity of the compound has not been published. The doses given correspond to 2 to 10 mg Hg, i.e., 0.03 to 0.15 mg/kg body weight in a 70 kg adult man. No symptoms were reported after the administration.

Chapter 9

GENETIC EFFECTS

Claes Ramel

9.1 Introduction

The genetic activity of mercury compounds has been known since 1937, when Sass reported that a fungicide containing ethyl mercury phosphate caused disturbances of mitosis and polyploidy in plant cells. This effect of ethyl mercury on mitosis was verified and analyzed further by Kostoff, 1939 and 1940. Levan, 1945, reported a similar effect of inorganic mercury. A comparative analysis of the cytological effects on plant cells of several organic and inorganic mercury compounds was further performed at Levan's laboratory (Fahmy, 1951).

A series of investigations of the cytological effects of phenyl mercury on plant material was made by Macfarlane and her collaborators. The investigations included effects on the mitotic spindle mechanism (Macfarlane and Schmoch, 1948, and Macfarlane, 1953), as well as chromosome breakage and somatic mutations (Macfarlane, 1950 and 1951, and Macfarlane and Messing, 1953). Other studies on the effect of organic mercury on plants were performed by Bruhin, 1955, using the fungicide Agrimax M, which contains phenyl mercury dinaphtylmethane-disulphonate.

Apart from these early studies, several investigations on the genetic effects of mercurials have been performed in the last years in Sweden. These investigations will be summarized below.

9.2 Effects on Cell Division
9.2.1 Mitotic Activity

In order to evaluate quantitatively the observed effects of a chemical treatment on chromosomes, it is important to know the effect on the mitotic activity. Bruhin, 1955, made a comparative investigation of the effect of colchicine and of phenyl mercury dinaphtylmethane-disulphonate on the mitotic activity in germinating seeds of *Crepis capillaris*. In spite of the fact that both substances had the same pronounced effect on the spindle mechanism, they had markedly different effects upon the mitotic activity. While colchicine caused a distinct initial increase of dividing cells, followed by a subsequent decrease, the mercurial had a slight effect on the mitotic activity, compared to the control.

In connection with the study of the c-mitotic effect of methyl mercury hydroxide on *Vicia faba*, as reported in a following section, the effect of treatment on the mitotic activity was analyzed (Ramel and Ahlberg, unpublished data). The time of the treatment was 24 hours. As can be seen in Table 9:1, no pronounced effect of the treatment occurred, although a slight increase of dividing cells at low doses and a slight decrease at high doses may be traced.

TABLE 9:1

Percent Dividing Cells in *Vicia Faba* Root Tip Cells after Treatment with CH_3HgOH in Different Concentrations for 24 Hours.* (From Ramel and Ahlberg, unpublished data.)

	Concentration in substrate (10^{-6} Mol/l)							
	0	0.1	0.2	0.4	0.8	1.6	3.2	6.4
	10.6	18.2	11.4	17.0	12.4	6.2	8.2	9.4
	20.2	11.0	17.2	18.4	9.8	20.6	14.4	8.8
	8.4	12.0	15.8	11.4	14.0	10.2	10.8	10.6
	13.2	14.8	14.8	19.8	14.8	10.8	12.2	11.6
	11.8	10.2	9.4	12.4	11.8	12.0		11.6
	15.8	19.6	18.8	18.6	9.0	13.0		8.8
Total mean	13.3	14.3	14.6	16.3	12.0	12.1	11.4	10.1

*Each number represents the mean of 5 roots of one bean and 100 cells per root

9.2.2 C-Mitosis

The most striking effect of mercury compounds on the genetic material concerns the distribution of chromosomes and the induction of polyploidy and other deviating chromosome numbers in the cell. Most of the published work therefore deals with this aspect of genetic effects of mercury compounds. In particular, tests on root tip cells of *Allium cepa* have been used (Levan, 1945, and in press, Macfarlane and Schmoch, 1948, Fahmy, 1951, Ramel, 1967 and 1969a, and Fiskesjö, 1969).

All mercury compounds studied cause c-mitosis, an inactivation of the spindle fiber mechanism at cell divisions, similar to the well-known effect of colchicine.

Some differences between the effect of colchicine and of mercury compounds can be observed, however. With increasing dosage, a complete block of the spindle fiber mechanism is acquired very rapidly with colchicine, while such a dose-response relationship, at least with organic mercury compounds, involves a more gradual series of transitions between normal and c-mitotic cell divisions (Ramel, 1969a). These stages include multipolar cell divisions, defect distributions of single or a few chromosomes, and other types of incomplete c-mitosis. The result will be that mercury compounds evidently tend to cause more variable chromosome numbers than colchicine. It may be pointed out that this circumstance has some practical relevance. Such deviations from the normal chromosome number, which only involves single chromosomes, constitute more of a genetic risk than defects concerning the whole chromosome set. These latter defects are almost invariably lethal at an early stage.

In investigations of the effect of hexyl mercury bromide on mitosis in *Allium*, Levan (in press) found a deviating kind of c-mitosis in treatments at high concentrations. The chromosomal mitotic cycle proceeded through telophase without the nuclear membrane disappearing. Levan suggested the name *endonuclear c-mitosis* for this variant of c-mitosis. Whether it gave rise to tetraploid nuclei could not be decided because of the toxic influence of the mercurial.

C-mitotic effects similar to the ones found in plants have been observed in animal cells. Umeda et al., 1969, treated tissue cultures of HeLa-cells with phenyl and ethyl mercury chloride, ethyl mercury cysteine, and *n*-butyl mercury chloride, and reported c-mitotic effects. Similar effects on human leucocytes, treated in vitro with methyl mercury chloride, were found by Fiskesjö, 1970. Okada and Oharazawa, 1967, found significantly increased frequency of polyploidy in tissue cultures from mice treated in vivo with subcutaneous injections of ethyl mercury phosphate.

9.2.3 Dose-Response Relationships of C-Mitosis

The ability to induce c-mitosis is by no means limited to colchicine and mercury compounds. On the contrary, it has been shown by Levan and Östergren, 1943, that organic substances in general have this property. The lowest dosage necessary to cause c-mitosis varies widely, however, among different substances. Östergren and Levan, 1943, could demonstrate with *Allium* tests that a close negative correlation exists between the "threshold" value for c-mitosis and the water solubility of the substances. Thus the more soluble a substance is, the higher is the concentration value at which c-mitosis is induced. In Figure 9:1 (from Ramel, 1969a) this threshold concentration for c-mitosis in *Allium* has been plotted in relation to the water solubility according to Östergren, 1951. The corresponding values for methyl mercury hydroxide, methyl mercury dicyandiamide, phenyl mercury hydroxide, and colchicine are indicated in the Figure. It can be seen that they fall entirely outside the main regression line. In spite of the fact that they have a fairly high solubility in water, they act at extremely low concentrations.

Considering the actual dose-response relationships, the lowest dose which causes c-mitosis is of a particular interest from a practical point of view. Comparative information on this point is available from *Allium* tests. These data are shown in Table 9:2 for various organic and inorganic mercury compounds, as well as for colchicine. It is clear that particularly organic mercury compounds are exceedingly effective c-mitotic agents. In fact, they even act at considerably lower concentrations than colchicine.

Although there are no corresponding comparative data for organisms other than *Allium*, the experimental results on other plant species point to a similar sensitivity toward mercury compounds (Kostoff, 1940, Macfarlane and Messing, 1953, and Bruhin, 1955).

In order to study the dose-response relation-

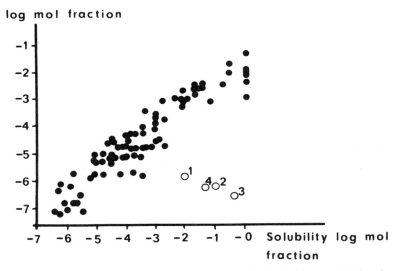

FIGURE 9:1. Correlation between solubility and threshold concentration for c-mitosis for different organic substances (from Östergren, 1951) and for colchicine (1), methyl mercury dicyandiamide (2), methyl mercury hydroxide (3), and phenyl mercury hydroxide (4). (From Ramel, 1969a.)

TABLE 9:2

Comparison of Approximate Threshold Values (in the substrate) for C-Mitosis in *Allium Cepa* of Mercury Compounds and Colchicine. (From Fahmy, 1951.)

Colchicine	$200 \cdot 10^{-6}$ Mol
Hg Br$_2$	$200 \cdot 10^{-6}$ Mol
Methyl HgBr	$0.5 \cdot 10^{-6}$ Mol
Ethyl HgBr	$0.2 \cdot 10^{-6}$ Mol
Butyl HgBr	$0.1 \cdot 10^{-6}$ Mol

ships of MeHg treatment in a species unrelated to *Allium*, an experiment was performed on *Vicia faba* (Ramel and Ahlberg, unpublished data). The treatment was applied for 24 hours at different concentrations of MeHg, as indicated in Figure 9:2. A significantly increased frequency of c-mitosis could be observed already at $0.1 \cdot 10^{-6}$ M in the substrate. The reason for the decline of the percentage of c-mitosis at the highest concentrations is not clear. It does not seem likely that it is related to the slight change of mitotic activity, reported in Table 9:1.

Observations by Fiskesjö, 1970, on human leucocytes treated in vitro with methyl mercury chloride gave a lowest concentration level for c-mitosis between 1 and $2 \cdot 10^{-6}$ M. This indicates a similar order of magnitude to that of plant cells.

Umeda et al., 1969, observed an inhibition of cell growth in treated HeLa cells at 0.32 ppm with phenyl mercury chloride, ethyl mercury chloride, and butyl mercury chloride, and the corresponding value for Hg Cl$_2$ was 3.2 ppm.

In the investigations outlined above, the dose-response relationships do not refer to the concentration of the mercury compounds in the tissues studied, but only to the concentration in the substrate. Without any knowledge of at least the gross uptake of the compounds in the actual tissues, a comparison of the cytological effects of the various compounds inevitably will suffer from some uncertainty. Thus it is difficult to know to what extent a difference in effect can be attributed to a real difference in biological effect or to a difference in the uptake of the tissue.

In order to elucidate this problem, some analyses have been made of the uptake of methyl mercury hydroxide, Hg (NO$_3$)$_2$ and colchicine in the root tissue of *Allium* (Ramel, Ahlberg, and Webjörn, unpublished data). The mercury compounds were labeled with ^{203}Hg and the concentration in the root tips was analyzed with gamma-spectrometry. A study of the corresponding uptake of colchicine was made with tritium labeled colchicine and liquid scintillation analyses. Table 9:3 gives the average uptakes of the 3 compounds in the roots, measured as dry

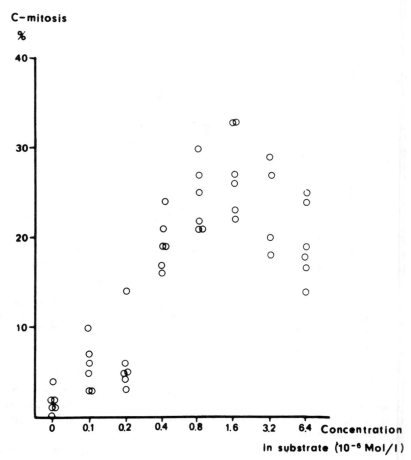

FIGURE 9:2. Percent c-mitosis in *Vicia faba* after 24 hours' treatment with CH_3 Hg OH. (Ramel and Ahlberg, unpublished data.)

TABLE 9:3

Accumulation of Mercury Compounds and Colchicine in 5 mm Root Tips of *Allium Cepa*. (From Ramel, Ahlberg, and Webjörn, unpublished data.)

Compound	CH_3 Hg OH	$Hg(NO_3)_2$						Colchicine			
Conc. Mol/l.	$8 \cdot 10^{-6}$	$1 \cdot 10^{-6}$		$10 \cdot 10^{-6}$		$100 \cdot 10^{-6}$		$1 \cdot 10^{-6}$		$10 \cdot 10^{-6}$	
Treatment, hours	6	4	24	4	24	4	24	4	24	4	24
Mean accumulation	1,002	605	1,981	1,010	1,251	1,422	2,523	347	398	275	275

weight. The accumulation of the organic and inorganic mercury is evidently similar to and around 3 times larger than the accumulation of colchicine. It is obvious that the difference in the c-mitotic effect of Hg^{2+}, CH_3 Hg^+, and colchicine (see Table 9:2) does not bear any relationship to the uptake of the compounds in the root tissue.

This indicates a difference in the biological and biochemical action.

In Table 9:3 a difference may be pointed out between the inorganic mercury compound and colchicine. Increasing the time of treatment from 4 to 24 hours with colchicine does not lead to an increased accumulation as it does with the mercury compound. That the tissue becomes more rapidly saturated with colchicine presumably depends on the fact that the target molecules are more specific for colchicine than for mercury compounds, as will be further discussed below.

9.2.4 Mechanisms of C-Mitotic Action

The mitotic action of colchicine as well as of organic mercurials shows a dose-response relationship which places them beside most other organic substances, as mentioned above. The unspecific c-mitotic action of chemicals in general has been suggested by Östergren and Levan, 1943, to be related to narcosis. With regard to colchicine, the target molecule has been demonstrated by Borisy and Taylor, 1967, and Shelanski and Taylor, 1967. According to these authors, colchicine very specifically binds to a structural protein with a sedimentation coefficient of 6S. This protein constitutes the building block not only of the spindle fibers but of microtubules in general. The binding of colchicine to this protein explains the biological effects of colchicine. The formation of the spindle fibers involves the polymerization of this protein unit to long chains and experimental data indicate that this polymerization depends on the formation of hydrogen bonds between the protein molecules (Mazia, 1955 and 1961). Presumably colchicine acts at this stage by binding to the protein molecules and thus preventing the hydrogen bonding.

On the other hand, however, it has long been known that sulfhydryl groups play an essential role in the formation of spindle fibers. Rapkine, 1931, found on sea urchin eggs that the occurrence of acid soluble sulfhydryl groups shows a cyclic variation in close phase with the mitotic cycle. According to Mazia, 1955 and 1961, sulfhydryl groups of protein units involved in the formation of the spindle fibers are oxidized to intermolecular disulfide bonds. It seems that a tetramere protein molecule is formed in this way and that this large unit constitutes the above mentioned protein of 6S involved in the formation of microtubules. The well-known reactivity of mercury and mercury compounds to sulfhydryl groups in protein (cf. Boyer, 1959) is *a priori* likely to interfere with the sulfhydryl-disulfide cycle of the proteins involved in the formation of the spindle fibers.

The importance of the sulfhydryl groups in this connection can be concluded from experimental data. It has been shown that mercaptans such as cysteine, glutathione, and dimercaptopropanol (BAL) act as efficient inhibitors against the c-mitotic action of organomercury compounds (Macfarlane, 1953, and Ramel, 1969a).

The effect of combined treatment with phenyl mercury hydroxide, BAL, and recovery in water upon *Allium* root cells is shown in Table 9:4 (Ramel, 1969a). The treatment was given in successive six-hour periods. It can be seen that simultaneous treatment with the mercurial and BAL as well as after-treatment with BAL almost eliminated the c-mitotic effect of the mercury compound. The simplest explanation of these data would be that the binding of the mercurial to BAL prevents it from reacting with the SH-groups of the protein units involved in the spindle fiber formation. That BAL and methyl mercury form such a complex is indicated by the observation of Berlin and Ullberg, 1963d, that the distribution of methyl mercury in the bodies of mice becomes altered by a simultaneous introduction of BAL. It could be pointed out, however, that the mercurials also may act in a more indirect way. For instance, they could inhibit enzymes involved in the formation of the spindle fibers or they could interfere with the ribosomes, where sulfhydryl groups are essential for the protein synthesis (Tamaoki and Miyazawa, 1967). It may be mentioned that an effect at the enzyme level by methyl mercury has not been indicated in experiments on crossing over or chromosome repair, as reported below.

9.3 Radiomimetic Effects

Organomercury compounds act not only on the

TABLE 9:4

C-Mitosis in *Allium Cepa* after Combined Treatments with $1.25 \cdot 10^{-6}$ M Phenyl Mercury Hydroxide, 10^{-4} M 2,3 Dimercapto-1-Propanol (BAL) and Recovery in Water. (From Ramel, 1969a.)

Treatment in successive 6-hour periods	Percent c-mitosis	Total mitosis
Control	0.4	2,518
Hg + Recov.	78.7	2,506
BAL + Recov.	0.6	2,492
Hg + BAL	2.4	2,511
Hg / BAL + Recov.	1.8	2,532
BAL + Hg + Recov.	52.8	2,422

distribution of the chromosomes, as dealt with above, but also directly upon the genetic material. Macfarlane, 1950 and 1951, reported the occurrence of somatic mutations, pollen sterility, and chromosome fragmentation in plants after treatment with phenyl mercury compounds. In experiments with *Allium*, chromosome breakages were observed by Ramel, 1969a, both with phenyl and methyl mercury. The chromosome breakage was considerably stronger with phenyl than with methyl mercury, as indicated in Table 9:5.

Levan (in press) found a strong chromosome breakage in *Allium* cells brought on by hexyl mercury bromide. The breakages were mostly of the chromatid type and a high frequency of them was localized to the centromeres.

The frequency of chromosome fragmentation produced by different mercury compounds seems to be independent of the c-mitotic effects. This is evident from comparison of methyl, hexyl, and phenyl mercury.

Skerfving, Hansson, and Lindsten, 1970, performed chromosome analyses on 13 human subjects, 9 of whom had remarkably elevated mercury levels in their blood, due to an extreme diet of mercury contaminated fresh water fish. The chromosome analyses were made on lymphocytes grown in vitro. The results are shown in Figure 9:3. Although the variation obviously is large, the authors found a significant rank correlation between chromosome breaks and the mercury concentration in the blood.

9.4 Effects on Meiosis
9.4.1 Cytological Observations

Preliminary observations on the effect of methyl mercury hydroxide on meiosis were performed on microsporogenesis in *Tradescantia* (Ramel and Engström, unpublished data). The treatment was given to cut stems, which were put in water containing MeHg. The transportation through the stem and the uptake of the mercurial in the flower buds were established by the use of methyl mercury, labeled with ^{203}Hg and by gamma-spectrometric analyses of flower buds. Cytological observations revealed that methyl mercury induced inactivation of the spindle fibers during meiosis (c-meiosis), resulting in chromosomal effects corresponding to the ones induced at mitosis.

9.4.2 Nondisjunction in Drosophila

The cytological observations evidently indicate that meiosis as well as mitosis is affected by the treatment with mercurials. It is, however, of interest also to know to what extent the effects on meiotic cells correspond to an actual transmission of gametes with altered chromosome sets from one generation to the next. A series of investigations of this problem has been performed on *Drosophila melanogaster* (Ramel and Magnusson, 1969 and 1971, and Ramel, 1969b and 1970). The advanced genetic technic with *Drosophila* and the large number of genetic markers and chromosomal

TABLE 9:5

Effect on Chromosome Fragmentation after Treatment with Phenyl and Methyl Mercury Hydroxide for 24 Hours, Followed by 48 Hours of Recovery in Water. (From Ramel, 1969a.)

Compound	Conc. mol $1 \cdot 10^{-6}$	Bridges	Fragm.	Bridges and fragm.	% Anaphases with bridges	% Anaphases with fragm.	Total anaphases
Phenyl Hg OH	2.5	78	49	18	19.8	13.8	486
	1.2	84	36	22	13.2	7.2	805
	0.25	33	10	2	2.2	0.7	1,627
	0	1	2	0	0.1	0.3	752
Methyl Hg OH	2.5	29	7	2	6.1	1.8	507
	1.2	4	8	0	0.6	1.2	688
	0.25	2	4	0	0.3	0.7	582
	0	4	0	0	0.2	0	1,750

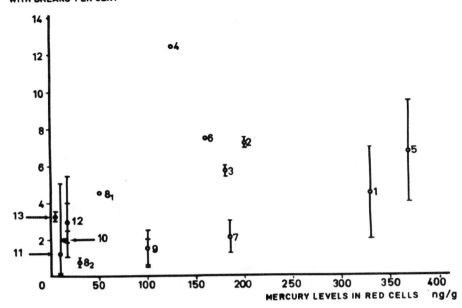

FIGURE 9:3. Chromosome breaks in relation to mercury concentration in red cells in Swedish consumers of fish containing methyl mercury. (From Skerfving, Hansson, and Lindsten, 1970.) The numbers in the diagram refer to the individual subjects investigated.

aberrations available make this organism particularly suitable for a detailed analysis of this kind.

The experimental procedure used for these investigations on *Drosophila* has been described by Ramel and Magnusson, 1969, and only a few points will be dealt with here. The treatment for the flies was made by mixing the mercury compounds in the corn agar substrate. Toxicity tests were performed and such doses were chosen which gave a delayed larval development without causing excessive lethality. Most of the experiments were made with methyl mercury hydroxide and the dosages used were 0.25 mg/l. substrate for treatment of larvae and 5 mg/l. for adults.

The actual genetic system used in the experiments was based on the meiotic distribution of the sex chromosomes. The X chromosomes were marked with the recessive mutant yellow which makes the bodies of the flies yellowish. The Y chromosome was marked with a translocated piece carrying the wild type allele of yellow. The genetic marking of the sex chromosomes enables their meiotic distribution to be followed. As shown in Figure 9:4, regular females carry the gene yellow in both of their X chromosomes, and consequently have a yellow body color. The regular males carry yellow in the X chromosome and the wild type allele in Y and they will be wild type. A wrong distribution of the sex chromosomes through nondisjunction during meiosis gives rise to gametes with 2 or no sex chromosomes. As shown in Figure 9:4, this will be manifested as exceptional offspring — wild type (XXY) females carrying an extra Y chromosome and yellow (XO) males with only 1 X chromosome.

9.4.2.1 Standard X Chromosomes

The results from the treatment of larvae with methyl mercury hydroxide are shown in Table 9:6. It is obvious that the frequency of exceptional offspring is increased by the mercury treatment. The difference between the Hg series and the control is highly significant for the time period of 4 to 10 days after the treatment. When more time had elapsed since the treatment, however, no effect could be traced. The frequency of exceptional offspring is identical in the Hg treatment series and in the control for the time period of 11 to 17 days after treatment. The lack of genetic effects of methyl mercury later than 10 days after treatment is in good agreement with the fate of this mercurial in the flies. Gas-liquid chromatographic studies of the mercury content of flies treated with methyl mercury have shown a

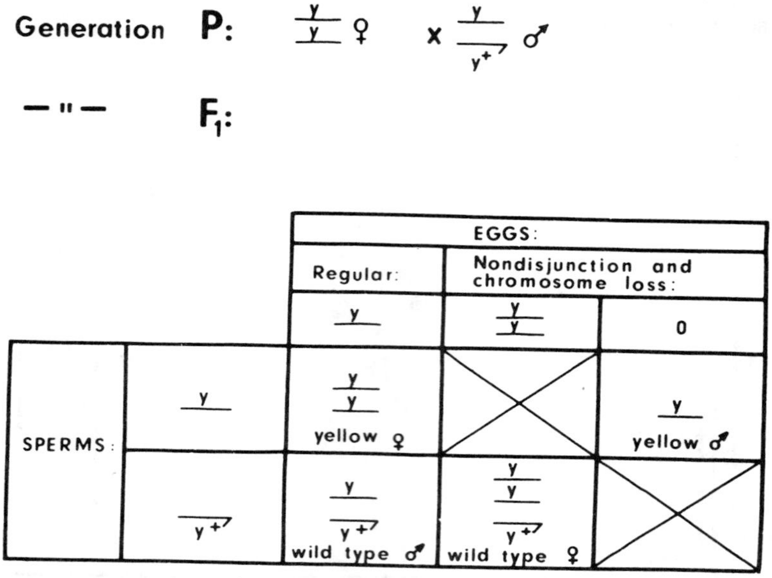

Generation P: $\frac{y}{y}$ ♀ × $\frac{y}{y^{+?}}$ ♂

— " — F₁:

		EGGS:		
		Regular:	Nondisjunction and chromosome loss:	
		$\frac{y}{}$	$\frac{y}{y}$	0
SPERMS:	$\frac{y}{}$	$\frac{y}{y}$ yellow ♀	×	$\frac{y}{}$ yellow ♂
	$\overline{y^{+?}}$	$\frac{y}{y^{+?}}$ wild type ♂	$\frac{y}{\frac{y}{y^{+?}}}$ wild type ♀	×

FIGURE 9:4. Test system with *Drosophila melanogaster* for screening nondisjunction and chromosome loss (a standard procedure).

TABLE 9:6

Effect of Methyl Mercury Hydroxide on *Drosophila melanogaster* (given as a larval treatment to the parental cross ywsn/ywsn x ywsn/ with 0.25 mg/l. Substrate). (From Ramel and Magnusson, 1969.)

Days after treatment	Hg			Control		
	% Exceptions		Total number	% Exceptions		Total number
	XXY-♀♀	XO-♂♂		XXY-♀♀	XO-♂♂	
4-6	0.32*	0.31	23,030	0.16	0.45	38,007
6-8	0.24**	0.43	44,405	0.11	0.38	40,935
8-10	0.18	0.38	33,299	0.12	0.45	34,730
Sum 4-10	0.24***	0.38	100,734	0.13	0.42	113,672
11-13	0.15	0.47	18,525	0.15	0.57	27,279
13-15	0.17	0.58	36,327	0.19	0.47	37,838
15-17	0.27	0.49	20,941	0.22	0.56	20,659
Sum 11-17	0.19	0.52	75,793	0.19	0.52	85,776

Significant differences vs. the controls:
*p=0.05-0.01
** p = 0.01 - 0.001
***p<0.001

pronounced decrease after about 10 days (Jensen, Magnusson, and Ramel, unpublished data).

It should be added that the effect of methyl mercury on the distribution of the sex chromosomes is restricted almost entirely to treatments of females. A substantial effect after treatment of males can only be obtained with the use of chromosomal aberrations, which cause a high spontaneous nondisjunction (Ramel, 1970).

The effect of methyl mercury on nondisjunction of chromosomes in *Drosophila* confirms at a genetic level the cytological observations of the c-mitotic effect of mercurials. The increased frequency of nondisjunctional gametes induced by the mercury treatment most likely results from an inactivation of the spindle fiber mechanism. This effect of the mercurial in *Drosophila* is, however, of an unusual kind. Theoretically, nondisjunction of the sex chromosomes should give rise to 2 types of gametes, half with 2 sex chromosomes and half with no sex chromosomes. An inspection of the results after mercury treatment in Table 9:6 clearly shows that only exceptional gametes with 2 sex chromosomes, giving rise to exceptional female offspring, were increased. No effect whatsoever can be traced on the reciprocal class of exceptional offspring, that is, males with only 1 sex chromosome. This result has been confirmed in other experimental series.

Just as concluded by Ramel and Magnusson, 1969, the data suggest that the mercury induced nondisjunction of chromosomes is connected with a nonrandom distribution of the chromosomes in such a way that the two X chromosomes are preferentially distributed to that pole which gives rise to the egg cell. Such a preferential distribution of chromosomes, "meiotic drive," is known in other connections, but apparently organomercury compounds constitute the first example of a chemically induced meiotic drive. It may be added that 1 experiment was made also with phenyl mercury. The result was the same as with methyl mercury:. that is, only an increase of XX-gametes occurred.

9.4.2.2 Inversion Heterozygotes

The effect of methyl mercury in *Drosophila* was not only tested with standard X chromosomes, but also with inversions which affect the normal meiotic pairing of chromosomes. It turned out that the effect on X chromosomes heterozygous for inversions was entirely different from the corresponding effect reported above on standard X chromosomes. Some experimental data on standard X (y w sn) and the complex inversion Muller 5 (M5) are given in Table 9:7 (from Ramel and Magnusson, 1969). While the effect on structurally homozygous chromosomes (y w sn/y w sn and y M5/y M5) primarily concerns XX exceptional gametes, the opposite is true for heterozygous X (y ec ct v f/y M5) which shows the strongest effect on the reciprocal O-gametes.

TABLE 9:7

Effect of Methyl Mercury Hydroxide on X Chromosomes With and Without Inversions (Muller-5) in *Drosophila Melanogaster*. (From Ramel and Magnusson, 1969.)

Genotype of mothers	Treatment	Offspring		Total number
		% Exceptions		
		XX -♀♀	XO-♂♂	
ywsn/ywsn	MeHg	0.24**	0.38	100,734
ywsn/ywsn	Control	0.13	0.42	113,672
yecctvf/yM5	MeHg	0.28**	0.84**	56,561
yecctvf/yM5	Control	0.08	0.31	81,338
yM5/yM5	MeHg	0.28**	0.12*	17,540
yM5/yM5	Control	0.02	0.02	25,805

* For further explanation, see text.
** Indications of statistical analysis, as in Table 9:6.

In order to analyze this difference in response to the mercury treatment of structurally homozygous as compared to structurally heterozygous X chromosomes, an extensive series of experiments was performed with various X chromosome inversions (Ramel and Magnusson, 1971, and unpublished data). The inversions were chosen in such a way that the influence of the heterochromatin could be established.

The results show that the effect of methyl mercury is almost identical with all heterozygous X chromosome inversions and independent of the heterochromatin balance as well as of the spontaneous frequency of nondisjunction. Concerning the predominant effect of methyl mercury on exceptional O-gametes the data are in agreement with the following explanation:

Under standard conditions, nondisjunction of chromosomes depends on a lack of pairing during meiosis, which is shown by the fact that the chromosomes involved almost invariably lack crossing over. The induction of nondisjunction by mercury, on the other hand, presumably depends on an entirely different mechanism, an inactivation of the spindle fiber mechanism. Consequently, the meiotic pairing, which is not influenced by the formation of the spindle fibers, remains normal. The chromosomes involved in mercury induced nondisjunction therefore have gone through normal meiotic crossing over, which will not affect the viability of structurally homozygous chromosomes. Crossing over in structurally heterozygous chromosomes will, on the contrary, drastically affect the viability of the chromosomes. Dicentric and acentric chromosome fragments will be produced through crossing over within the inverted segment, invariably leading to an elimination of the chromosomes involved. Potential XX-gametes will, therefore, lose the X chromosomes and be converted to O-gametes.

This hypothesis, that the high incidence of mercury induced O-gametes in heterozygous inversions is caused by an elimination of crossovers, is supported by other experimental data. An increase of crossing over by genetic means (interchromosomal effects of autosomal inversions) also significantly increases the mercury induced O-gametes in heterozygous M5 (Ramel and Magnusson, 1969 and 1971).

Finally, it should be emphasized that the fact that the introduction of heterozygous inversions in the test system causes a shift from XX toward O-gametes serves as a strong indication that the mercurial actually acts on the spindle mechanism and not on other mechanisms involved in chromosomal segregation.

9.4.3 Effects on Crossing Over and Chromosome Repair

Mercury compounds, very active as enzyme inhibitors, might have an indirect genetic effect through the inhibition of enzymes involved in different genetic processes, as mentioned above. In order to investigate this possibility, the effect of methyl mercury was analyzed on crossing over and chromosome repair in *Drosophila* (Ramel and Magnusson, unpublished data).

In these experiments, methyl mercury was distributed to the flies in the same way and in the same concentrations as in the experiments on nondisjunction dealt with above. A detailed presentation of the experimental procedure and results of the crossing over experiments will be published elsewhere and for the present purpose it is, therefore, sufficient to summarize some of the main points and the general conclusion.

Meiotic crossing over was studied in the X chromosome and chromosome 2. In the X chromosome crossing over was analyzed in the following intervals: y-ec-ct-v-f-car, covering nearly the whole euchromatic part of X. In chromosome 2, the intervals were b-cn-vg-bw, covering the centromere region and most of the right arm. Although the material was fairly extensive (over 40,000 flies analyzed), no influence on the meiotic recombination process of the mercury treatment could be established.

The repair mechanism after radiation-induced chromosome lesions constitutes another experimental system which can be used to study an effect on repair enzymes. An inhibition of these enzymes by mercurials would lead to a synergistic effect between radiation and the mercurial. The consequence would be an increased radiation-induced effect on the chromosomes.

In the experiments dealing with this problem, an experimental system equivalent to the one used for the nondisjunction test was employed. Radiation-induced chromosome breaks lead to elimination of the chromosomes and can be scored as an increased frequency of O-gametes. This is a well-known effect of radiation in *Drosophila*. The use of a Y chromosome with translocated pieces of X, marked y^+ and B, enabled the scoring of loss of either of the markers as another indication of chromosome breakage.

In the present experiments, 1100r of x-ray were given to males, half of which had received treatment with methyl mercury during their larval development. It can be seen in Table 9:8 that the mercury treatment does not have any noticeable effect on the frequency of radiation-induced chromosome loss.

From the lack of effect of methyl mercury on crossing over and radiation-induced chromosome breaks, it can be concluded that the mercurial does not have any appreciable effect on the enzyme systems of chromosome repair and DNA synthesis. Inasmuch as methyl mercury causes chromosome breakage, this action presumably emanates from a direct effect on the chromosomes rather than from an indirect enzymatic effect. Such a direct effect on the chromosomes would be in accordance with in vitro studies of methyl mercury and DNA. Gruenwedel and Davidson, 1966, have shown that methyl mercury binds to DNA and causes a denaturation of DNA in vitro.

9.4.4 Point Mutations

The mutagenic effect of mercury compounds is of importance to establish, considering the ability of methyl mercury to react with DNA and the chromosome breaking action, particularly of phenyl mercury. With regard to phenyl mercury, a mutagenic activity is indicated by the observations of Macfarlane and Messing, 1953, on somatic mutations in various plant materials.

In order to study the mutagenic effect of methyl mercury, experiments on sex-linked recessive lethals were made on *Drosophila melanogaster* with the standard Muller-5 technique (Ramel, 1969b, and unpublished data). The mercury treat-

TABLE 9:8

Effect of Irradiation With or Without Treatment with Methyl Mercury on *Drosophila* y/y⁺ YB^s ♂♂ Mated to y/y ♀♀. Exceptional XO-♂♂ and Loss of the Y Chromosome markers y⁺ and B Indicate Chromosome Breakage (Ramel and Magnusson, unpublished data).

	Days after irradiation	Treatment	% Exceptions		Total number
			XO-♂♂	Loss of y⁺ or B in Y	
Exp. 1	5-6	Hg + 1100 r	0.88	0.53	2,275
		1100 r	0.47	0.50	4,215
	6-7	Hg + 1100r	0.72	0.42	3,597
		1100 r	0.69	0.17	4,766
	7-8	Hg + 1100 r	0.88	0.28	2,169
		1100 r	0.62	0.27	3,710
	8-9	Hg + 1100 r	0.54	0.48	1,680
		1100 r	0.96	1.09	2,291
Exp. 2	5-6	Hg + 1100 r	0.76	0.15	5,896
		1100 r	0.67	0.20	6,132
	6-7	Hg + 1100 r	0.98	0.25	3,167
		1100 r	0.96	0.45	4,489
	7-8	Hg + 1100 r	1.24	0.24	1,691
		1100 r	1.05	0.31	3,241
	8-9	Hg + 1100 r	0.82	0.15	1,334
		1100 r	0.81	0.11	1,861

ment and the dose were the same as in the nondisjunction experiments on *Drosophila* reported above. The analysis also included 2 series on phenyl mercury hydroxide. The result of the experiments is presented in Table 9:9. The sizes of the separate series are small and no significant difference occurred between the treated series and their respective controls. There is, however, a clear tendency toward an increased frequency of recessive lethals after mercury treatment. A statistical analysis of the combined series according to Fisher, 1950, shows a significant difference between the series treated with methyl mercury and the control at a p-level of 0.025. The tendency is the same in the series with phenyl mercury, but the material is too small for a statistical significance.

Although methyl mercury evidently has a mutagenic effect, this effect is small — less than twice the spontaneous rate. The data point to a similar effect of phenyl mercury. The reason for this comparatively small mutagenic activity in vivo may be that the majority of the mercury molecules entering the cell gets "trapped" and inactivated before reaching the nucleus and the chromosomes by various proteins and polypeptides, such as the ones forming the spindle mechanism.

It should be added that the mutagenic effects of mercury compounds in mammalian systems have only been studied by means of dominant lethals, which do not discriminate among different kinds of genetic effects. In an experiment by Frölén and Ramel, briefly reported by Ramel, 1967, male CBA mice were treated with methyl mercury diacyandiamide, at a dose of 3 mg Hg/kg I.P. Ten injected males were mated immediately to 4 females each. Each week for 6 weeks after treatment new females were given to the males in order to cover the whole spermatogenesis. No significant increase of dominant lethals was obtained, although a significant reduction of pregnant females was found as compared to the control, treated with NaCl.

A dominant lethal experiment with rats was performed by Khera (1971). Male rats were treated orally with 0, 1, 2.5, and 5 mg/Hg/kg of

TABLE 9:9

Recessive Lethals with Methyl and Phenyl Mercuric Compounds (Ramel, unpublished data).

Exp. No.	Treatment	Sex treated	Tested chromosomes	% Lethals	P-value vs. control
1	Me Hg Control	♀♀	2,396 2,381	0.38 0.13	0.09
2	Me Hg Control	♀♀	3,073 3,103	0.30 0.23	0.65
3	Me Hg Control	♂♂	637 583	0.63 0.00	0.15
4	Me Hg Control	♂♂	2,415 2,298	0.50 0.30	0.29
5	Me Hg Control	♂♂	3,210 3,093	0.12 0.16	0.69
6	Ph Hg Control	♀♀	3,149 3,072	0.29 0.23	0.29
7	Ph Hg Control	♂♂	3,194 3,196	0.25 0.19	0.59

Combined P-values: Me Hg $0.025 > P > 0.01$
Ph Hg $0.50 > P > 0.30$

methyl mercury chloride for 7 days. For 5 days each, 14 sequential matings to untreated females were performed. From matings during the first 20 days following the treatment a reduction in the average number of viable embryos per litter was observed in all the Hg-treated groups. These results are somewhat unexpected as the highest sensitivity must have involved gametes treated as spermatozoa or late spermatids. In another experiment, males continuously received 0.1, 0.5, and 1 mg/kg/day. The treatment with the highest dose caused a significant depression in the average litter size after 25 days. No such effect was observed with the lower doses.

9.5 Concluding Remarks

It is obvious that mercury compounds have various effects on the genetic material. Apparently all compounds are active as c-mitotic agents, although the effectiveness is considerably higher for organic mercury compounds like alkyl and phenyl compounds than for inorganic ones. At least alkyl and phenyl mercury compounds also cause chromosome breakage and, to a minor extent, point mutations.

The question naturally arises as to what significance these observations have for evaluating the genetic risks of mercury pollution. Because mercury released in the aquatic environment becomes methylated through the action of microorganisms (Jensen and Jernelöv, 1969), the interest in this respect focuses on the genetic effect of methyl mercury. A wealth of data has been accumulated on the behavior of methyl mercury in different biological systems, including the human system. The high biological stability of methyl mercury and its long retention in the body are well-known and constitute matters of great concern from the point of view of environmental pollution. These circumstances are, of course, also highly relevant from a genetic point of view.

The intake of methyl mercury inevitably will result in an exposure also of the tissues and of cells to methyl mercury — in human beings as well as in *Drosophila* and *Allium*. The fact that the mercurials act on basic genetic systems like the spindle fiber mechanism and DNA makes it furthermore justified to assume the same effect in different organisms as long as the compound reaches the target molecules. That this in fact does occur is supported by the experimental evidence on widely different organisms. The data also indicate a uniform reaction of the genetic material to mercurials. The analysis by Skerfving, Hansson, and Lindsten, 1970, on lymphocytes from mercury-exposed human subjects, as reported above, is in accordance with this. Furthermore, their data indicate that the mercury pollution has reached a level at which genetic effects on human beings actually do take place, although little is known of the medical significance of chromosomal defects in blood cells.

The genetic risk of mercury exposure may involve somatic as well as germ cells. With regard to somatic cells, the consequences of genetic changes in postnatal life are quite obscure, although a connection with carcinogenesis may be suspected. In prenatal tissues, however, the action of mercury compounds is of more immediate concern. It is a well-known fact that methyl mercury readily passes through the placenta and may cause intra-uterine intoxication, as during the Minamata catastrophe, when 22 such cases were reported. Although there is no evidence that any of these cases originated from a chormosomal disorder, such an effect certainly must be taken into consideration. The dosage at which methyl mercury interferes particularly with chromosomal segregation is evidently very small. This should be considered in view of the fact that chromosomal disorders usually are estimated to cause around a third of all spontaneous abortions.

Concerning the effect of mercurials on germ cells, the c-mitotic action also appears to be of importance. As discussed above, mercurials to a large extent cause irregular c-mitosis with only a partially inactivated spindle fiber mechanism. This leads to errors in the distribution of single chromosomes, and cells with a far greater chance of survival than more completely polyploid cells are produced. It is possible that the result will be an increase of congenital disorders like mongolism, which depend on such an erroneous distribution of a single chromosome.

Finally, it should be pointed out that the direct mutagenic effect of methyl mercury, as revealed by the recessive lethal tests in *Drosophila*, is of comparatively small magnitude. This aspect of the genetic hazards from mercury pollution therefore seems to be less serious.

Chapter 10

GENERAL DISCUSSION AND CONCLUSIONS – NEED FOR FURTHER RESEARCH

Lars Friberg and Jaroslav Vostal

In previous chapters the metabolism and toxicity of mercury and different mercury compounds have been treated separately and systematically. Conclusional sections have been included in each chapter and a comprehensive summary is not necessary. However, some main conclusions will be emphasized and a comparison will be made among various mercury compounds.

There is no doubt that mercury can constitute a serious health problem. Within industry injurious exposure to metallic mercury vapors as well as to both inorganic and organic compounds may occur. In the general population exposure to methyl mercury, particularly via fish, is by far the most dangerous form of exposure to mercury. This does not mean that contamination of the environment with other forms of mercury is of no importance. It has been made obvious from several studies that a microbiological methylation of other forms of mercury takes place in the bottom sediment in water. As a result, mercury in fish is found almost exclusively as methyl mercury independent of which form originally contaminated the water. Lakes and rivers can be contaminated primarily via sewage and via contaminated air and rainwater.

From the toxicological point of view all of the alkyl mercury compounds must be considered first. Both methyl mercury and ethyl mercury (here and in the following are meant mono-methyl and mono-ethyl mercury compounds) are highly toxic, giving rise to severe damage of the central nervous system, with sensory disturbances, ataxia, visual disturbances, and deafness. The prognosis is poor and in severe cases the fatality rate is high. These injuries are often called the Minamata disease, after the place in Japan where the first epidemic in a general population was identified. Prenatal poisoning with methyl mercury has been reported in human beings. The symptoms are those of an unspecific infantile cerebral palsy with mental retardation and motor disturbances. Other organic compounds such as aryl and alkoxyalkyl compounds including phenyl and methoxyethyl mercury have a much lower toxicity. Few poisonings have been reported and the clinical manifestations are not well-known.

After exposure to inorganic mercury, particularly metallic mercury vapors, symptoms from the central nervous system with tremor and unspecified neurasthenic symptoms dominate. Renal damage may occur. The prognosis is much more favorable than that for alkyl mercury compounds.

Organic mercury compounds, particularly methyl mercury and phenyl mercury, are highly active genetically as shown for c-mitosis and chromosome breakage in onion root cells. Data from one study tend to show a higher frequency of chromosome breakage in lymphocytes in human beings exposed to methyl mercury via fish. The medical consequences of such findings are not known, however. The mutagenic effect as studied on *Drosophila* seems to be fairly low for methyl mercury. In rats, a positive dominant lethal test has been observed after exposure to methyl mercury. In view of the genetic findings and the stability of alkyl mercury compounds in the body, the possibility of significant genetic effects of methyl mercury must be borne in mind.

The differences in toxicity among the various mercury compounds are explained to a great extent by differences in metabolism. Methyl mercury and to some extent also ethyl mercury have considerable stability in the body, while other forms of mercury are sooner or later transformed into mercuric mercury.

Vapors of metallic mercury are rapidly and almost completely absorbed via inhalation. No quantitative data are available on the systemic absorption of mercury compounds after inhalation. Clinical evidence indicates a high absorption after exposure to alkyl mercury vapors, however. Absorption of metallic mercury via ingestion is negligible. Soluble mercuric mercury salts are absorbed to a limited extent. Methyl mercury, and with all probability ethyl mercury, are almost completely absorbed. Phenyl mercury is probably absorbed to a considerable degree when taken into the body by the peroral route. Skin penetration may occur after contact with several mercury compounds.

Animal and human data have shown that methyl mercury and ethyl mercury easily pass the placenta and accumulate in the fetus. For other mercury compounds the placenta constitutes a relatively effective barrier against penetration. Inhaled mercury vapor exists in vapor form in the blood for a short period, which allows the mercury to penetrate the brain membranes rapidly. As a result, the concentration of mercury in the brain after exposure to mercury vapor is about 10 times higher than after administration of a corresponding dose of mercuric mercury. This explains the higher toxicity for the central nervous system after exposure to mercury vapor.

The distribution of mercury within the body is affected by biotransformation. For methyl and ethyl mercury the distribution is much more even than that after exposure to other compounds. The highest levels of mercury are found in liver, kidneys, the central nervous system, and blood cells. In a human tracer dose experiment with methyl mercury, about 10% of the total body burden was found in the head, probably mainly in the brain, and about 5% in the blood.

The distribution of inorganic mercury shows a different picture. It changes with time so that relatively more mercury is found in the kidneys and the brain some time after the exposure. Generally, the kidney contains the highest concentration, the liver comes next, and then the spleen and brain. Also, within the organs the distribution is uneven. The blood contains a high concentration immediately after exposure but the concentration decreases rapidly with time. Much less is known on the distribution of aryl and alkoxyalkyl mercury but when some time has elapsed, the pattern tends to resemble that of inorganic mercury.

The distribution within the blood is one point of interest when discussing mercury in blood as an index of exposure. In man, methyl mercury has a ratio of about 10:1 between cells and plasma. After exposure to metallic mercury the ratio of mercury in red cells to plasma is about 1:1. The same holds true some time after exposure to phenyl mercury, while in this case relatively more mercury is found initially in the cells.

Data on retention and excretion for different mercury compounds during differing exposure situations are rather scanty. For methyl mercury, however, investigations in a number of animal species and in man indicate a monophasic exponential elimination. The biological half-life differs among species; it has been found to be between 70 and 90 days in human tracer dose experiments. Excretion occurs via urine, feces, and hair. In man the excretion via feces is about 10 times greater than that via urine.

The elimination of inorganic mercury is probably similar for exposure to mercuric mercury and mercury vapor. The biological half-life after single peroral tracer doses followed up to 3 to 4 months has been found to be about 30 to 60 days in human beings. Urinary and fecal excretions of inorganic mercury are about equal. Animal data show that the excretion follows not a monophasic exponential curve, but 2 to 3 consecutive exponential curves with increasing half-lives. Further interpretations of half-lives are difficult due to the time-related redistributions within the body, with an uneven distribution among and within organs in combination with a slow excretion from, for example, kidneys and the central nervous system. The data taken together indicate a risk of high accumulation in critical organs at prolonged exposure.

Due to the high degree of biotransformation of aryl and alkoxyethyl organic mercury compounds, interpretation of half-lives is difficult. However, animal data indicate that after phenyl and methoxyethyl mercury exposure, mercury is eliminated faster than after exposure to short chain alkyl mercury compounds but more slowly than after exposure to mercuric mercury. The excretion occurs through both feces and urine.

There is reason to consider the central nervous system the critical organ in chronic exposure to both inorganic and organic mercury, even if the toxic manifestations differ considerably for different compounds. In certain cases, the kidneys may be critical organs in chronic exposure to inorganic mercury and phenyl mercury. For alkyl mercury compounds genetic effects may be of considerable importance.

Dose-response relationships are not known for most exposure situations. For inhalation of vapors of metallic mercury and peroral exposure to methyl mercury data are available, however, which make it possible to evaluate risks to some extent. Experience with mercury vapor comes exclusively from animal experiments and industrial exposure. Prolonged exposure in an industrial environment to about 0.1 mg Hg/m^3 involves a risk for mercury intoxication. Recent data from studies in the chlorine industries in the U.S. as well as some industrial and animal data from Russia show, however, that some effects may be seen after exposure to lower, in fact considerably lower, concentrations. The significance of these findings is difficult to evaluate.

It is not possible to state a lowest concentration which may give rise to some medical manifestations. Even as little as 0.01 to 0.05 mg Hg/m^3 could not be considered for certain a no-effect level for industrial exposure according to the data at hand from the U.S. and U.S.S.R. Based on animal data from the U.S.S.R., it seems that still lower concentrations may give rise to certain effects. Without knowledge of the accumulation rate of mercury in different parts of the central nervous system, of the effects of continuous long-term exposure, and of the nature of particularly sensitive groups, it is not possible to make a realistic estimation of the concentration to which the industrial concentrations given would correspond within the general population. Taking only differences in exposure over a one-year period into consideration (365 vs. 225 days, a lung ventilation of 20 m^3 per day vs. 10 m^3) would give a reduction with a factor of about 3. This means that a concentration in industry of 0.01 mg Hg/m^3 would correspond to a concentration in the general population of about 0.003 mg Hg/m^3. With a lung ventilation of 20 m^3/day and an absorption of 80%, this corresponds to a daily absorption of about 50 μg.

It should be pointed out that the above mentioned calculations do not refer to oral intake of inorganic mercury. The concentration of mercury in CNS after inhalation of elemental mercury vapor will be much higher than that after exposure via ingestion. This will occur partly because the absorption rate of mercury is higher after inhalation and partly because a substantially higher portion of the absorbed amount gets into the CNS after inhalation of mercury vapors.

No good biological indicator is available for evaluating the risk of mercury intoxication through inhalation of mercury vapor. Neither mercury in blood nor in urine is satisfactory. It is true that on a group basis mercury levels in blood and urine will parallel exposure, but probably mainly recent exposure. There is no evidence that the concentrations in blood and urine during exposure will reflect concentrations in critical organs, and intoxications may occur at low levels of mercury in urine while high mercury levels are not necessarily accompanied by signs of intoxication. For evaluating recent exposure, blood and urinary mercury levels may be of importance. An exposure to about 0.1 mg Hg/m^3 of air seems to correspond on an average to 200 to 250 μg Hg/l. of urine.

Dose-response relationships in regard to methyl mercury are based primarily on data from the Niigata epidemic in Japan. The lowest mercury level in whole blood which gave rise to clinical intoxication was about 0.2 μg Hg/g or about 0.4 μg Hg/g red cells. In this report it has been considered reasonable to assume that this is the lowest level at which intoxication (in this case, irreversible changes) was observed. At the same time, it must be emphasized that several people in Japan as well as in Scandinavia are known to have had higher concentrations without clinical symptoms of methyl mercury intoxication. But then it should also be appreciated that the diagnosis of intoxication was made with rather crude clinical methods and subclinical effects of intoxication may well have occurred at lower exposure levels. Furthermore, there is evidence that the fetus may be more sensitive to methyl mercury than a pregnant woman/mother. Possible genetic effects were not studied, which complicates further the interpretation of the dose-response curve.

Empirical data from exposed persons as well as from animal experiments, together with knowledge of the metabolism of methyl mercury, show that methyl mercury and even total mercury in red cells or in whole blood are good indices of the concentration of mercury in the critical organ. If external contamination can be excluded, hair can also be used as an index. Mercury determinations in urine are of very limited value as indices of exposure to methyl mercury or indices for evaluating risks of intoxication.

The concentration of mercury in hair in relation to whole blood in man is about 300:1 corresponding to about 60 μg/g hair as a critical concentration. The critical levels in blood and hair mentioned correspond roughly to a daily exposure of 0.3 mg Hg as methyl mercury in a 70 kg man, or 4 μg/kg body weight. If a "safety factor" of 10 is applied (as was done in Sweden by Berglund et al., 1971) to allow for differences in sensitivity, including the possible greater sensitivity of the fetus, and for genetic and subclinical effects, this would mean values for whole blood, red cells, and hair of 0.02, 0.04, and 6 µg/g, respectively. The corresponding daily intake of mercury as methyl mercury would then be 0.03 mg for a 70 kg man, corresponding to about 0.4 µg/kg body weight.

The critical levels in blood mentioned above may be compared with levels found in "non-exposed" people from Scandinavia. Among such people the mercury content in whole blood is below or about 0.005 µg/g.

The figure given above for daily intake of methyl mercury, 300 µg, assumed to be the lowest level at which clinical intoxication has been observed in adults, is in contrast to the figure 50 µg discussed for the daily absorption of metallic mercury vapors, a result of a continuous exposure to 0.003 mg Hg/m^3. It should be pointed out strongly that for methyl mercury we are dealing with severe, irreversible damage. Furthermore, there is reason to believe that damage to the fetus may occur already at levels of daily intake lower than 300 µg Hg/m^3. Concerning the effects of metallic vapor especially, the criteria used were subtle, reversible effects. The difference between the 2 figures discussed, then, does not seem unreasonable.

What has been mentioned above concerning methyl mercury probably is valid to a considerable extent for ethyl mercury.

As for further research, there is an immediate need for more epidemiological studies on dose-response relationships with regard to all the mercury compounds. Particularly subclinical effects should be looked for. For methyl mercury only fairly gross effects have been studied and differences in sensitivity among individuals and subgroups of the population are not known. Such differences might well be substantial and thus important for evaluating acceptable exposure.

Data from studies in both the U.S. and U.S.S.R. indicate that exposure to metallic mercury may give rise to effects at considerably lower concentrations than have been recognized before. There is a need, however, to repeat and to extend these studies with due caution against potential analytical and epidemiological errors.

One major drawback with the epidemiological studies carried out in industry to date is the lack of coordination among them. This disadvantage is not unique for mercury, but there do seem to be excellent possibilities to study the toxicity of several mercury compounds by modern epidemiological techniques in a much better fashion than has been done hitherto. This presupposes cooperative efforts among several industrial groups as well as between state and independent researchers.

The evidence of fetal lesions in human beings after exposure to methyl mercury calls for intensive studies. Very little is known about dose-response relationships in this context. Genetic effects after exposure to different mercury compounds should also be investigated in more depth. Results from studies in fruit flies and plants up to now prompt investigations in higher animal species.

Though valuable information concerning uptake, biotransformation, and excretion is already available for several compounds, many more data are needed. This is true not only for compounds like alkyl and aryl mercury but also for inorganic compounds and metallic mercury. Despite the fact that exposure to, for example, metallic mercury has occurred for very long times, the biological half-life and accumulation risk in human beings in different organs are not known in any detail.

When reviewing the toxicological literature, particularly that dealing with metallic mercury vapors, it becomes obvious that widely different methods are used in different countries for studying effects. The differences in the approaches in the U.S.S.R. as compared with those in western countries are particularly apparent. Of special importance would be to study effects at very low exposure levels. Investigations should not be limited to conditions inside factories, but should also include populations living in the vicinity of the mercury emitting source. To coordinate international efforts in this field is a challenge for intergovernmental and other international health agencies.

REFERENCES

AIHA, Analytical guides, *Amer. Industr. Hyg. Ass. J.*, 30, 327, 1969.

Abbott, D. C. and Tatton, J. O'G., Pesticide residues in the total diet in England and Wales, 1966 to 1967. IV. Mercury content of the total diet, *Pestic. Sci.* 1, 99, 1970.

Aberg, B., Ekman, L., Falk, R., Greitz, U., Persson, G., and Snihs, J.-O., Metabolism of methyl mercury (^{203}Hg) compounds in man. *Arch. Environ. Health*, 19, 478, 1969.

Abramowicz, I., Deposition of mercury in the eye, *Brit. J. Opthal.*, 30, 696, 1946.

Abrams, J. D. and Majzoub, U., Mercury content of the human lens, *Brit. J. Opthal.*, 54, 59, 1970.

Adam, K. R., The effects of dithiols on the distribution of mercury in rabbits, *Brit. J. Pharmacol.*, 6, 483, 1951.

Ahlborg, G. and Ahlmark, A., Alkyl mercury compound poisoning: clinical aspect and risks of exposure, *Nord. Med.*, 41, 503, 1949, (Swedish with English summary).

Ahlborg, U., Berglund, F., Berlin, M., Grant, C., Von Haartman, U., and Nordberg, G., to be published, F. Berglund, National Institute of Public Health, S-104 01 Stockholm 60.

Ahlmark, A., Poisoning by methyl mercury compounds, *Brit. J. Industr. Med.*, 5, 117, 1948.

Ahlmark, A. and Tunblad, B., Toxic hazards of organic mercury compounds, Proc. 2nd. Int. Congr. Crop Protection, London, 1951, 1.

Aho, I., The occurrence of mercury in Åland pike, *Husö Biol. Stat. Medd.*, 13, 5, 1968, (Swedish).

Aikawa, J. K., Blumberg, A. J., and Catterson, D. A., Distribution of Hg^{203}-labeled mercaptomerin in organs of normal rabbits, *Proc. Soc. Exp. Biol. Med.*, 89, 204, 1955.

Aikawa, J. K. and Fitz, R. H., The distribution of Hg^{203}-labeled mercaptomerin in human tissues, *J. Clin. Invest.*, 35, 775, 1956.

Air Quality Criteria for Particulate Matter, U. S. Dept. Health, Education and Welfare Nat. Air Pollut. Contr. Admin. Pub. no. AP-49, 1969.

Albanus, L., Berglund, F., Berlin, M., Frankenberg, L., Grant, C., Von Haartman, U., Jernelöv, A., Rydälv, M., Skerfving, S., and Sundwall, A., Pilot experiments in a comparative investigation of toxicity on cats, between methyl mercury accumulated in vivo in Swedish fish and methyl mercuric hydroxide added to a homogenate of fish, S. Skerfving, National Institute of Public Health, S-104 01 Stockholm 60, Stencils, 1969, (Swedish).

Albanus, L., Berglund, F., Berlin, M., Frankenberg, L., Grant, C., Von Haartman, U., Jernelöv, A., Nordberg, G., Rydälv, M., Skerfving, S., and Sundwall, A., to be published, S. Skerfving, National Institute of Public Health, S-104 01 Stockholm 60.

Albanus, L., Frankenberg, L., and Sundwall, A., Mercury poisoning in cat fed with Swedish freshwater fish, FOA 1-Report 1970, FOA 1, Fack, S-172 04 Sundbyberg 4, (Swedish).

Almkvist, J., Experimentelle studien über die Lokalisation des Quecksilbers bei Quecksilbervergiftung, *Nord. Med. Arch.*, 36(2), 1, 1903.

Almkvist, J., Quecksilberchädigungen, in *Handbuch der Haut- und Geschlechtskrankheiten*, Jadassohn, J. Ed., Springer Verlag, Berlin, 1928, 18, 178.

Analytical Methods Committee, The determination of small amounts of mercury in organic matter, *Analyst*, 90, 515, 1965.

Anas, R. E., Mercury in fur seals, in Mercury in the Western Environment, abstract of papers presented at the Environmental Health Center Workshop, Oregon State University, Corvallis, Oregon, February 25 to 26, 1971.

Andersson, A., Mercury in the soil, *Grundförbättring*, 3-4, 95, 1967, (Swedish with English Summary).

Andersson, A. and Wiklander, I., Something about mercury in nature, *Grundförbättring*, 18, 171, 1965.

Anghileri, L. J., Absorption and excretion of 3-chloromercuri-2-methoxypropylurea labeled with Hg^{203}, *Nucl. Med.*, 4, 193, 1964.

Aomine, S. and Inoue, K., Retention of mercury by soils. II. Adsorption of phenyl mercuric acetate by soil colloids, *Soil Sci. Plant Nutr.*, 13, 195, 1967.

Aomine, S., Kawasaki, H., and Inoue, K., Retention of mercury by soils. I. Mercury residue of paddy and orchard soils, *Soil Sci. Plant Nutr.*, 13, 186, 1967.

Arighi, L. S., Mercury level studies in migratory waterfowl, in Mercury in the Western Environment, abstract of papers presented at the Environmental Health Center Workshop, Oregon State University, Corvallis, Oregon, February 25 to 26, 1971.

Armeli, G. and Cavagna, G., The value of the determination of urinary mercury in the prevention of mercury poisoning, *Med. Lavoro*, 57, 524, 1966, (Italian with English summary).

Armstrong, R. D., Leach, L. J., Belluscio, P. R., Maynard, E. A., Hodge, H. C., and Scott, J. K., Behavorial changes in the pigeon following inhalation of mercury vapor, *Amer. Industr. Hyg. Ass. J.*, 24, 366, 1963.

Artagaveytia, D., Degrossi, O. J., and Pecorini, V., Thyroid tumour scanning with ^{197}HgCl$_2$, *Nucl. Med.*, 9, 350, 1970.

Ashe, W. F., Largent, E. J., Dutra, F. R., Hubbard, D. M., and Blackstone, M., Behavior of mercury in the animal organism following inhalation, *A.M.A. Arch. Industr. Hyg. Occup. Med.*, 7, 19, 1953.

Atkinson, W. S., A colored reflex from the anterior capsule of the lens which occurs in mercurialism, *Amer. J. Opthal.*, 26, 685, 1943.

Atkinson, W. and von Sallman, L., Mercury in lens (hydrargyrosis lentis), *Trans. Amer. Opthal. Soc.*, 44, 65, 1946.

Axelrod, J. and Tomschick, R., Enzymatic O-methylation of epinephrine and other catechols, *J. Biol. Chem.*, 233, 702, 1958.

Bache, C. A., Gutenmann, W. H., and Lisk, D. J., Residues of total mercury and methylmercuric salts in lake trout as a function of age, *Science*, 172, 951, 1971.

Bäckström, J., Distribution studies of mercury pesticides in quail and some fresh water fishes, *Acta Pharmacol.*, 27, suppl. 3, 1, 1969a.

Bäckström, J., Differences in distribution of mercury in various species: Mouse — bird — fish, *Nord. Hyg. T.*, 50, 130, 1969b, (Swedish).

Baker, J. R., Ranson, R. M., and Tynen, J., A new chemical contraceptive, *Lancet*, 2, 882, 1938.

Bakulina, A. V., The effect of a subacute granosan poisoning on the progeny, *Sovet Med.*, 31, 60, 1968.

Baldi, G., Hyperthyroidism and chronic mercurialism, *Med. Lavoro*, 40, 113, 1949, (Italian with English summary).

Baltrukiewicz, Z., Contamination by ^{203}Hg of newborn rats fed with milk of mothers receiving intravenously ^{203}Hg-labeled Neohydrin®, *Acta Physiol. Pol.*, 20, 669, 811, 1969.

Baltrukiewicz, Z., Trials of evaluating the dose of ionizing radiation in kidneys after administration of Neohydrin® labeled with ^{197}Hg or ^{203}Hg, *Acta Physiol. Pol.*, 21, 90, 1970.

Bank, N., Mutz, B. F., and Aynedjian, H. S., The role of "leakage" of tubular fluid in anuria due to mercury poisoning, *J. Clin. Invest.*, 46, 695, 1967.

Barber, J., Beauford, W., and Shieh, Y. J., Some aspects of mercury uptake by plant, algal and bacterial systems in relation to its biotransformation and volatilization, in *Mercury, Mercurials, and Mercaptans*, Miller, M. W. and Clarkson, T. W., Eds., from the 4th Rochester Conf. on Environmental Toxicity, June 17 to 19, 1971, Charles C Thomas, Springfield, Ill., in press.

Barnes, E. C., The determination of mercury in air, *J. Industr. Hyg. Toxic.*, 28, 257, 1946.

Barringer, A. R., Interference-free spectrometer for high-sensitivity mercury analyses of soils, rocks, and air, *Inst. Mining Met. Trans.*, Sect. B., 75, 120, 1966.

Bass, M. H., Idiosyncrasy to ammoniated mercury ointment, *J. Mount Sinai Hosp.*, 10, 199, 1943.

Bate, L. C. and Dyer, F. F., Trace elements in human hair, *Nucleonics*, 23, 74, 1965.

Baumann, T., Akrodynie und Polyradiculitis — Quecksilber und Coxsackie-Virus, *Mod. Probl. Paediat.*, 1, 773, 1954.

Beani, L., Contribution to the study of chronic mercury poisoning, Med. Lavoro, 46, 633, 1955, (Italian with English summary).

Beliles, R. P., Clark, R. S., and Yuile, C. L., The effects of exposure to mercury vapor on behavior of rats, *Toxic. Appl. Pharmacol.*, 12, 15, 1968.

Benning, D., Outbreak of mercury poisoning in Ohio, *Industr. Med. Surg.*, 27, 354, 1958.

Berg, W., Johnels, A., Sjöstrand, B., and Westermark, T., Mercury content in feathers of Swedish birds from the past 100 years, *Oikos*, 17, 71, 1966.

Berglund, F., Experiments with rats relating to the toxicity of methyl mercury compounds, *Nord. Hyg. T.*, 50, 118, 1969, (Swedish).

Berglund, F., Berlin, M., Birke, G., von Euler, U., Friberg, L., Holmstedt, B., Jonsson, E., Ramel, C., Skerfving, S., Swensson, A., and Tejning, S., Methyl mercury in fish, a toxicologic-epidemiologic evaluation of risks, report from an expert group, *Nord. Hyg. T.*, suppl. 4, 1971 (published in *Nord. Hyg. T.*, Suppl. 3, Swedish, 1970).

Berglund, F., Berlin, M., Grant, C., and von Haartman, U., to be published, Nat. Inst. Public Health, S-104 01 Stockholm 60.

Bergman, T., Ekman, L., and Östlund, K., Retention and distribution of methylradiomercury in pigs, to be published, K. Östlund, Royal Veterinary College, S-104 05 Stockholm 50.

Bergstrand, A., Friberg, L., Mendel, L., and Odeblad, E., The localization of subcutaneously administered radioactive mercury in the rat kidney, *J. Ultrastruct. Res.*, 3, 234, 1959a.

Bergstrand, A., Friberg, L., Mendel, L., and Odeblad, E., Studies on the excretion of mercury in the kidneys, Transact. 12th Scand. Cong. Path. Microbiol., 1959b, 115.

Bergstrand, A., Friberg, L., and Odeblad, E., Localization of mercury in the kidneys after subcutaneous administration, *A.M.A. Arch. Industr. Health*, 17, 253, 1958.

Berkhout, P. G., Paterson, N. J., Ladd, A. C., and Goldwater, L. J., Treatment of skin burns due to alkyl mercury compounds, *Arch. Environ. Health*, 3, 106, 1961.

Berlin, M., Advantages and disadvantages of whole-organ assay and whole-body section autoradiography as revealed in studies of the body distribution of mercury and cadmium, Proc. 2nd. Symp. Int. Limites Tolérables, Paris, 1963a, 223.

Berlin, M., On estimating threshold limits for mercury in biological material, *Acta Med. Scand.*, Suppl. 396, 1, 1963b.

Berlin, M., Renal uptake, retention and excretion of mercury. II. A study in the rabbit during infusion of methyl- and phenylmercury compounds, *Arch. Environ. Health*, 6, 626, 1963c.

Berlin, M., Recent progress in mercury toxicology research and its consequences for current occupational mercury problems, *Proc. 15th Int. Congr. Occup. Health*, Wien, 1966, Vol. 3, 107.

Berlin, M., Fazackerly, J., and Nordberg, G., The uptake of mercury in the brains of mammals exposed to mercury vapour and to mercuric salts, *Arch. Environ. Health*, 18, 719, 1969.

Berlin, M. and Gibson, S., Renal uptake retention and excretion of mercury. I. A study in the rabbits during infusion of mercuric chloride, *Arch. Environ. Health*, 6, 617, 1963.

Berlin, M., Jerksell, L. -G., and Nordberg, G., Accelerated uptake of mercury by brain caused by 2,3-dimercaptopropanol (BAL) after injection into the mouse of a methylmercuric compound, *Acta Pharmacol.*, 23, 312, 1965.

Berlin, M., Jerksell, L. -G., and von Ubisch, H., Uptake and retention of mercury in the mouse brain, *Arch. Environ. Health*, 12, 33, 1966.

Berlin, M. and Johansson, L. G., Mercury in mouse brain after inhalation of mercury vapour and after intravenous injection of mercury salt, *Nature*, 209, 85, 1964.

Berlin, M., Nordberg, G., and Hellberg, J., The uptake and distribution of methylmercury in the brain of *Saimiri Sciureus* in relation to behavioral and morphological changes, in *Mercury, Mercurials and Mercaptans,* Miller, M. W. and Clarkson, T. W., Eds., from the 4th Rochester Conference on Environmental Toxicity, June 17 to 19, 1971, Charles C Thomas, Springfield, Ill., in press.

Berlin, M. H., Nordberg, G. F., and Serenius, F., On the site and mechanism of mercury vapor resorption in the lung, *Arch. Environ. Health*, 18, 42, 1969.

Berlin, M., Ramel, C., and Swensson, A., Poisoning by the consumption of fish containing a methyl mercury compound. A report from a study trip to Japan in 1968, Stencils, 1969, (Swedish).

Berlin, M. and Ullberg, S., Accumulation and retention of mercury in the mouse. I. An autoradiographic study after a single intravenous injection of mercuric chloride, *Arch. Environ. Health*, 6, 589, 1963a.

Berlin, M. and Ullberg, S., Accumulation and retention of mercury in the mouse. II. An autoradiographic comparison of phenylmercuric acetate with inorganic mercury, *Arch. Environ. Health*, 6, 602, 1963b.

Berlin, M. and Ullberg, S., Accumulation and retention of mercury in the mouse. III. An autoradiographic comparison of methylmercuric dicyandiamide with inorganic mercury, *Arch. Environ. Health*, 6, 610, 1963c.

Berlin, M. and Ullberg, S., Increased uptake of mercury in mouse brain caused by 2,3-dimercaptopropanol (BAL), *Nature*, 197, 84, 1963d.

Bertilsson, L. and Neujahr, H. Y., Methylation of mercury compounds by methylcobalamin, *Biochemistry*, 10, 2805, 1971.

Bertine, K. K. and Goldberg, E. D., Fossil fuel combustion and the major sedimentary cycle, *Science*, 173, 233, 1971.

Biber, T. U. L., Mylle, M., Baines, A. D., Gottschalk, C. W., Oliver, J. R., and MacDowell, M. C., A study by micropuncture and microdissection of acute renal damage in rats, *Amer. J. Med.*, 44, 644, 1968.

Bidstrup, P. L., Bonnell, J. A., Harvey, D. G., and Locket, S., Chronic mercury poisoning in men repairing direct-current meters, *Lancet*, 251, 856, 1951.

Birke, G., Johnels, A. G., Plantin, L. -O., Sjöstrand, B., and Westermark, T., Mercury poisoning through eating fish?, *Läkartidningen*, 64, 3628, 3654, 1967, (Swedish with English summary).

Birke, G., Hagman, D., Johnels, A. G., Plantin, L. -O., Sjöstrand, B., Skerfving, S., Westermark, T., and Österdahl, B., to be published, in *Arch. Environ. Health,* 1972, King Gustaf V Research Institute, S-104 01 Stockholm 60.

Birkhaug, K. E., Phenyl-mercuric-nitrate, *J. Infect. Dis.*, 53, 250, 1933.

Biskind, L. H., Phenyl mercury nitrate. Its clinical uses in gynecology, a preliminary report, *Surg. Gynec. Obstet.*, 57, 261, 1933.

Biskind, L. H., The therapeutic application of phenylmercuric salts, *Lancet*, 2, 1049, 1935.

Bivings, L., Acrodynia: a summary of BAL therapy reports and a case report of calomel disease, *J. Pediatr.*, 34, 322, 1949.

Bivings, L. and Lewis, G., Acrodynia: a new treatment with BAL, *J. Pediatr.*, 32, 63, 1948.

Blau, M. and Bender, M. A., Radiomercury (Hg^{203}) labeled neohydrin: a new agent for brain tumor localization, *J. Nucl. Med.*, 3, 83, 1962.

Bligh, E. G., Mercury levels in Canadian fish, in Mercury in Man's Environment, Proc. Roy. Soc. Canad., Symp. February 15 to 16, 1971, 73.

Boley, L. E., Morrill, C. C., and Graham, R., Evidence of mercury poisoning in feeder calves, *N. Amer. Vet.*, 22, 161, 1941.

Bonnevie, P., Dalgaard-Mikkelsen, S., Hansen, S. C., Riber Petersen, B., Poulsen, E., Somer, E., and Uhl, E., Mercury investigations on Danish eggs, pig's liver and fish, *Fra Sundhedsstyr.*, 8, 81, 1969, (Danish).

Bonnin, M., Organic mercury dust poisoning, *Roy. Adelaide Hosp. Rep.*, 31, 11, 1951.

Booer, J. R., The behavior of mercury compounds in soil, *Ann. Appl. Biol.*, 31, 340, 1944.

Borg, K., Effects of dressed seed on game birds, VIII, *Nord. Veterinärmötet i Helsingfors*, 398, 1958, (Swedish with English summary).

Borg, K., Swedish studies of game that had been found dead as well as specimens that had been shot, *Oikos*, Suppl. 9, 38, 1967.

Borg, K., The terrestrial fauna, and pesticides containing mercury, *Fauna Flora*, 63, 186, 1968, (Swedish).

Borg, K., The mercury problem as related to the terrestrial fauna, discovery and development, *Nord. Hyg. T.*, 50, 9, 1969a.

Borg, K., Annual report on game research at the National Veterinary Institute, S-104 05 Stockholm 50 Stencils, 1, 1969b, (Swedish).

Borg, K., Erne, K., Hanko, E., and Wanntorp, H., Experimental secondary methyl mercury poisoning in the goshawk (*Accipiter g. gentilis L.*), *Environ. Pollut.*, 1, 91, 1970.

Borg, K., Wanntorp, H., Erne, K., and Hanko, E., Mercury poisonings in Swedish wildlife, National Veterinary Institute, S-104 05 Stockholm 50 Stencils, 1965, (Swedish with English summary).

Borg, K., Wanntorp, H., Erne, K., and Hanko, E., Mercury poisoning in Swedish wildlife, *J. Appl. Ecol.*, Suppl. 3, 171, 1966, (English with French and German summary).

Borg, K., Wanntorp, H., Erne, K., and Hanko, E., Alkyl mercury poisoning in terrestrial Swedish wildlife, *Viltrevy*, 6, 299, 1969a.

Borg, K., Wanntorp, H., Erne, K., and Hanko, E., Mercury poisoning in Swedish wildlife, National Veterinary Institute, S-104 05 Stockholm 50 Stencils, 1, 1969b, (Swedish).

Borghgraef, R. R. M., Kessler, R. H., Pitts, R. F., Parks, M. E., Van Woert, W., and MacLoed, M. B., Plasma regression, distribution and excretion of radiomercury in relation to diuresis following the intravenous administration of Hg^{203} labelled chlormerodrin to the dog, *J. Clin. Invest.*, 35, 1055, 1956.

Borghgraef, R. R. M. and Pitts, R. F., The distribution of chlormerodrin (Neohydrin®) in tissues of the rat and dog, *J. Clin. Invest.*, 35, 31, 1956.

Borinski, P., Das Vorkommen Kleinster Hg-Mengen in Harn Und Faeces, *Klin. Wochenschr.*, 10, 149, 1931a.

Borinski, P., Die Herkunft der Quecksilbers in den Ausscheidungen, *Zahnärztliche Rundsch.*, 40, 221, 1931b.

Borisy, G. G. and Taylor, E. W., The mechanism of action of colchicine. Binding of colchicine - ^3H to cellular protein, *J. Cell. Biol.*, 34, 525, 1967.

Bornmann, G., Henke, G., Alfes, H., and Möllmann, H., Über die Enterale Resorption von Metallischem Quecksilber, *Arch. Toxikol.*, 26, 203, 1970.

Bouveng, H. O., Organo-mercurials in pulp and paper industry, *Oikos*, Suppl. 9, 18, 1967.

Bowen, H. J. M., Standard materials and intercomparisons, in *Advances in Activation Analysis*, Vol. 1, Lenihan, J. M. A. and Thomson, S. I., Eds., Academic Press, London, 1969, 101.

Boyer, P. D., Sulfhydryl and disulfide groups of enzymes, in *The Enzymes*, Boyer, P. D., Lardy, H., and Myrbäck, K., Eds., Academic Press, New York, 1959, I, 511.

Brar, S. S., Nelson, D. M., Kanabrocki, E. L., Moore, C. E., Burnham, C. D., and Hattori, D. M., Thermal neutron activation analysis of airborne particulate matter in Chicago metropolitan area, Nat. Bur. Stand. (V.S.), spec. pub. 43, 1969.

Brock, D. W., An analysis of mercury residues in Idaho pheasants, in Mercury in the Western Environment, abstract of papers presented at the Environmental Health Center Workshop, Oregon State University, Corvallis, Oregon, February 25 to 26, 1971.

Brown, E. A., Reactions to the organomercurial compounds, *Ann. N.Y. Acad. Sci.*, 65, 545, 1957.

Brown, I. A., Chronic mercurialism, *Arch. Neurol. Psychiat.*, 72, 674, 1954.

Brown, J. R. and Kulkarni, M. V., A review of the toxicity and metabolism of mercury and its compounds, *Med. Serv. J. Canada*, 23, 786, 1967.

Bruhin, A., Über polyploidisierende Wirkung eines Samenbeizmittels, *Phytopathol. Z.*, 23, 381, 1955.

Brune, D., Low temperature irradiation applied to neutron activation analysis of mercury in human whole blood, *Acta Chem. Scand.*, 20, 1200, 1966.

Brune, D., Aspects of low-temperature irradiation in neutron activation analysis, *Anal. Chem. Acta*, 44, 15, 1969, (English with French and German summary).

Brune, D. and Jirlow, K., Determination of mercury in aqueous samples by means of neutron activation analysis with an account of flux disturbances, *Radiochem. Acta*, 8, 161, 1967.

Buhler, D. R., Claeys, R. R., and Rayner, H. J., The mercury content of Oregon pheasants, in Mercury in the Western Environment, abstract of papers presented at the Environmental Health Center Workshop, Oregon State University, Corvallis, Oregon, February 25 to 26, 1971.

Buhler, D. R., Claeys, R. R., and Shanks, W. E., Mercury in aquatic species from the Pacific Northwest, in Mercury in the Western Environment, abstract of papers presented at the Environmental Health Center Workshop, Oregon State University, Corvallis, Oregon, February 25 to 26, 1971.

Burch, G., Ray, T., Threefoot, S., Kelly, F. J., and Svedberg, A., The urinary excretion and biologic decay periods of radiomercury labeling a mercurial diuretic in normal and diseased man, *J. Clin. Invest.*, 29, 1131, 1950.

Bureau, Y., Boiteau, H., Barriere, H., Litoux, P., and Bureau, B., Acrodynie d'origine mercurielle. Action de la penicillamine, *Bull. Soc. Franc. Derm. Syph.*, 77, 184, 1970.

Burns, R. A., Mercurialentis, *Proc. Roy. Soc. Med.*, 55, 322, 1962.

Burston, J., Darmady, E. M., and Stranack, F., Nephrosis due to mercurial diuretics, *Brit. Med. J.*, 1, 1277, 1958.

Butt, E. M. and Simonsen, D. G., Mercury and lead storage in human tissues, *Amer. J. Clin. Path.*, 20, 716, 1950.

Bürgi, E., Grösse und Verlauf der Quecksilberausscheidung durch die Nieren bei den verschiedenen üblichen Kuren, *Arch. Derm. Syph.*, 79, 3, 1906.

Cadigan, R. A., Mercury in sedimentary rocks of the Colorado plateau region, in *Mercury in the Environment*, U.S. Geological Survey professional paper, no. 713, Washington, D.C., U.S.G.P.O., 1970, 17.

Cammarota, V. A., Jr., Mercury, in Minerals Yearbook 1970, Bureau of Mines, Washington, D.C., in press.

Campbell, E. E. and Head, B. M., The determination of mercury in urine — single extraction method, *Industr. Hyg. Quart.*, 16, 275, 1955.

Campbell, J. S., Acute mercurial poisoning by inhalation of metallic vapour in an infant, *Canad. Med. Ass. J.*, 58, 72, 1948.

Cantor, M. O., Mercury lost in the gastrointestinal tract, *J.A.M.A.*, 146, 560, 1951.

Cassano, G. B., Amaducci, L., and Viola, P. L., Distribution of mercury (Hg^{203}) in the brain of chronically intoxicated mice (autoradiographic study), *Riv. Pat. Nerv. Ment.*, 87, 214, 1966.

Cassano, G. B., Amaducci, L., and Viola, P. L., Étude autoradiographique de la distribution du mercure (Hg^{203}) inhalé, dans l'encéphale du rat et de la souris, *Acta Neurol. Belg.*, 67, 1099, 1967.

Cassano, G. B., Viola, P. L., Ghetti, B., and Amaducci, L., The distribution of inhaled mercury (Hg^{203}) vapors in the brain of rats and mice, *J. Neuropath. Exp. Neurol.*, 28, 308, 1969.

Cember, H., The influence of the size of the dose on the distribution and elimination of inorganic mercury $Hg(NO_3)_2$ in the rat, *Amer. Industr. Hyg. Ass. J.*, 23, 304, 1962.

Cember, H., A model for the kinetics of mercury elimination, *Amer. Industr. Hyg. Ass. J.*, 30, 367, 1969.

Cember, H. and Donagi, A., The influence of dose level and chemical form on the dynamics of mercury elimination, *Excerpta Med. Int. Cong. Ser.*, 62, 440, 1964.

Cember, H., Gallagher, P., and Faulkner, A., Distribution of mercury among blood fractions and serum proteins, *Amer. Industr. Hyg. Ass. J.*, 29, 233, 1968.

Center for Disease Control, Neurotropic Diseases Surveillance, report CC-13 submitted to U.S. Dept. of Health, Education, and Welfare, March 15, 1971, no. 1.

Cheek, D. B., Bondy, R. K., and Johnson, L. R., The effect of mercurous chloride (calomel) and epinephrine (sympathetic stimulation) on rats. The importance of the findings to mechanisms in infantile acrodynia, *Pediatrics*, 23, 302, 1959.

Cheek, D. B. and Hicks, C. S., Pink disease or infantile acrodynia: its nature, prevention, and cure, *Med. J. Aust.*, 1, 107, 1950.

Cholak, J., The nature of atmospheric pollution in a number of industrial communities, Proc. 2nd Nat. Air Pollut. Symp., Pasadena, California, 1952.

Cholak, J. and Hubbard, D. M., Microdetermination of mercury in biological material, *Industr. Engin. Chem. Analyt.*, 18, 149, 1946.

Clarkson, T. W., Biochemical aspects of mercury poisoning, *J. Occup. Med.*, 10, 351, 1968.

Clarkson, T. W., Isotope exchange methods in studies of the biotransformation of organomercurial compounds in experimental animals, in *Chemical Fallout, Current Research on Persistent Pesticides*, Miller, M. W. and Berg, G. G., Eds., Charles C Thomas, Springfield, Ill., 1969, 274.

Clarkson, T. W., Epidemiological and experimental aspects of lead and mercury contamination of food, *Food Cosmet. Toxicol.*, 9, 229, 1971.

Clarkson, T. W., Gatzy, J., and Dalton, C., Studies on the equilibration of mercury vapor with blood, Division of Radiation Chemistry and Toxicology, University of Rochester Atom. Ener. Project, Rochester, N.Y., UR-582, 1961.

Clarkson, T. W. and Greenwood, M., The mechanism of action of mercurial diuretics in rats; the renal metabolism of p-chloromercuribenzoate and its effects on urinary excretion, *Brit. J. Pharmacol.*, 26, 50, 1966.

Clarkson, T. W. and Greenwood, M. R., Simple and rapid determination of mercury in urine and tissues by isotope exchange, *Talanta*, 15, 547, 1968.

Clarkson, T. W. and Greenwood, M. R., Selective determination of inorganic mercury in the presence of organomercurial compounds in biological material, *Anal. Biochem.*, 37, 236, 1970.

Clarkson, T. W. and Rothstein, A., The excretion of volatile mercury by rats injected with mercuric salts, *Health Phys.*, 10, 1115, 1964.

Clarkson, T. W., Rothstein, A., and Sutherland, R., The mechanism of action of mercurial diuretics in rats: the metabolism of ^{203}Hg-labelled chlormerodrin, *Brit. J. Pharmacol.*, 24, 1, 1965.

Clarkson, T. W. and Shapiro, R. E., The absorption of mercury from food, its significance and new methods of removing mercury from the body, in Mercury in Man's Environment, Proc. Roy. Soc. Canad., Symp. February, 15 to 16, 1971, 124.

Clarkson, T. W., Small, H., and Norseth, T., The effect of a thiol containing resin on the gastrointestinal absorption and fecal excretion of methylmercury compounds in experimental animals, *Fed. Proc.*, 30, 543 Abs., 1971.

Clegg, D. J., Embryotoxicity of mercury compounds, in Mercury in Man's Environment, Proc. Roy. Soc. Canad., Symp., February 15 to 16, 1971, 141.

Clennar, G. and Lederer, H., Mercurial diuretics and nephrosis, *Brit. Med. J.*, 1, 1544, 1958.

Cohen, M. M., An avocational dermatitis, *Md. Med. J.*, 7, 236, 1958.

Coldwell, B. B. and Platonow, N., The effect of methylmercuric acetate on the rate of disappearance of ethanol from the blood of swine, *Toxicol. Appl. Pharmacol.*, 14, 368, 1969.

Cole, H. N., Gericke, A. J., and Sollmann, T., The treatment of syphilis by mercury inhalations, *Arch. Derm. Syph.*, 5, 18, 1922.

Cole, H. N., Schreiber, N., and Sollmann, T., Mercurial ointments in the treatment of syphilis, *Arch. Derm.*, 21, 372, 1930.

Coleman, R. F., Cripps, F. H., Stimson, A., Scott, H. D., and Aldermaston, A. W. R. E., The trace element content of human head hair in England and Wales and the application to forensic science, *Atom Monthly Info. Bull. U. K. At. Energy Auth.*, 123, 12, 1967.

Cooke, N. E. and Beitel, A., Some aspects of other sources of mercury in the environment, in Mercury in Man's Environment, Proc. Roy. Soc. Canad., Symp., February 15 to 16, 1971, 53.

Corneliussen, P. E., Residues in food and feed. Pesticide residues in total diet samples (IV), *Pestic Monit. J.*, 2, 140, 1969.

Cotter, L. H., Hazard of phenylmercuric salts, *J. Occup. Med.*, 4, 305, 1947.

Cowder, P. J., The effect on human life of toxic agricultural chemicals, H.M.S.O. 6th rep. from the Estimates Committee Maff, 1961, 298.

Curley, A., Sedlak, V. A., Girling, E. F., Hawk, R. E., Barthel, W. F., Pierce, P. E., and Likosky, W. H., Organic mercury identified as the cause of poisoning in humans and hogs, *Science*, 172, 65, 1971.

Dahhan, S. S. and Orfaly, H., Electrocardiographic changes in mercury poisoning, *Amer. J. Cardiol.*, 14, 178, 1964.

Dalgaard-Mikkelsen, S., The occurrence of mercury in the Danish environment, *Nord. Hyg. T.*, 50, 34, 1969, (Danish).

Dall'Aglio, M., The abundance of mercury in 300 natural water samples from Tuscany and Latium, in *Origin and Distribution of Elements*, Ahrens, L. H , Ed., Pergamon Press, New York, 1968, 1065.

Dams, R., Robbins, J. A., Rahn, K. A., and Winchester, J. W., Nondestructive neutron activation analysis of air pollution particulates, *Anal. Chem.*, 42, 861, 1970.

Daniel, J. W. and Gage, J. C., The metabolism by rats of phenylmercury acetate, *Biochem. J.*, 122, 24p, 1971.

Daniel, J. W., Gage, J. C., and Lefevre, P. A., The metabolism of methoxyethylmercury salts, *Biochem. J.*, 121, 411, 1971.

Davies, D. J. and Kennedy, A., The excretion of renal cells following necrosis of the proximal convoluted tubule, *Brit. J. Exp. Pathol.*, 48, 45, 1967.

Dencker, I. and Schütz, A., Mercury content of food, *Läkartidningen*, 68, 4031, 1971, (Swedish).

Dérobert, L. and Marcus, O., Intoxication professionelle par inhalation de composé organique mercurielle antiparasitaire, *Ann. Med. Leg.*, 36, 294, 1956.

Dillon Weston, W. A. R. and Booer, J. R., Seed disinfection. I. An outline of an investigation on disinfectant dusts containing mercury, *J. Agric. Sci.*, 25, 628, 1935.

Dinman, B. D., Evans, E. E., and Linch, A. L., Organic mercury: Environmental exposure, excretion, and prevention of intoxication in its manufacture, *Arch. Industr. Health*, 18, 248, 1958.

D'Itri, F. M., The environmental mercury problem, report to the Michigan House of Representatives resulting from H. R.-424 Great Lakes Contamination (Mercury) Comm., Inst. Water Res., Michigan State University, June, 1971, CRC Press, in press, 1972.

Dreisbach, R. H. and Taugner, R., Renale "Stapelung" und Ausscheidung von ^{203}Hg-Sublimat bei der Ratte, *Nucl. Med.*, 5, 421, 1966.

Drogtjina, E. A. and Karimova, A. K., Poisoning with granozan, *Gig. Sanit.*, 21, 31, 1956.

Dustman, E. H., Stickel, L. F., and Elder, J. B., Mercury in wild animals in the Lake St. Clair region, in *Environmental Mercury Contamination*, Hartung, R. and Dinman, B. D., Eds., Ann Arbor Science Publishers, Ann Arbor, Michigan, in press.

Eades, J. F., Pesticide residues in the Irish environment, *Nature*, 210, 650, 1966.

Eastman, N. J. and Scott, A. B., Phenylmercuric acetate as a contraceptive, *Hum. Fertil.*, 9, 33, 1944.

Eberle, H., Experimentelle Untersuchungen über die unterschiedliche Empfindlichkeit von ein- und zweinierigen Tieren gegen Sublimat, Urannitrat und Kaliumbichromat, *Z. Unfallmed. Berufskr.*, 3, 196, 1951.

Edelstam, C., Johnels, A. G., Olsson, M., and Westermark, T., Ecological aspects of the mercury problem, *Nord. Hyg. T.*, 50, 14, 1969, (Swedish).

Edwards, C. M., Mercury poisoning in a horse as a result of eating treated oats, *Vet. Rec.*, 54, 5, 1942.

Edwards, G. N., Two cases of poisoning by mercuric methide, *St. Barts Hosp. Rep.*, 1, 141, 1865.

Edwards, G. N., Note on the termination of the second case of poisoning by mercuric methide, *St. Barts Hosp. Rep.*, 2, 211, 1866.

Edwards, J. G., The renal tubule (nephron) as affected by mercury, *Am. J. Pathol.*, 18, 1011, 1942.

Ekman, L., Greitz, U., Persson, G., and Aberg, B., Metabolism and retention of methyl-^{203}-mercurynitrate in man, *Nord. Med.*, 79, 450, 1968a, (Swedish with English summary).

Ekman, L., Grietz, U., Magi, A., Snihs, J. O., and Aberg, B., Distribution of ^{203}Hg in volunteers after administration of methyl-^{203} mercury-nitrate, *Nord. Med.*, 79, 456, 1968b, (Swedish with English summary).

Ekman, L., Aberg, B., Greitz, U., Persson, G., Falk, R., and Snihs, J. O., Metabolism of ^{203}Hg administered as CH$_3$^{203}HgCl perorally to man, *Nord. Hyg. T.*, 50, 116, 1969, (Swedish).

Elkins, H. B. and Pagnotto, L. D., Is the 24-hour urine sample a fallacy?, *Amer. Industr. Hyg. Ass. J.*, 26, 456, 1965.

Ellis, F. A., The sensitizing factor in Merthiolate®, *J. Allergy*, 18, 212, 1947.

Ellis, F. A. and Robinson, H. M., Cutaneous sensitivity to Merthiolate® and other mercurial compounds, *Arch. Derm. Syph.*, 46, 425, 1942.

Ellis, R. W. and Fang, S. C., Elimination, tissue accumulation, and cellular incorporation of mercury in rats receiving an oral dose of ^{203}Hg-labelled phenylmercuric acetate and mercuric acetate, *Toxicol. Appl. Pharmacol.*, 11, 104, 1967.

El-Sadik, Y. M. and El-Dakhakhny, A-A., Effects of exposure of workers to mercury at a sodium hydroxide producing plant, *Amer. Industr. Hyg. Ass. J.*, 31, 705, 1970.

Enders, A. and Noetzel, H., Spezifische Veränderungen im Kleinhirn bei chronischer oraler Vergiftung mit Sublimat, *Arch. Exp. Path. Pharmakol.*, 225, 346, 1965.

Engleson, G. and Herner, T., Alkyl mercury poisoning, *Acta Paediat. Scand.*, 41, 289, 1952, (English with French, German, and Spanish summary).

Environmental Mercury Contamination Conference, Ann Arbor, Michigan, September 30 to October 2, 1970. Proceedings in *Environmental Mercury Contamination*, Hartung, R. and Dinman, B. D., Eds., Ann Arbor Science Publishers, Ann Arbor, Michigan, in press; also in *Environ. Sci. Technol.*, 4, 890, 1970.

Epstein, E., Rees, W. J., and Maibach, H. I., Recent experience with routine patch test screening, *Arch. Derm.*, 98, 18, 1968.

Esbo, H. and Fritz, T., Annual report of the Swedish State Seed Testing Institute for the Fiscal Year 1968–1969, *Medd. Statens Centrala Frökontrollanstalt*, 45, 3, 1970, (Swedish with English summary).

Eyl, T. B., Alkylmercury contamination of foods, *J.A.M.A.*, 215, 287, 1971.

Fagerström, T. and Jernelöv, A., Formation of methyl mercury from pure mercuric sulphide in aerobic organic sediment, *Water Res.*, 5, 121, 1971.

Fahmy, F. Y., Cytogenetic analysis of the action of some fungicide mercurials, Thesis, University of Lund, Sweden, 1951.

Falk, R., Snihs, J. O., Ekman, L., Greitz, U., and Åberg B., Whole-body measurements on the distribution of mercury-203 in humans after oral intake of methylradiomercury nitrate, *Acta. Radiol.*, 9, 55, 1970, (English with German and French summary).

Fanconi, G., Botsztejn, A., and Schenker, P., Überempfindlichkeitsreaktionen auf Quecksilbermedikation im Kindesalter mit bedonderer Berücksichtigung der Calomel-krankheit, *Helv. Pediat. Acta*, 2 (suppl. 4), 3, 1947.

Fanconi, G. and von Murait, G., Die Feersche Krankheit (Akrodynie), eine seltsame Krankheit, *Deut. Med. Wochenschr.*, 78, 20, 1953.

Farquahar, H. G., Mercurial poisoning in early childhood, *Lancet*, 2, 1186, 1953.

Farquahar, J. W., Crawford, T. B. B., and Law, W., Urinary sympathin excretion of normal infants and of infants with pink diseases, *Br. Med. J.*, 276, 1956.

Farvar, M. A. and Cember, H., Difference between in vitro and in vivo distribution of mercury in blood proteins, *J. Occup. Med.*, 11, 11, 1969.

Fernandez, M.De M., Catalan, P. A., and Murias, B. S. F., A case of atmospheric contamination by mercury, *Rev. Sanid. Hig. Publica*, 40, 1, 1966, (Spanish).

Filby, R. H., Davis, A. I., Shah, K. R., and Haller, W. A., Determination of mercury in biological and environmental materials by instrumental neutron activation analysis, *Mikrochim. Acta*, 6, 1130, 1970.

Fimreite, N., Mercury uses in Canada and their possible hazards as sources of mercury contamination, *Environ. Pollut.*, 1, 119, 1970a.

Fimreite, N., Effects of methyl mercury treated feed on the mortality and growth of leghorn cockerels, *Can. J. Anim. Sci.*, 50, 387, 1970b.

Fimreite, N., Effects of dietary methyl mercury on ring-necked pheasant (*Phasianus colchicus*) with special reference to reproduction, Canadian Wildlife Service Occasional Papers Series, in press.

Fimreite, N., Fyfe, R. W., and Keith, J. A., Mercury contamination of Canadian prairie seed eaters and their avian predators, *Can. Field Naturalist*, 84, 269, 1970.

Fimreite, N., Holsworth, W. N., Keith, J. A., Pearce, P. A., and Gruchy, I. M., Mercury in fish and fish-eating birds near sites of industrial contamination, *Can. Field Naturalist*, in press.

Fischer, J., Mundschenk, H., and Wolf, R., Milzszintigraphie mit 1 Bromomercuri (^{197}Hg)-2-hydroxypropane (BMHP), *Fortshr. Roentgenstr.*, 103, 349, 1965.

Fishbein, L., Chromatographic and biological aspects of organomercurials, *Chromatogr. Rev.*, 13, 83, 1970.

Fisher, R. A., *Statistical Methods for Research Workers*, 11th ed., Oliver & Boyd, London, 1950.

Fiskesjö, G., Some results from *Allium* tests with organic mercury halogenides, *Hereditas*, 62, 314, 1969.

Fiskesjö, G., The effect of two organic mercury compounds on human leucocytes in vitro, *Hereditas*, 64, 142, 1970.

Fitzhugh, O. G., Nelson, A. A., Laug, E. P., and Kunze, F. M., Chronic oral toxicities of mercuri-phenyl and mercuric salts, *Industr. Hyg. Occup. Med.*, 2, 433, 1950.

Flanigan, W. J. and Oken, D. E., Renal micropuncture study of the development of anuria in the rat with mercury-induced acute renal failure, *J. Clin. Invest.*, 44, 449, 1965.

Flewelling, F. J., Loss of mercury to the environment from chloralkali plants, in Mercury in Man's Environment, Proc. Roy. Soc. Canad. Symp., February 15 to 16, 1971, 34.

Forbes, G. and White, J., Chronic mercury poisoning in latent finger-print development, *Br. Med. J.*, 1, 899, 1952.

Fowler, B. A., Ultrastructural evidence for nephropathy induced by long-term exposure to small amounts of methyl mercury, *Science*, 175, 780, 1972.

Fraser, A. M., Melville, K. I., and Stehle, R. L., Mercury-laden air: the toxic concentration, the proportion absorbed, and the urinary excretion, *J. Industr. Health*, 16, 77, 1934.

Fregert, S. and Hjorth, N., Increasing evidence of mercury sensitivity. The possible role of organic mercury compounds, *Contact. Derm. Newsletter*, 5, 88, 1969.

Friberg, L., Aspects of chronic poisoning with inorganic mercury based on observed cases, *Nord. Hyg. T.*, 32, 240, 1951, (Swedish with English summary).

Friberg, L., Studies on the accumulation, metabolism and excretion of inorganic mercury (Hg^{203}) after prolonged subcutaneous administration to rats, *Acta Pharmacol.*, 12, 411, 1956.

Friberg, L., Studies on the metabolism of mercuric chloride and methyl mercury dicyandiamide, *A.M.A. Arch. Industr. Health*, 20, 42, 1959.

Friberg, L., On the value of measuring mercury and cadmium concentrations in urine, *Pure Appl. Chem.*, 3, 289, 1961.

Friberg, L., Hammarström, S., and Nyström, A., Kidney injury after chronic exposure to inorganic mercury, *A.M.A. Arch. Industr. Hyg. Occup. Med.*, 8, 149, 1953.

Friberg, L., Odeblad, E., and Forssman, S., Distribution of two mercury compounds in rabbits after a single subcutaneous injection, *A.M.A. Arch. Industr. Health*, 16, 163, 1957.

Friberg, L., Piscator, M., and Nordberg, G., *Cadmium in the Environment*, Chemical Rubber Company, Cleveland, 1971, 27.

Friberg, L., Skog, E., and Wahlberg, J. E., Resorption of mercuric chloride and methyl mercury dicyandiamide in guinea-pigs through normal skin and through skin pre-treated with acetone, alkylarylsulphonate and soap, *Acta Derm. Venereol.*, 41, 40, 1961, (English with French, German, and Spanish summary).

Friedman, H. L., Relationship between chemical structure and biological activity in mercurial compounds, *Ann. N.Y. Acad. Sci.*, 65, 461, 1956.

Frithz, A., Cellular changes in the psoriatic epidermis. IX. Neutron activation analysis of mercury in patients topically treated with ammonium mercuric chloride, *Acta Derm. Venereol.*, 50, 345, 1970.

Frithz, A. and Lagerholm, B., Cellular changes in the psoriatic epidermis. VI. The submicroscopic intracellular distribution of mercury compound in the normal epidermis in comparison with that in the psoriatic epidermis, *Acta Derm. Venereol.*, 48, 403, 1968.

Frölen, H. and Ramel, C., to be published, Swedish Defence Research Institute, S-172 04 Sundbyberg 4.

Fujimoto, Y., Ohshima, K., Satoh, H., and Ohta, Y., Pathological studies on mercury poisoning in cattle, *Jap. J. Vet. Res.*, 4, 17, 1956.

Fujimura, Y., Studies on the toxicity of mercury. Urinary excretion of mercury and its deposition in organs in non-exposed man, *Jap. J. Hyg.*, 19, 33, 1964, (Japanese with English summary).

Fukuda, K., Metallic mercury induced tremor in rabbits and mercury content of the central nervous system, *Br. J. Ind. Med.*, 28, 308, 1971.

Fukunaga, K. and Tsukano, Y., Pesticide regulations and residue problems in Japan, *Residue Rev.*, 26, 1, 1969.

Furukawa, K., Suzuki, T., and Tonomura, K., Decomposition of organic mercurial compounds by mercuryresistant bacteria, *Rep. Agr. Biol. Chem.*, 33, 128, 1969.

Furukawa, K. and Tonomura, K., Enzyme system involved in the decomposition of phenyl mercuric acetate by mercury-resistant *Pseudomonas*, *Agr. Biol. Chem.*, 35, 604, 1971.

Furutani, S. and Osajima, Y., Studies on residual components of agricultural chemicals in food. III. Value of mercury in wheat, some vegetables and paddy field soil, *Kyushu Daigaku Nogakubu Gakugei Zasshi*, 22, 45, 1965.

Furutani, S. and Osajima, Y., Studies on residual components of agricultural chemicals in food. V. Mercury content in rice and other foods, *Nippon Shokuhin Kogyo Gakkaishi*, 14, 15, 1967, (Japanese).

Gage, J. C., The distribution and excretion of inhaled mercury vapour, *Br. J. Ind. Med.*, 18, 287, 1961a.

Gage, J. C., The trace determination of phenyl- and methylmercury salts in biological material, *Analyst*, 86, 457, 1961b.

Gage, J. C., Distribution and excretion of methyl- and phenylmercury salts, *Br. J. Ind. Med.*, 21, 197, 1964.

Gage, J. C., The metabolism of methoxyethylmercury and phenylmercury in the rat, in *Mercury, Mercurials and Mercaptans*, Miller, M. W. and Clarkson, T. W. Eds., from the 4th Rochester Conf. on Environmental Toxicity, June 17 to 19, 1971, Charles C Thomas, Springfield, Ill., in press.

Gage, J. C. and Swan, A. A. B., The toxicity of alkyl- and arylmercury salts, *Biochem. Pharmacol.*, 8, 77, 1961.

Gage, J. C. and Warren, J. M., The determination of mercury and organic mercurials in biological samples, *Am. Occup. Hyg.*, 13, 115, 1970.

Gaul, L. E., Preliminary and short report. Sensitizing component in thiosalicylic acid, *J. Invest. Derm.*, 3, 91, 1958.

Gayer, J., Graul, H., and Hundeshagen, H., Die Lokalisierung des Transportes von Hg^{++}-Ionen in der Niere durch Stop-flow-Analyse, *Klin. Wochenschr.*, 40, 953, 1962.

Gebhards, S. V., Mercury residue in Idaho fishes, in Mercury in the Western Environment, abstract of papers presented at the Environmental Health Center Workshop, Oregon State University, Corvallis, Oregon, February 25 to 26, 1971.

Gibbs, O. S., Pond, H., and Hansmann, G. A., Toxicological studies on ammoniated mercury, *J. Pharmacol.*, 72, 16, 1941.

Gibel, H. and Kramer, B., Idiosyncrasy to mercury preparations in childhood, *Am. J. Dis. Child.*, 66, 155, 1943.

Glomski, C. A., Brody, H., and Pillay, S. K. K., Distribution and concentration of mercury in autopsy specimens, *Nature*, 232, 200, 1971.

Goldberg, A. A. and Shapero, M., Toxicological hazards of mercurial paints, *J. Pharm. Pharmacol.*, 9, 469, 1957.

Goldberg, A. A., Shapero, M., and Wilder, E., The penetration of phenylmercuric dinaphtylmethane disulphonate into skin and muscle tissue, *J. Pharm. Pharmacol.*, 2, 89, 1950.

Goldblatt, M. W., Vesication and some vesicants, *Br. J. Ind. Med.*, 2, 183, 1945.

Goldman, H. S. and Freeman, L. M., Radiographic and radioisotopic methods of evaluation of the kidneys and urinary tract, *Pediatr. Clin. North Am.*, 18, 409, 1971.

Goldwater, L. J., Kidney injury after chronic exposure to inorganic mercury, *A. M. A. Arch. Ind. Hyg.*, 8, 588, 1953.

Goldwater, L. J., Occupational exposure to mercury, Harben Lectures, *J. Roy. Inst. Public Health Hyg.*, 27, 279, 1964.

Goldwater, L. J., Jacobs, M. B., and Ladd, A. C., Absorption and excretion of mercury in man. I. Relationship of mercury in blood and urine, *Arch. Environ. Health*, 5, 537, 1962.

Goldwater, L. J. and Joselow, M. M., Absorption and excretion of mercury in man. XIII. Effects of mercury exposure on urinary excretion of coproporphyrin and delta-aminolevulinic acid, *Arch. Environ. Health*, 15, 327, 1967.

Goldwater, L. J., Ladd, A. C., Berkhout, P. G., and Jacobs, M. B., Acute exposure to phenylmercuric acetate, *J. Occup. Med.*, 6, 227, 1964.

Goldwater, L. J., Ladd, A. C., and Jacobs, M. B., Absorption and excretion of mercury in man. VII. Significance of mercury in blood, *Arch. Environ. Health*, 9, 735, 1964.

Gościńska, Z., Durch Alkyle Quecksilberverbindung Hervorgerufenes Degenerationssyndrom des Gehirns ("Minamata-Krankheit"), *Helv. Paediatr. Acta,* 15, 216, 1965.

Grant, C. A., Pathology of experimental methyl mercury intoxication: some problems of exposure and response, in *Mercury, Mercurials, and Mercaptans,* Miller, M. W., and Clarkson, T. W., Eds., (from the 4th Rochester Conf. on Environmental Toxicity, June 17 to 19, 1971) Charles C Thomas, Springfield, Ill., in press.

Grant, C., Moberg, A., and Westöö, G., to be published, National Board of Health and Welfare, Department of drugs, S-104 01 Stockholm 60.

Greaves, F. C., Phenyl mercuric nitrate in the treatment of otitis externa and of the dermatophytoses, *U. S. Naval Med. Bull.,* 34, 527, 1936.

Greenlaw, R. H. and Quaife, C. M., Retention of Neohydrin®-Hg^{203} as determined with a total-body scintillation counter, *Radiology,* 78, 970, 1962.

Greif, R. L., Sullivan, W. J., Jacobs, G. S., and Pitts, R. F., Distribution of radiomercury administered as labelled chlormerodrin (Neohydrin®) in the kidneys of rats and dogs, *J. Clin. Invest.,* 35, 38, 1956.

Griffith, W. H., Mercury contamination in California's fish and wildlife, in Mercury in the Western Environment, abstract of papers presented at the Environmental Health Center Workshop, Oregon State University, Corvallis, Oregon, February 25 to 26, 1971.

Griffith, G. C., Butt, E. M., and Walker, J., The inorganic element content of certain human tissues, *Ann. Intern. Med.,* 41, 504, 1954.

Gritzka, T. L. and Trump, B. F., Renal tubular lesions caused by mercuric chloride, *Am. J. Pathol.,* 52, 1225, 1968.

Gross, E. R., *Dermatitis venenata* (mercury and its salts), *Arch. Derm. Syph.,* 37, 689, 1938.

Grossman, J., Weston, R. E., Lehman, R. A., Halperin, J. P., Ullmann, T. D., and Leiter, L., Urinary and fecal excretion of mercury in man following administration of mercurial diuretics, *J. Clin. Invest.,* 30, 1208, 1951.

Gruenwedel, D. W. and Davidson, N., Complexing and denaturation of DNA by methylmercuric hydroxide. I. Spectrophotometric studies, *J. Mol. Biol.,* 21, 129, 1966.

Gurba, J. B., Use of mercury in Canadian agriculture, in Mercury in Man's Environment, Proc. Roy. Soc. Can., Symp. February 15 to 16, 1971, 44.

Haber, M. H. and Jennings, R. B., Sex differences in renal toxicity of mercury in the rat, *Nature,* 201, 1235, 1964.

Haddad, J. K. and Sternberg, E., Jr., Bronchitis due to acute mercury inhalation, *Am. Rev. Resp. Dis.,* 88, 543, 1963.

Hagen, U., Toxikologie organischer Quecksilberverbindungen, *Arch. Exp. Pathol. Pharmakol.,* 224, 193, 1955.

Halldin, A., Industrial sources, *Nord. Hyg. T.,* 50, 154, 1969, (Swedish).

Hallee, T. J., Diffuse lung disease caused by inhalation of mercury vapor, *Am. Rev. Resp. Dis.,* 99, 430, 1969.

Hamaguchi, H., Rokuro, K., and Hosohara, K., Photometric determination of traces of mercury in sea water, *J. Chem. Soc. Jap.,* 82, 347, 1961, (Japanese).

Hamilton, A., *Industrial Poisons in the United States,* The MacMillan Co., New York 1925, 234.

Hanko, E., Erne, K., Wanntorp, H., and Borg, K., Poisoning in ferrets by tissues of alkyl mercury-fed chickens, *Acta Vet. Scand.,* 11, 268, 1970.

Hannerz, L., Experimental investigations on the accumulation of mercury compounds in water organisms, *Rep. Inst. Freshwater Res., Sweden,* 48, 120, 1968.

Hansen, H. -J., Mercury and game, *Sven. Vet. T.,* 20, 1, 1965a, (Swedish).

Hansen, H. -J., Mercury poisonings among game in Sweden, in *The Mercury Problem in Sweden,* Royal Swedish Ministry of Agriculture, Stockholm, 1965b, 11, (Swedish).

Hanson, A., Man-made sources of mercury, in Mercury in Man's Environment, Proc. Roy. Soc. Can., Symp., February 15 to 16, 1971, 22.

Hansson, H. and Möller, H., Patch test reactions to Merthiolate® in healthy young subjects, *Br. J. Derm.,* 83, 349, 1970.

Hansson, H. and Möller, H., Cutaneous reactions to Merthiolate® and their relationship to vaccination with tetanus toxoid, *Acta. Allergol.,* 26, 150, 1971a.

Hansson, H. and Möller, H., Intracutaneous test reactions to tuberculin containing Merthiolate® as a preservative, *Scand. J. Infect. Dis.,* 3, 169, 1971b.

Hapke, H. -J., Hinweise auf zentralnervöse Wirkungen geringer Quecksilberdosen bei Ratten, *Naunyn-Schmiedelberg Arch. Pharmacol.,* 226, 348, 1970.

Haq, I. U., Agrosan poisoning in Man, *Br. Med. J.,* 5335, 1579, 1963.

Harada, M., Neuropsychiatric disturbances due to organic mercury poisoning during the prenatal period, *Psychiatr. Neurol. Jap.,* 66, 429, 1964, (Japanese with English summary).

Harada, Y., Clinical investigations on Minamata disease. B. Infantile Minamata disease, in *Minamata Disease,* Kutsuna, M., Ed., study group of Minamata disease, Kumamoto University, Japan, 1968a, 73.

Harada, Y., Clinical investigations on Minamata disease. C. Congenital (or fetal) Minamata disease, in *Minamata Disease,* Kutsuna, M., Ed., study group of Minamata disease, Kumamoto University, Japan, 1968b, 93.

Harada, Y., Miyamoto, Y., Nonaka, I., Ohta, S., and Ninomiya, T., Electroencephalographic studies of Minamata disease in children, *Dev. Med. Child Neurol.,* 10, 257, 1968.

Harper, P., Idiosyncrasy to ammoniated mercury ointment in childhood, *J. Pediatr.,* 5, 794, 1934.

Hartung, J., Phenyl-Quecksilberacetat und phenyl-Quecksilberoleat in Textilien, *Berufsdermatosen,* 13, 116, 1965.

Häsänen, E. and Sjöblom, V., Mercury content of fish in Finland in 1967, *Suom. Kalatalous (Finlands fiskerier)*, 36, 5, 1968, (Finnish with English summary).

Hasselrot, T. B., Report on current field investigations concerning the mercury content in fish, bottom sediment, and water, *Rep. Inst. Freshwater Res.*, 48, 102, 1968.

Hasselrot, T. B., Field investigations concerning the occurrence of mercury in fish, water, bottom sediment and bottom organisms, Report to the Research Board of the National Environmental Protection Board, Stencils, 1969, (Swedish).

Hay, W. J., Rickards, A. G., McMenemey, W. H., and Cumings, J. N., Organic mercurial encephalopathy, *J. Neurol. Neurosurg. Psychiatry*, 26, 199, 1963.

Hayes, A. D. and Rothstein, A., The metabolism of inhaled mercury vapor in the rat studied by isotope techniques, *J. Pharmacol.*, 138, 1, 1962.

Hayes, H., Muir, J., and Whitby, L. M., A rapid method for the determination of mercury in urine, *Am. Occup. Hyg.*, 13, 235, 1970.

Heide, F., Lerz, H., and Böhm, G., Gehalt des Saale-wassers an Blei und Quecksilber, *Naturwissenschaften*, 44, 441, 1957.

Helminen, M., Karppanen, E., and Koivisto, I., Mercury content of the ringed seal of Lake Saimaa, *Suom. Eläinlääkärilehti*, 74, 87, 1968, (Finnish).

Henderson, C. and Shanks, W. E., Mercury concentrations in fish, in Mercury in the Western Environment, abstract of papers presented at the Environmental Health Center Workshop, Oregon State University, Corvallis, Oregon, February 25 to 26, 1971.

Hengst, W., von der Ohe, M., and Kienle, G., Untersuchungen über die Möglichkeit einer beschleunigten Elimination von ^{203}Hg-Ionen aus der Niere nach Schädelszintigraphie mit ^{203}Hg-Chlormerodrin, *J. Nucl. Med.*, 6, 378, 1967.

Henriksson, K., Karppanen, E., and Helminen, M., High residues of mercury in Finnish white-tailed eagles, *Ornis Fenn.*, 43, 38, 1966, (English with Finnish summary).

Henriksson, K., Karppanen, E., and Helminen, M., The amounts of mercury in seals from lakes and sea, *Nord. Hyg. T.*, 50, 54, 1969, (Swedish).

Hepp, P., Über Quecksilberäthylverbindungen und über das Verhältniss der Quecksilberäthyl – zur Quecksilbervergiftung, *Arch. Exp. Pathol. Pharmakol.*, 23, 91, 1887.

Herberg, W. W., Mercury poisoning in a dairy herd, *Vet. Med.*, 49, 401, 1954.

Herdman, R. C., Statement of Roger C. Herdman, M.D., Deputy Commissioner for Research and Development, New York State Dept. of Health, before the Subcommittee on the Environment of the Committee on Commerce of the U. S. Senate, May 20, 1971.

Herner, T., Poisoning from organic compounds of mercury, *Nord. Med.*, 26, 833, 1945, (Swedish with English summary).

Hesse, E., Versuche zur Therapie der Quecksilbervergiftung. II. Mitteilung: Die parenterale Hg-Vergiftung, *Arch. Exp. Pathol. Pharmakol.*, 177, 266, 1926.

Hill, H. A. O., Pratt, J. M., Ridsdale, S., Williams, F. R., and Williams, R. J. P., Kinetics of substitution of co-ordinated carbanions in cobalt (III) corrinoids, *Chem. Commun.*, 6, 341, 1970.

Hill, W. H., A report on two deaths from exposure to the fumes of a di-ethyl mercury, *Can. J. Public Health*, 34, 158, 1943.

Hiroshi, K., Hiroshi, K., Masao, C., and Masaharu, T., Diagnosis of mercury poisoning. III. Investigations concerning histopathology and histochemistry of the organic mercury poisoning observed around the Agano River, in Report on the cases of mercury poisoning in Niigata, Ministry of Health and Welfare, Tokyo, Stencils, 1967, 47, (Japanese with Swedish translation).

Hirschman, S. Z., Feingold, M., and Boylen, G., Mercury in house paint as a cause of acrodynia, *New Eng. J. Med.*, 269, 889, 1963.

Holzel, A. and James, T., Mercury and pink disease, *Lancet*, 1, 441, 1952.

Höök, O., Lundgren, K. -D., and Swensson, Å., On alkyl mercury poisoning: with a description of two cases, *Acta Med. Scand.*, 1954, 150, 131.

Hopmann, A., Acute poisoning from mercury vapour, *Bull. Hyg. (London)*, 3, 585, 1928.

Hoshino, O., Tanzawa, K., Hasegawa, Y., and Ukita, T., Differences in mercury content in the hairs of normal individuals depending on their home environment, *J. Hyg. Chem. Soc. Jap.*, 12, 90, 1966, (Japanese with English summary and English translation).

Hosohara, K., Mercury contents in water from deep seas, *J. Chem. Soc. Jap.*, 82, 1107, 1961, (Japanese).

Hosohara, K., Kozuma Hirotaka, Kawasaki Katsuhiko, and Tokumatsu, T., Total mercury content in sea water, *J. Chem. Soc. Jap.*, 82, 1479, 1961, (Japanese).

Howie, R. A. and Smith, H., Mercury in human tissue, *J. Forensic Sci. Soc.*, 7, 90, 1967.

Hughes, W. L., A physicochemical rationale for the biological activity of mercury and its compounds, *Ann. N.Y. Acad. Sci.*, 65, 454, 1957.

Hull, E. and Monte, L. A., Bichloride of mercury poisoning, a statistical study of 302 cases, *South. Med. J.*, 27, 918, 1934.

Hunter, D., Bomford, R. R., and Russell, D. S., Poisoning by methyl mercury compounds, *Quart. J. Med.*, 33, 193, 1940.

Hunter, D. and Russell, D. S., Focal cerebral and cerebellar atrophy in a human subject due to organic mercury compounds, *J. Neurol. Neurosurg. Psychiatry*, 17, 235, 1954.

Hutcheon, D. E., Diuretics, in *Drill's Pharmacology in Medicine*, DePalma, J. R., Ed., McGraw-Hill, New York, 1965, 659.

IUPAC, Analytical methods for use in occupational hygiene incorporating second replacement – addition issue, up-to-date to 1965, Butterworths, London, No. 30: *Mercury in Air*, No. U1: *Mercury in Urine*, 1969.

Imura, N., Sukegawa, E., Pan, S.-K., Nagao, K., Kim, J.-Y., Kwan, T., and Ukita, T., Chemical methylation of inorganic mercury with methylcobalamin, a vitamin B 12 analog, *Science*, 172, 1248, 1971.

Irukayama, K., Kai, F., Kondo, T., Ushikusa, S., Fujiki, M., and Tajima, S., Toxicity and metabolism of methyl mercury compounds in animals — especially in relation to Minamata disease, *Jap. J. Hyg.*, 20, 11, 1965, (Japanese with English summary and Swedish translation).

Itsuno, Y., Toxicologic studies on organic mercury compounds. Poisoning of rats with organic mercury compounds, in *Minamata Disease*, Kutsuna, M., Ed., study group of Minamata disease, Kumamoto University, Japan, 1968, 267.

Jackson, M. H., A new chemical contraceptive, *Lancet*, 2, 1030, 1938.

Jacobs, M. B. and Goldwater, L. J., Absorption and excretion of mercury in man. VIII. Mercury exposure from house paint — a controlled study on humans, *Arch. Environ. Health*, 11, 582, 1965.

Jacobs, M. B., Goldwater, L. J., and Gilbert, H., Ultramicrodetermination of mercury in blood, *Am. Ind. Hyg. Ass. J.*, 21, 276, 1961.

Jacobs, M. B., Ladd, A. C., and Goldwater, L. J., Absorption and excretion of mercury in man. III. Blood mercury in relation to duration of exposure, *Arch. Environ. Health*, 5, 634, 1963.

Jacobs, M. B., Ladd, A. C., and Goldwater, L. J., Absorption and excretion of mercury in man. VI. Significance of mercury in urine, *Arch. Environ. Health*, 9, 454, 1964.

Jacobs, M. B., Yamaguchi, S., Goldwater, L. J., and Gilbert, H., Determination of mercury in blood, *Am. Ind. Hyg. Ass. J.*, 21, 475, 1960.

Jakubowski, M., Piotrowski, J., and Trojanowska, B., Binding of mercury in the rat: studies using $^{203}HgCl_2$ and gel filtration, *Toxicol. Appl. Pharmacol.*, 16, 743, 1970.

Jalili, M. A. and Abbasi, A. H., Poisoning by ethyl mercury toluene sulphonanilide, *Br. J. Ind. Med.*, 18, 303, 1961.

James, C. H. and Webb, J. S., Sensitive mercury vapour meter for use in geochemical prospecting, *Bull. Inst. Mining Metallurg.*, 691, 633, 1964.

Janson, H., Vergiftung mit einem Modernen Saatbeizmittel, *Med. Welt.*, 3, 1767, 1929.

Järvenpää, T., Tillander, M., and Miettinen, J. K., Methylmercury: half-time of elimination in flounder, pike and eel, *Suom. Kemistilehti*, 43, 439, 1970b.

Jenne, E. A., Atmospheric and fluvial transport of mercury, in *Mercury in the Environment*, U.S. Geological Survey Professional Paper, No. 713, Washington, D.C., U.S.G.P.O., 1970.

Jenne, E. A., Mercury concentrations in waters of the United States, in *Mercury in the Western Environment*, abstract of papers presented at the Environmental Health Center Workshop, Oregon State University, Corvallis, Oregon, February 25 to 26, 1971, 40.

Jensen, S. and Jernelöv, A., Biological methylation of mercury in aquatic organisms, *Nature*, 223, 753, 1969.

Jernelöv, A., The metabolism and turnover of mercury in nature and what we can do to affect it, *Nord. Hyg. T.*, 50, 174, 1969a, (Swedish).

Jernelöv, A., Conversion of mercury compounds, in *Chemical Fallout. Current Research on Persistent Pesticides*, Miller, M. W. and Berg, G. G., Eds., Charles C Thomas, Springfield, Ill., 1969b, 68.

Jernelöv, A., Mercury levels in the fish from river Delånger, Swedish Water and Air Pollution Research Laboratory, Drottning Kristinas väg 47, S-114 28 Stockholm, Stencils, 1969c, (Swedish).

Jernelöv, A., A new biochemical pathway for the methylation of mercury and some ecological implications, in *Mercury, Mercurials and Mercaptans*, from the 4th Rochester Conference on Environmental Toxicity, Miller, M. W. and Clarkson, T. W., Eds., June 17 to 19, 1971, Charles C Thomas, Springfield, Ill., in press.

Jervis, R. E., Debrun, D., LePage, W., and Tiefenbach, B., Mercury residues in Canadian foods, fish, wildlife, summary of progress, National Health Grant Project No. 605-7-510, "Trace Mercury in Environmental Materials" for the period September, 1969 to May, 1970, Dept. of Chemical Engineering and Applied Chemistry, University of Toronto, Toronto 181, Canada.

Joensuu, O. I., Fossil fuels as a source of mercury pollution, *Science*, 172, 1027, 1971.

Johansson, F., Ryhage, R., and Westöö, G., Identification and determination of methylmercury compounds in fish using combination gas chromatograph-mass spectrometer, *Acta Chem. Scand.*, 24, 2349, 1970.

Johnels, A. G., Edelstam, C., Olsson, M., and Westermark, T., Mercury as an environmental poison in Sweden, *Fauna Flora*, 63, 172, 1968, (Swedish).

Johnels, A. G., Olsson, M., and Westermark, T., Mercury in fish: investigations on mercury levels in Swedish fish, *Vår Föda*, 7, 67, 1967, (Swedish).

Johnels, A. G., Olsson, M., and Westermark, T., *Esox Lucius* and some other organisms as indicators of mercury contamination in Swedish lakes and rivers, *Bull. Off. Int. Epizoot.*, 69, 1439, 1968.

Johnels, A. G. and Westermark, T., Mercury and old feathers, *Ronden*, 22, 255, 1968, (Swedish).

Johnels, A. G. and Westermark, T., Mercury contamination of the environment in Sweden, in *Chemical Fallout. Current Research on Persistent Pesticides*, Miller, M. W. and Berg, G. G., Eds., Charles C Thomas, Springfield, Ill., 1969, 221.

Johnels, A. G., Westermark, T., Berg, W., Persson, P. I., and Sjöstrand, B., Pike (*Esox Lucius L.*) and some other aquatic organisms in Sweden as indicators of mercury contamination in the environment, *Oikos*, 18, 323, 1967, (English with Russian summary).

Johnson, J. E. and Johnson, J. A., A new value for the long component of the effective half-retention time of ^{203}Hg in the human, *Health Phys.*, 14, 265, 1968.

Jonasson, I. R., Mercury in the natural environment. A review of recent work, Geological Survey of Canada, paper no. 70-57, 1970.

Jonasson, I. R. and Boyle, R. W., Geochemistry of mercury, in Mercury in Man's Environment, Proc. Roy. Soc. Can., Symp., February 15 to 16, 1971, 5.

Jonek, J., Histochemische Untersuchungen über das Verhalten einiger Enzyme im Herzmuskel nach experimenteller Vergiftung mit Quecksilberdämpfen, *Int. Arch. Gewerbepath.*, 21, 1, 1964.

Jonek, J. and Grzybek, H., Untersuchungen über das Verhalten der gamma-Glutamyl-Transpeptidase in einigen Organen nach chronischer Vergiftung mit Quecksilberdämpfen, *Int. Arch. Gewerbepath.*, 20, 572, 1964.

Jonek, J. and Kośmider, S., Verhalten der Atmungsfermente und der alkalischen Phosphatase in der Leber bei experimenteller Vergiftung mit Quecksilberdämpfen, *Int. Arch. Gewerbepath.*, 20, 496, 1964.

Jonek, J., Pacholek, A., and Jez, W., Histochemische Untersuchungen über das Verhalten der Adenosintriphosphatase, 5-Nucleotidase, sauren Phosphatase, sauren Desoxyribonuclease II u. der unspezifischen Esterasen in der Leber bei experimenteller Vergiftung mit Quecksilberdämpfen, *Int. Arch. Gewerbepath.*, 20, 562, 1964.

Jones, H. R., Mercury pollution control, Pollution Control Review No. 1, Noyes Data Corporation, Park Ridge, N. J., 1971.

Jordi, A., Quecksilbervergiftungen bei munitionsarbeitern, *Schweiz. Med. Wochenschr.*, 77, 621, 1947.

Joselow, M. M. and Goldwater, L. J., Absorption and excretion of mercury in man. XII. Relationship between urinary mercury and proteinuria, *Arch. Environ. Health*, 15, 155, 1967.

Joselow, M. M., Goldwater, L. J., and Weinberg, S. B., Absorption and excretion of mercury in man. XI. Mercury content of "normal" human tissues, *Arch. Environ. Health*, 15, 64, 1967.

Joselow, M. M., Ruiz, R., and Goldwater, L. J., Absorption and excretion of mercury in man. XIV. Salivary excretion of mercury and its relationship to blood and urine mercury, *Arch. Environ. Health*, 17, 35, 1968.

Juliusberg, F., Experimentelle Untersuchungen über die Quecksilberresorption bei der Schmierkur, *Arch. Derm. Syph.*, 56, 5, 1901.

Kai, F., Metabolism of mercury compounds in animals administered with the shellfish from Minamata bay and methylmercuric compounds, *J. Kumamoto Med. Soc.*, 37, 673, 1963, (Japanese with English summary).

Kantarjian, A. D., A syndrome clinically resembling amyotrophic lateral sclerosis following chronic mercurialsim, *Neurology*, 11, 639, 1961.

Katsunuma, H., Suzuki, T., Nishi, S., and Kashima, T., Four cases of occupational organic mercury poisoning, *Rep. Inst. Sci. Labour*, 61, 33, 1963.

Kawasaka, I., Tanabe, H., Hosomi, Y., Kondo, T., Takeda, M., Kanoda, K., Tatsuno, T., Amano, T., Matsui, K., Kobata, T., Ukita, C., Hoshino, O., Tanzawa, K., Ueda, K., Nishimura, M., Aoki, H., Kondo, T., Nakazawa, Y., and Ishikura, T., State of pollution due to mercury compounds, in *Report on the cases of mercury poisoning in Niigata*, Part II. Ministry of Health and Welfare, Tokyo, Stencils, 1967, 75, (Japanese with Swedish translation).

Kazantzis, G., The Measurement of tremor in the early diagnosis of chronic mercurialism. Working paper for Int. Symp. on MAC values of mercury, Stockholm, 1968.

Kazantzis, G., Schiller, F. R., Asscher, A. W., and Drew, R. G., Albuminuria and the nephrotic syndrome following exposure to mercury and its compounds, *Quart. J. Med.*, 31, 403, 1962.

Keckes, S. and Miettinen, J. K., Mercury as marine pollutant, FAO Technical Conference on Marine Pollution and Its Effects on Living Resources and Fishing, Rome, December 9 to 18, 1970, 1.

Keith, J. A. and Gruchy, I. M., Mercury residues in Canadian wildlife, in Mercury in Man's Environment, Proc. Roy. Soc. Can., Symp., February 15 to 16, 1971, 91.

Kesic, B. and Haeusler, V., Hematological investigation on workers exposed to mercury vapor, *Ind. Med. Surg.*, 20, 485, 1951.

Kessler, R. H., Lozano, R., and Pitts, R. F., Studies on structure diuretic activity relationships of organic compounds of mercury, *J. Clin. Invest.*, 36, 656, 1957.

Khera, K. S., The effect of methyl mercury on male fertility: a comparative study in rats and mice, presented at 2nd Ann. Meet. Environmental Mutagen Soc., Washington, D.C., March 21 to 24, 1971.

Kim, C. K. and Silverman, J., Determination of mercury in wheat and tobacco leaf by neutron activation analysis using mercury-197 and a simple exchange separation, *Anal. Chem.*, 37, 1616, 1965.

Kimura, Y. and Miller, U. L., Vapor phase separation of methyl or ethyl mercury compounds and metallic mercury, *Anal. Chem.*, 32, 420, 1960.

Kimura, Y. and Miller, V. L., The degradation of organomercury fungicides in soil, *J. Agric. Food Chem.*, 12, 253, 1964.

King, G. W., Acute pneumonitis due to accidental exposure to mercury vapor, *Ariz. Med.*, 11, 335, 1954.

Kitagawa, T., Clinical investigations on Minamata disease. E. Rehabilitation in Minamata disease, in *Minamata Disease*, Kutsuna, M. Ed., study group of Minamata Disease, Kumamoto University, Japan, 1968, 127.

Kitamura, S., Determination on mercury content in bodies of inhabitants, cats, fishes and shells in Minamata district and the mud of Minamata bay, in *Minamata Disease*, Kutsuna, M., Ed., study group of Minamatata disease, Kumamoto University, Japan, 1968, 257.

Kitamura, S., Tsukamoto, T., Hayakawa, K., Sumino, K. and Shibata, T., Application of gas chromatography to the analysis of alkylmercury compounds, *Med. Biol.*, 72, 274, 1966, (Japanese).

Kiwimäe, A., Swensson, Å., and Ulfvarson, U., Long-term experiments on the effect of feeding white leghorn hens wheat treated with different mercury compounds, *Stud. Laboris Salutis*, 7, 1, 1970.

Kiwimäe, A., Swensson, Å., Ulfvarson, U., and Westöö, G., Methyl mercury compounds in eggs from hens after oral administration of mercury compounds, *Agr. Food Chem.*, 17, 1014, 1969.

Kloss, G., Strahlenbelastung durch Hg^{203} — markierte Testpräparate zur Nierendiagnostik, *J. Nucl. Med.*, 3, 82, 1962.

Koelsch, F., Gesundheitsschädigungen durch organische Quecksilberverbindungen, *Int. Arch. Gewerbepath.*, 8, 113, 1937.

Koeman, J. H., Vink, J. A. J., and Goeij, J. J. M., Causes of mortality in birds of prey and owls in the Netherlands in the winter of 1968-1969, *Ardea*, 57, 67, 1969, (English with Dutch summary).

Kolmer, J. A. and Lucke, B., A study of the histologic changes produced experimentally in rabbits by mercurial compounds, *Arch. Derm. Syph.*, 3, 531, 1921.

Korst, D. R., Nixon, J. C., Boblitt, D. E., and Quirk, J., Studies on the selective splenic sequestration of erythrocytes labeled with radioactive mercurihydroxypropane (MHP), *J. Lab. Clin. Invest.*, 66, 788, 1965.

Kosmider, S., Mineralhaushaltstörungen im Blutserum und in den Geweben bei experimenteller Vergiftung mit Quecksilberdämpfen, *Int. Arch. Gewerbepath.*, 21, 60, 1964.

Kosmider, S., Untersuchungen über den toxischen Wirkungsmechanismus des metallischen Quecksilbers, *Int. Arch. Gewerbepath.*, 21, 282, 1965.

Kosmider, S., Zusammenhänge zwischen Schwermetallvergiftungen und Atheromatose bzw. Herz- und Kreislaufstörungen, *Z. Gesamte. Hyg.*, 14, 355, 1968.

Kosmider, S., Kossmann, S., and Zajaczkowski, S., Das Verhalten der Eiweisskörper der Muco- und lipoproteide sowie der alkalischen Phosphatase im Blutserum bei experimenteller akuter Vergiftung mit Quecksilbersalzen, *Int. Arch. Gewerbepath.*, 20, 171, 1963.

Kosmider, S., Wocka-Marek, T., and Kujawska, A., Beurteilung der Brauchbarkeit biochemischer und enzymatischer Teste in der Frühdiagnostik der Quecksilbervergiftung, *Int. Arch. Gewerbepath.*, 25, 232, 1969.

Kostoff, D., Effect of the fungicide "Granosan®" on atypical growth and chromosome doubling in plants, *Nature*, 144, 334, 1939.

Kostoff, D., Atypical growth, abnormal mitosis and polyploidy induced by ethyl-mercury-chloride, *Phytopathol. Z.*, 23, 90, 1940.

Kournossov, V. N., Some evidence to form a basis for a threshold value for mercury in air, *Gig. Sanit.*, 1, 7, 1962, (Russian, with Swedish translation); and/or Kournossov, V. N., Basic data for the hygienic determination of limits of allowable mercury vapor concentrations, in *Atmospheric Pollutants*, Ryazanov, V. A., Ed., Book 5, Moscow, 1962, 39. (Translated by B. Levine, U.S. Dept. of Commerce, Washington, D.C.)

Krylova, A. N., Naumov, V. M., Rubtsov, S. F., Yakovleva, V. I., Distribution of mercury in the internal organs of rats poisoned with ethylmercuric chloride, *Chem. Abstr.*, (110221s) 72, 286, 1970.

Kulczycka, B., Resorption of metallic mercury by the conjunctiva, *Nature*, 206, 943, 1965.

Kurland, L. T., Faro, S. N., and Siedler, H., Minamata Disease: The outbreak of a neurologic disorder in Minamata, Japan, and its relationship to the ingestion of seafood contaminated by mercuric compounds, *World Neurol.*, 1, 370, 1960.

Kussmaul, A., *Untersuchungen über den konstitutionellen Mercurialismus und sein Verhältniss zur konstitutionellen Syphilis*, Stahel, Würzburg, 1861.

Kuwahara, S., Toxicity of mercurials, especially methyl mercury compound in chick embryo. Report I. Toxicity of mercurials injected in fertilized chick egg, *J. Kumamoto Med. Soc.*, 44, 81, 1970a, (Japanese with English summary).

Kuwahara, S., Toxicity of mercurials, especially methyl mercury compound, in chick embryo. Report II. Transfer of mercurials, which were administered to hens subcutaneously, into the eggs and toxicity of mercurials in chick-embryo, *J. Kumamoto Med. Soc.*, 44, 90, 1970b, (Japanese with English summary).

Ladd, A. C., Goldwater, L. J., and Jacobs, M. B., Absorption and excretion of mercury in man. II. Urinary mercury in relation to duration of exposure, *Arch. Environ. Health*, 6, 480, 1963.

Ladd, A. C., Goldwater, L. J., and Jacobs, M. B., Absorption and excretion of mercury in man. V. Toxicity of phenylmercurials, *Arch. Environ. Health*, 9, 43, 1964.

Ladd, A. C., Zuskin, E., Valic, F., Almonte, J. B., and Gonzales, T. V., Absorption and excretion of mercury in miners, *J. Occup. Med.*, 8, 127, 1966.

Landner, L., Biochemical model for the biological methylation of mercury suggested from methylation studies in vivo with *Neurospora crassa*, *Nature*, 230, 452, 1971.

Lane, C. G., Cutaneous disturbances caused by therapeutic measures, *J. Omaha. Mid-West Clin. Soc.*, 6, 45, 1945.

Lapp, H. and Schafé, K., Morphologische, Histochemische und Speicherungs-Untersuchungen über den Verlauf der Sublimatnephrose bei der Ratte, *Beitr. Pathol. Anat.*, 123, 77, 1960.

Lauckhart, J. B., Mercury in Washington wildlife, in Mercury in the Western Environment, abstract of papers presented at the Environmental Health Center Workshop, Oregon State University, Corvallis, Oregon, February 25 to 26, 1971.

Laug, E. P. and Kunze, F. M., The absorption of phenylmercuric acetate from the vaginal tract of the rat, *J. Pharmacol. Exp. Ther.*, 95, 460, 1961.

Laug, E. P., Vos, E. A., Kunze, F. M., and Umberger, E. J., A study of certain factors governing the penetration of mercury through the skin of the rat and the rabbit, *J. Pharmacol. Exp. Ther.*, 89, 52, 1947.

Lecomte, J. and Bacq, Z. M., Action Vésicante des Dérivés Organiques Mercuriels, *C. R. Soc. Biol.*, 143, 1296, 1949.

Ledergerber, E., Einiges zu den Todesfällen und über die zum Tode führenden Erkrankungen der Arbeiter der Zündkapselfabrikation, *Schweiz. Med. Wochenschr.*, 30, 263, 1949.

Lee, D. F. and Roughan, J. A., Pesticide residues in foodstuffs in Great Britain, XIV. Mercury residues in potatoes, *Pestic. Sci.*, 1, 150, 1970.

Leff, W. A. and Nussbaum, H. E., Renal tolerance to long-term administration of organomercurial diuretics, *Ann. N.Y. Acad. Sci.*, 65, 520, 1957.

Lehman, A. J., Chemicals in foods: a report to the Association of Food and Drug Officials on current developments. II. Pesticides, *Quart. Bull. Ass. Food Drug. Officials U.S.*, 15, 122, 1951.

Lehotzy, K. and Bordas, S., Study on the subacute neurotoxic effect of methoxy-ethyl mercury chloride (MEMC) in rats, *Med. Lavoro*, 59, 241, 1968.

Leites, R. G., The limit of allowable concentration of mercury vapors in the air of inhabited areas, in *Limits of Allowable Concentrations of Atmospheric Pollutants*, Ryazanov, V. A., Ed., Book 1, Translated by B. S. Levine, U.S. Department of Commerce, Washington, D.C., 1952, 74.

Levan, A., Cytological reactions induced by inorganic salt solutions, *Nature*, 156, 751, 1945.

Levan, A., Cytogenetic effects of hexyl mercury bromide in the allium test, *J. Indian Bot. Soc.*, Golden Jubilee Vol. 50.

Levan, A. ánd Östergren, G., The mechanism of c-mitotic action, *Hereditas*, 29, 381, 1943.

Levine, B., Use of phenylmercuric nitrate in tinea and yeast infections of the skin, *J.A.M.A.*, 101, 2109, 1933.

Lidums, V. and Ulfvarson, U., Mercury analysis in biologic material by direct combustion in oxygen and photometric determination of the mercury vapour, *Acta Chem. Scand.*, 22, 2150, 1968a.

Lidums, V. and Ulfvarson, U., Preparation of average sample solution of heterogeneous organic materials for determination of microquantities of mercury using purified sodium hydroxide, *Acta Chem. Scand.*, 22, 2379, 1968b.

Lieb, C. C. and Goodwin, G. M., The excretion of mercury by the gastric mucous membrane, *J.A.M.A.*, 64, 2041, 1915.

Lihnell, D., Seed treatment in the Nordic countries 1968, *Nord. Hyg. T.*, 50, 147, 1969, (Swedish).

Lihnell, D. and Stenmark, A., Mercury in small rodents, *Vaxtskyddsnotiser*, 31, 36, 1967, (Swedish).

Lihnell, D. and Stenmark, A., On the occurrence of mercury in small rodents, *Statens Växtskyddsanstalt Meddelanden* (National Institute for Plant Protection Contributions), 13 (110), 361, 1969, (Stockholm), (Swedish with English summary).

Linch, A. L., Stalzer, R. F., and Lefferts, D. T., Methyl and ethyl compounds-recovery from air and analysis, *Am. Ind. Hyg. Ass. J.*, 29, 77, 1968.

Lindstedt, G., A rapid method for the determination of mercury in urine, *Analyst*, 95, 264, 1970.

Lindstedt, G. and Skare, I., Microdetermination of mercury in biologic samples. II. An apparatus for rapid atomic determination of Hg in digested samples. *Analyst*, 96, 223, 1971.

Lindström, O., Rapid microdetermination of mercury by spectrophotometric flame combustion, *Anal. Chem.*, 31, 461, 1959.

Linfield, W. M., Sherrill, J. C., Casely, R. E., Noel, D. R., and Davis, G. A., Studies in the development of antibacterial surfactants: I. Institutional use of antibacterial fabric softeners, *J. Am. Oil Chem. Soc.*, 37, 248, 1960.

Ling, C., Sensitive simple mercury photometer using mercury resonance lamp as a monochromatic source, *Anal. Chem.*, 39, 798, 1967.

Lippman, R. W., Finkle, R. D., and Gillette, D.D., Effect of proteinuria on localization of radiomercury in rat kidney, *Proc. Soc. Exp. Biol. Med.*, 77, 68, 1951.

Ljunggren, K., Sjöstrand, B., Hagman, D., and Westermark, T., Recent experience in activation analysis, *Nord. Hyg. T.*, 50, 75, 1969, (Swedish).

Ljunggren, K., Sjöstrand, B., Johnels, A. G., Olsson, M., Otterlind, G., and Westermark, T., Activation analysis of mercury and other environmental pollutants in water and aquatic egosystems, Proc. Symp. "Nuclear Techniques in Environmental Pollution," Salzburg, October 26 to 30, 1970, IAEA-SM-142a/22, Vienna, 1971, 373.

Locket, S. and Nazroo, I.A., Eye changes following exposure to metallic mercury, *Lancet*, 1, 528, 1952.

Löfroth, G., Methylmercury. A review of health hazards and side effects associated with the emission of mercury compounds into natural systems, *Ecol. Res. Comm. Bull.*, 4, 1, 1969, Stencils.

Loken, M. K., Bugby, R. D., and Lowman, J. T., Evaluation of splenic function using [127]Hg-mercurihydroxypropane, *J. Nucl. Med.*, 10, 615, 1969.

Lomholt, S., Quecksilber. Theoretisches, Chemisches und Experimentelles, in Jadassohn, J., Ed., *Handbuch der Haut- und Geschlechtskrankheiten*, Springer Verlag, Berlin, 1928, Vol. 18, 1.

Loosmore, R. M., Harding, J. D. J., and Lewis, G., Mercury poisoning in pigs, *Vet. Rec.*, 81, 268, 1967.

Lundgren, K.-D. and Swensson, Å., On alkyl mercuric compounds causing occupational illness, *Nord. Hyg. T.*, 29, 1, 1948, (Swedish with English summary).

Lundgren, K.-D. and Swensson, Å., Occupational poisoning by alkyl mercury compounds, *J. Ind. Hyg.*, 31, 190, 1949.

Lundgren, K.-D. and Swensson, Å., Hygienic problems concerning the industrial use of phenyl mercury compounds, *Nord. Hyg. T.*, 31, 207, 1950, (Swedish with English summary).

Lundgren, K.-D. and Swensson, Å., A survey of results of investigations on some organic mercury compounds used as fungicides, *Am. Ind. Hyg. Ass. J.*, 21, 308, 1960a.

Lundgren, K.-D. and Swensson, Å., On the control and protection of workers handling organic mercury compounds, *Ark. Hig. Rad. Toksikol.*, 11, 27, 1960b, (English with Yugoslavian summary).

Lundgren, K.-D., Swensson, Å., and Ulfvarson, U., Studies in humans on the distribution of mercury in the blood and the excretion in urine after exposure to different mercury compounds, *Scand. J. Clin. Lab. Invest.*, 20, 164, 1967.

McAfee J. G. and Wagner, H. N., Visualization of renal parenchyma by scintiscanning with Hg^{203} Neohydrin®, *Radiology*, 75, 820, 1960.

McCarthy, J. H., Meuschke, J. L., Ficklin, W. H., and Learned, R. E., Mercury in the atmosphere, in *Mercury in the Environment*, U.S. Geological Survey professional paper, No. 713 Washington, D.C., U. S. G. P. O., 1970, 37.

McCord, C. P., Meek, S. F., and Neal, T. A., Phenyl mercuric oleate skin irritant properties, *J. Ind. Hyg.*, 23, 466, 1941.

McDuffie, B., Tentative findings on mercury levels in selected persons from the Binghampton area, State University of New York, *Binghampton News*, Office of University and Community Relations, January 13, 1971.

McDuffie, B., Discussion, in *Mercury, Mercurials and Mercaptans*, Miller, M. W. and Clarkson, T. W. Eds., from the 4th Rochester Conference on Environmental Toxicity, June 17 to 19, 1971, Charles C Thomas, Springfield, Ill., in press.

McEntee, K., Mercurial poisoning in swine, *Cornell Vet.*, 40, 143, 1950.

Macfarlane, E. W. E., Somatic mutations produced by organic mercurials in flowering plants, *Genetics*, 35, 122, 1950.

Macfarlane, E. W. E., Effects of water source on toxicity of mercurial poisons. II: Reactions of allium root tip cell on indicators of penetration, *Growth*, 15, 241, 1951.

Macfarlane, E. W. E., Cytological conditions in root tip meristem after gross antagonism of phenylmercuric poisoning, *Exp. Cell Res.*, 5, 375, 1953.

Macfarlane, E. W. E. and Messing, Sr. A. M., Shoot chimeras after exposure to mercurial compounds, *Bot. Gaz.*, 115, 66, 1953.

Macfarlane, E. W. E. and Schmoch, N. G., The colchicine and colchicine-like reaction as possible response to enzymic poisoning, *Science*, 108, 712, 1948.

Magos, L., Mercury-blood interaction and mercury uptake by the brain after vapor exposure, *Environ. Res.*, 1, 323, 1967.

Magos, L., Uptake of mercury by the brain, *Br. J. Ind. Med.*, 25, 315, 1968.

Magos, L. and Cernik, A. A., A rapid method for estimating mercury in undigested biological samples, *Br. J. Ind. Med.*, 26, 144, 1969.

Magos, L., Tuffery, A. A., and Clarkson, T. W., Volatilization of mercury by bacteria, *Br. J. Ind. Med.*, 21, 294, 1964.

Malaiyandi, M. and Barrette, J. P., Determination of submicro quantities of mercury in biological materials, *Anal. Lett.*, 3, 579, 1970.

Mambourg, A. M. and Raynaud, C., Etude a l'Aide d'Isotope Radioactifs du Mecanisme de l'Excretion Urinaire du Mercure chez le Lapin, *Rev. Fr. Etud. Clin. Biol.*, 10, 414, 1965.

Maruyama, Y., Komiya, K., and Manri, T., Determination of copper, arsenic and mercury in cigarettes by neutron activation analysis, *Radioisotopes*, 19, 250, 1970.

Massey, T. H. and Fang, S. C., A comparative study of the subcellular binding of phenylmercuric acetate and mercuric acetate in rat liver and kidney slices, *Toxicol. Appl. Pharmacol.*, 12, 7, 1968.

Massmann, W., Beobachtungen beim Umgang mit Phenylquecksilber-brenzkatechin, *Zbl. Arbeitsmed.*, 7, 9, 1957.

Mastromatteo, E. and Sutherland, R. B., Mercury in humans in the Great Lakes region, in *Environmental Mercury Contamination*, Hartung, R. and Dinman, B. D., Eds., Ann Arbor Science Publishers, Ann Arbor, Michigan in press.

Mathews, K. P. and Pan, P. M., Immediate type hypersensitivity to phenylmercuric compounds, *Am. J. Med.*, 44, 310, 1968.

Matsuda, S., Irukayama, K., Ueda, K., Kitano, H., Kitamura, M., Kosaka, T., Noriki, H., and Hirayama, T., III, Investigation on the epidemiology, of the mercury poisoning, in Report on the cases of mercury poisoning in Niigata, Ministry of Health and Welfare, Tokyo, Stencils, 1967, 266, (Japanese with Swedish translation).

Matsumoto, H., Koya, G., and Takeuchi, T., Fetal Minamata disease: A neuropathological study of two cases of intrauterine intoxication by methyl mercury compound, *J. Neuropathol. Exp. Neurol.*, 24, 563, 1965.

Matsumoto, H., Suzuki, A., Morita, C., Nakamura, K., and Saeki, S., Preventive effect of penicillamine on the brain defect of fetal rat poisoned transplacentally with methyl mercury, *Life Sci.*, 6, 2321, 1967.

Matsumura, F., Gotoh, Y., and Mallory Boush, G., Phenylmercuric acetate: metabolic conversion by microorganisms, *Science*, 173, 49, 1971.

Matsuoka, S., Sasaki, Y., Kaneko., and Tsuneyama, H., Diagnosis of mercury poisoning. II. Clinical observations 3. Studies on porphyrin metabolism, in Report on the cases of mercury poisoning in Niigata, Ministry of Health and Welfare, Tokyo, Stencils, 1967, 36, (Japanese with Swedish translation).

Matthes, F. T., Kirchner, R., Yow, M., and Brennan, J. C., Acute poisoning associated with inhalation of mercury vapor, *Pediatrics*, 22, 675, 1958.

Mazia, D., The organization of the mitotic apparatus, Symp., *Soc. Exp. Biol.*, 9, 335, 1955.

Mazia, D., Mitosis and the physiology of cell division, in *The Cell*, Bracket, D. and Mirsky, A. E. Eds., Academic Press, New York, 1961, Vol. 3, 77.

Maximum allowable concentrations of mercury compounds (MAC), *Arch. Environ. Health*, 19, 891, 1969.

Medved, L. I., Spynu, E. I., and Kagan, I. S., The method of conditioned reflexes in toxicology and its application for determining the toxicity of small quantities of pesticides, *Residue Rev.*, 6, 42, 1964.

Menten, M. L., Pathological lesions produced in the kidney by small doses of mercuric chloride, *J. Med. Res.*, 43, 315, 1922.

Mercury Contamination, in Man and His Environment, FAO-IAEA-ILO-WHO, to be published in IAEA Technical Report Series.

Mercury in Man's Environment, Proc. Symp., February 15 to 16, 1971, Ottawa, Canada, The Royal Society of Canada, Ottawa, 1971.

Mercury in the Western Environment, abstracts of papers presented at the Environmental Health Center Workshop, Oregon State University, Corvallis, Oregon, February 25 to 26, 1971.

Merewether, E. R. A., Industrial health, *Annual Report of the Chief Inspector of Factories for the year 1945*, H. M. S. O., London, 1946.

Meshkov, N. V., Glezer, I. I., and Panov, P. V., Effect of diethylmercury on the kidneys, *Arkh. Patol.*, 25, 519, 1963.

Miettinen, J. K., Absorption and elimination of dietary mercury (Hg^{2+}) and methylmercury in man, in *Mercury, Mercurials and Mercaptans*, Miller, M. W. and Clarkson, T. W. Eds. from the 4th Rochester Conference on Environmental Toxicity, June 17 to 19, 1971, Charles C Thomas, Springfield, Ill., in press.

Miettinen, J. K., Heyraud, M., and Keckes, S., Mercury as hydrospheric pollutant. II. Biologic half-time of methyl mercury in four Mediterranean species: a fish, a crab, and two molluscs, FAO FIR: MP:70:E-90. Rome, November 5, 1970.

Miettinen, J. K., Rahola, T., Hattula, T., Rissanen, K., and Tillander, M., Retention and excretion of ^{203}Hg-labelled methylmercury in man after oral administration of $CH_3^{203}Hg$ biologically incorporated into fish muscle protein — preliminary results, Fifth R I S (Radioactivity in Scandinavia) Symp., Department of Radiochemistry, University of Helsinki, Stencils, 1969b.

Miettinen, J. K., Rahola, T., Hattula, T., Rissanen, K., and Tillander, M., Elimination of ^{203}Hg-methyl mercury in man, *Ann. Clin. Res.*, 3, 116, 1971.

Miettinen, J. K., Tillander, M., Rissanen, K., Miettinen, V., and Ohmomo, Y., Distribution and excretion rate of phenyl- and methyl mercury nitrate in fish, mussels, molluscs and crayfish, in Proc. 9th Jap. Conf. Radioisotopes, B/II-17 pp. 474-478, Jap. Ind. Forum, Inc., Tokyo, 1969a.

Miettinen, V., Blankenstein, E., Rissanen, K., Tillander, M., Miettinen, J. K., and Valtonen, M., Preliminary study on the distribution and effects of two chemical forms of methylmercury in pike and rainbow trout, FAO FIR, MP/70/E-91, Rome, November 5, 1970.

Miller, H. C., Green, J. P., and Levine, J., The biologic fate of Neohydrin®-Hg^{203} in dogs and humans, *J. Nucl. Med. Abstr.*, 3, 208, 1962.

Miller, M. W. and Berg, G. G., *Chemical Fallout, Current Research on Persistent Pesticides*, Charles C Thomas, Springfield, Ill., 1969.

Miller, M. W. and Clarkson, T. W., *Mercury, Mercurials and Mercaptans*, Proc. 4th Rochester Interntl. Conf. Environmental Toxicity, Charles C Thomas, Springfield, Ill., in press.

Miller, V. L. and Csonka, E., Mercury retention in two strains of mice, *Toxicol. Appl. Pharmacol.*, 13, 207, 1968.

Miller, V. L., Gould, C. J., and Polley, D., Some chemical properties of phenylmercury acetate in relation to fungicidal performance, *Phytopathology*, 47, 722, 1957.

Miller, V. L., Klavano, P. A., and Csonka, E., Absorption, distribution and excretion of phenylmercuric acetate, *Toxicol. Appl. Pharmacol.*, 2, 344, 1960.

Miller, V. L., Klavano, P. A., Jerstad, A. C., and Csonka, E., Absorption, distribution and excretion of ethylmercuric chloride, *Toxicol. Appl. Pharmacol.*, 3, 459, 1961.

Miller, V. L., Lillis, D., and Csonka, E., Microestimation of intact phenylmercury compounds in animal tissue, *Anal. Chem.*, 30, 1705, 1958.

Milne, J., Christophers, A., and de Silva, P., Acute mercurial pneumonitis, *Br. J. Ind. Med.*, 27, 334, 1970.

Milnor, J. P., Binding of the mercury of an organic mercurial diuretic by plasma proteins, *Proc. Soc. Exp. Biol. Med.*, 75, 63, 1950.

Minamata Report, Kutsuna, M., Ed., Minamata disease, Study group of Minamata disease, Kumamoto University, Japan, 1968.

Minerals Yearbook 1970, Metals, Minerals and Fuels, Bureau of Mines, Washington, D. C., in press.

Miyakawa, T. and Deshimaru, M., Electron microscopical study of experimentally induced poisoning due to organic mercury compound. Mechanism of development of the morbid change, *Acta Neuropathol.* (Berlin), 14, 126, 1969.

Miyakawa, T., Deshimaru, M., Sumiyoshi, S., Teraoka, A., Udo, N., and Miyakawa, K., Electron microscopic studies on experimental Minamata disease — the changes caused by a long term administration of a small amount of organic mercury compound, *Psychiatr. Neurol. Jap.*, 71, 757, 1969, (Japanese with English summary).

Miyakawa, T., Deshimaru, M., Sumiyoshi, S., Teraoka, A., Udo, N., Hatteri, E., and Tatetsu, S., Experimental organic mercury poisoning — pathological changes in peripheral nerves, *Acta Neuropathol.*, 15, 45, 1970.

Miyakawa, T., Deshimaru, M., Sumiyoshi, S., Teraoka, A., and Tatetsu, S., Experimental organic mercury poisoning — regeneration of peripheral nerves, *Acta Neuropathol.*, 17, 6, 1971a.

Miyakawa, T., Deshimaru, M., Sumiyoshi, S., Teraoka, A., and Tatetsu, S., Experimental organic mercury poisoning. Pathological changes in muscles, *Acta Neuropathol.*, 17, 80, 1971b.

Miyama, T., Murakami, M., Suzuki, T., and Katsunuma, H., Retention of mercury in the brain of rabbit after intravenous or subcutaneous injection of sublimate, *Ind. Health*, 6, 107, 1968.

Mnatsakonov, T. S., Mamikonyan, R. S., Govorkyan, G. G., and Nazaretyan, K. L., Electrolyte and electrocardiographic changes in a chronic Granosan poisoning, *Gig. Tr. Prof. Zabol.*, 12, 39, 1968, (Russian with English summary).

Moffitt, A. E., Jr. and Kupel, R. E., A rapid method employing impregnated charcoal and atomic absorption spectrophotometry for the determination of mercury in atmospheric, biological, and aquatic samples, *Atom. Absorp. Newsletter*, 9, 113, 1970.

Molyneux, M. K., Observations on the excretion rate and concentration of mercury in urine, *Ann. Occup. Hyg.*, 9, 95, 1966.

Morikawa, N., Pathological studies on organic mercury poisoning. I. Experimental organic mercury poisoning in cats and its relation to the causative agent of Minamata disease, *Kumamoto Med. J.*, 14, 71, 1961a.

Morikawa, N., Pathological studies on organic mercury poisoning. II. Experimental production of congenital cerebellar atrophy by bis-ethylmercuric sulfide in cats, *Kumamoto Med. J.*, 14, 87, 1961b.

Moriyama, H., A study on the congenital Minamata disease, *Kumamoto Igakki Zasshi*, 41, 506, 1968, (Japanese with English summary, and Swedish translation).

Morris, G. E., Dermatose from phenylmercuric salts, *Arch. Environ. Health*, 1, 53, 1960.

Morrow, P. E., Gibb, F. R., and Johnson, L., Clearance of insoluble dust from the lower respiratory tract, *Health Phys.*, 10, 543, 1964.

Morsy, S. M. and El-Assaly, F. M., Body elimination rates of ^{134}Cs, ^{60}Co and ^{203}Pb, Hg, *Health Phys.*, 19, 769, 1970.

Mudge, G. H., Diuretics and other agents employed in the mobilization of edema fluid, in *The Pharmacological Basis of Therapeutics*, Goodman, L. S. and Gilman, A., Eds., 4th ed., New York, MacMillan, 1970, 839.

Mudge, G. H. and Weiner, I. M., The mechanism of action of mercurial and xanthine diuretics, *Ann. N. Y. Acad. Sci.*, 71, 344, 1958.

Murakami, U., Embryotoxic effect of some organic mercury compounds, *J. Jap. Med. Ass.*, 61, 1059, 1969, (Japanese, English translation).

Murakami, U., Embryo-fetotoxic effect of some organic mercury compounds, *Ann. Rep. Res. Inst. Environ. Med. Nagoya Univ.*, 18, 33, 1971.

Murakami, U., Kameyama, Y., and Kato, T., Effects of a vaginally applied contraceptive with phenylmercuric acetate upon developing embryos and their mother animals, *Ann. Rep. Res. Inst. Environ. Med. Nagoya Univ.*, 88, 1956.

Murone, I., Shirakawa, K., Sato, T., and Yamada, K., Diagnosis of mercury poisoning. II. Clinical observations. 2. Electromyographic examinations, in Report on the cases of mercury poisoning in Niigata, Ministry of Health and Welfare, Tokyo, Stencils, 1967, 30.

Mustakallio, K. K. and Telkkä, A., Selective inhibition patterns of succinic dehydrogenase and local necrobiosis in tubules of rat kidney induced by six mercurial diuretics, *Ann. Med. Exp. Biol. Fenn.*, 33, 1, 1955.

Nadkarni, R. A. and Ehmann, W. D., Neutron activation analysis of wheat flour samples, *Radiochem. Radioanal. Lett.*, 6, 89, 1971.

Nakamura, H., Studies on the decomposition of organomercury compounds. II. Decomposition of phenylmercury acetate in vivo, *J. Kumamoto Med. Soc.*, 43, 994, 1969.

Nakamura, K. and Suzuki, A., Expériences des malformations congénitales de l'encéphalie provoquées par intervention de certains agents, *Arch. Anat. Pathol.*, 15, 116, 1967.

Neal, P. A., Flinn, R. H., Edwards, T. I., Reinhart, W. H., Hough, J. W., Dallavalle, J. M., Goldman, F. H., Armstrong, D. W., Gray, A. S., Coleman, A. L., Postman, B. F., Mercurialism and its control in the felt-hat industry, Public Health Bull. No. 263, U.S. Public Health Service, Federal Security Agency, 1941, 1.

Neal, P. A. and Jones, R., Chronic mercurialism in the hatters' fur-cutting industry, *J.A.M.A.*, 110, 337, 1938.

Neal, P. A., Jones, R. R., Bloomfield, J. J., Dallavalle, J. M., and Edwards, T. I., A study of chronic mercurialism in the hatters' fur-cutting industry, Public Health Bull. No. 234, U.S. Treasury Department, Public Health Service, 1937, 1.

Nelson, N., Byerly, T., Kolbye, A., Kurland, L., Shapiro, R., Shibko, S., Stickel, W., Thompson, J., Van Den Berg, L., and Weissler, A., Hazards of mercury, Special Report to the U.S. H. E. W. Secretary's Pesticide Advisory Committee, *Environ. Res.*, 4, 1, 1971.

Newsome, W. H., Determination of methylmercury in fish and in cereal grain products, *J. Agric. Food Chem.*, 19, 567, 1971.

Nielsen Kudsk, F., Absorption of mercury vapour from the respiratory tract in man, *Acta Pharmacol.*, 23, 250, 1965a.

Nielsen Kudsk, F., The influence of ethyl alcohol on the absorption of mercury vapour from the lungs in man, *Acta Pharmacol.*, 23, 263, 1965b.

Nielsen Kudsk, F., Factors influencing the in vitro uptake of mercury vapour in blood, *Acta Pharmacol.*, 27, 161, 1969a.

Nielsen Kudsk, F., Uptake of mercury vapour in blood in vivo and in vitro from Hg-containing air, *Acta Pharmacol.*, 27, 149, 1969b.

Niigata Report, Report on the cases of mercury poisoning in Niigata, Ministry of Health and Welfare, Tokyo, Stencils, 1967, (Japanese with Swedish translation).

Nixon, G. S. and Smith, H., Hazard of mercury poisoning in the dental surgery, *J. Oral Ther. Pharmacol.*, 1, 512, 1965.

Nobel, S., Mercury in urine, in *Standards of Clinical Chemistry*, Vol. 3, Academic Press, New York, 1961, 176.

Nomura S., Epidemiology of Minamata disease, in *Minamata Disease*, Kutsuna, M., Ed., study group of Minamata disease, Kumamoto University, Japan, 1968, 5.

Nonaka, I., An electron microscopical study on the experimental congenital Minamata disease in rat, *Kumamoto Med. J.*, 22, 27, 1969.

Norberg, E. R., Brock, D. W., and Shields, F., A pheasant feeding study using methyl and phenyl mercury treated seed, in Mercury in the Western Environment, abstract of papers presented at the Environmental Health Center Workshop, Oregon State University, Corvallis, Oregon, February 25 to 26, 1971.

Nordberg, G. F., Berlin, M. H., and Grant, C. A., Methyl mercury in the monkey — autoradiographical distribution and neurotoxicity, *Proc. 16th Int. Cong. Occup. Health,* Tokyo, September 22 to 27, 1969, 1971, 234.

Nordberg, G. and Serenius, F., Deposition of inhaled mercury in the lung and brain — preliminary communication from an investigation of the guinea-pig, *Nord. Hyg. T.,* 47, 26, 1966, (Swedish with English summary).

Nordberg, G. F. and Serenius, F., Distribution of inorganic mercury in the guinea pig brain, *Acta Pharmacol.,* 27, 269, 1969.

Norden, Å., Dencker, I., and Schütz, A., Experiences from a food consumption survey based on the double-portion technique in the Dalby community in 1968 — 1969, *Näringsforskning,* 14, 40, 1970, (Swedish).

Nordiskt Symposium on the Problems of Mercury, Lidingö, Sweden, October 10 to 11, 1968, *Nord. Hyg. T.,* 50 (No. 2), 1969, (Swedish).

Norseth, T., The intracellular distribution of mercury in rat liver after methoxyethylmercury intoxication, *Biochem. Pharmacol.,* 16, 1645, 1967.

Norseth, T., The intracellular distribution of mercury in rat liver after a single injection of mercuric chloride, *Biochem. Pharmacol.,* 17, 581, 1968.

Norseth, T., Studies of intracellular distribution of mercury, in *Chemical Fallout. Current Research on Persistent Pesticides,* Miller, M. W. and Berg, G. G., Eds., Charles C Thomas, Springfield, Ill., 1969a, 408.

Norseth, T., Studies on the biotransformation of methylmercury salts in the rat, Thesis, Department of Radiation Biology and Biophysics, University of Rochester, Rochester, New York, Stencils, 1969b.

Norseth, T., Biotransformation of methyl mercuric salts in the mouse studied by specific determination of inorganic mercury, *Acta Pharmacol.,* 29, 375, 1971.

Norseth, T. and Brendeford, M., Intracellular distribution of inorganic and organic mercury in rat liver after exposure to methylmercury salts, *Biochem. Pharmacol.,* 20, 1101, 1971.

Norseth, T. and Clarkson, T. W., Biotransformation of methyl mercury salts in the rat studied by specific determination of inorganic mercury, *Biochem. Pharmacol.,* 19, 2775, 1970a.

Norseth, T. and Clarkson, T. W., Studies on the biotransformation of ^{203}Hg-labeled methyl mercury chloride in rats, *Arch. Environ. Health,* 21, 717, 1970b.

Okada, M. and Oharazawa, H., Diagnosis of mercury poisoning. Influence of ethylmercuric phosphate on pregnant mice and their fetuses, in Report on the cases of mercury poisoning in Niigata, Ministry of Health and Welfare, Tokyo, Stencils, 63, 1967, (Japanese, Swedish translation).

Okinaka, S., Yoshikawa, M., Mozai, T., Mizuno, Y., Terao, T., Watanabe, H., Ogihara, K., Hirai, S., Yoshino, Y., Inose, T., Anzai, S., and Tsuda, M., Encephalomyelopathy due to an organic mercury compound, *Neurology,* 14, 69, 1964.

Oliver, J., MacDowell, M. C., and Tracy, A., The pathogenesis of acute renal failure associated with traumatic and toxic injury. Renal ischemia, nephrotoxic damage and the ischemuric episode, *J. Clin. Invest.,* 30, 1307, 1951.

Oliver, W. T. and Platonow, N., Studies on the pharmacology by *N*-(ethylmercuri)-*p*-toluenesulfonanilide, *Am. J. Vet. Res.,* 21, 906, 1960, (English with Spanish summary).

Olsson, M., Discussion contribution: Disappearance of mercury from lakes, *Nord. Hyg. T.,* 50, 179, 1969, (Swedish).

Ordonez, J. V., Carrillo, J. A., Miranda, M., and Gale, J. L., Estudio epidemiologico de una enfermedad considerada como encefalitis en la region de los altos de Guatemala, *Bol. Ofic. Sani. Panamericana,* 60, 510, 1966, (Portuguese with English summary, partial Swedish translation).

Orton, S. T. and Bender, L., Lesions in the lateral horns of the spinal cord in acrodynia, pellagra and pernicious anemia, *Bull. Neurol. Inst.* (New York), 1, 506, 1931.

Oshiumi, Y., Matsuura, K., and Komaki, S., Effect of deoxidized glutathione (Tathion) on excretion of ^{203}Hg-MHP in kidney, *Nippon Acta Radiol.,* 29, 1038, 1969.

Östergren, G., Narcotized mitosis and the precipitation hypothesis of narcosis, Mecanisme de la narcose, *Colloq. Int. Centre Nat. Rech. Sci.,* 26, 77, 1951.

Östergren, G. and Levan, A., The connection between c-mitotic activity and water solubility in some monocyclic compounds, *Hereditas,* 29, 494, 1943.

Östlund, K., Separation of organic mercury compounds by thin-layer chromatography, *Nord. Hyg. T.,* 50, 82, 1969a, (Swedish).

Östlund, K., Studies on the metabolism of methyl mercury and dimethyl mercury in mice, *Acta Pharmacol.,* 27, Suppl. 1, 1, 1969b.

Otterlind, G. and Lennerstedt, I., The Swedish bird fauna, and biocide damages, *Vår Fågelvärld,* 23, 363, 1964, (Swedish).

Oyake, Y., Tanaka, M., Kubo, H., and Chichibu, M., Neuropathological studies on organo mercury intoxication with special reference to distribution of mercury granules, *Progr. Neurol. Res.,* 10, 744, 1966, (Japanese with English translation).

Paavila, H. D., Use of mercury in the Canadian pulp and paper industry, in Mercury in Man's Environment, Proc. Roy. Soc. Can., Symp., February 15 to 16, 1971, 40.

Palmer, J. S., Mercurial fungicidal seed protectant toxic for sheep and chickens, *J. Am. Vet. Ass.,* 142, 1385, 1963.

Parizek, J. and Ostádalová, I., The protective effect of small amounts of selenite in sublimate intoxication, *Experientia,* 23, 142, 1967.

Patterson, D. and Greenfield, J. G., Erythroedema polyneuritis, *Quart. J. Med.,* 16, 6, 1923.

Pekkanen, T., Clinical and neurologic findings in cats poisoned by methyl mercury, in Swedish-Finnish mercury symposium, Helsinki, November 7, 1969, Department of Radiochemistry, University of Helsinki, Stencils, 1969, 28, (Swedish).

Pentschew, A., Intoxikationen, Quecksilbervergiftung, in *Handbuch der Speziellen Pathologischen Anatomie und Histologie,* Uehlinger, E., Ed., Springer Verlag, Berlin, 1958, Vol. XIII/2, 2007, 2419.

Perkons, A. K. and Jervis, R. E., Hair individualization studies, Proc. 1965 Int. Conf. Modern Trends in Activation Analysis, Texas A and M University, 1965, 295.

Phillips, R. and Cember, H., The influence of body of radiomercury on radiation dose, *J. Occup. Med.,* 11, 170, 1969.

Piechocka, J., Chemical and toxicological studies in the fungicide, phenyl mercury acetate. Part I. Effects on rats fed with food contaminated with fungitox OR, *Rocz. Panstw. Zalk. Hig.,* 19, 389, 1968.

Pierce, A. P., Botbol, J. M., and Learned, R. E., Mercury content of rocks, soil and stream sediments, in *Mercury in the Environment,* U. S. Geological Survey professional paper, no. 713, Washington, D.C., U.S. G. P. O., 1970, 14.

Pieter Kark, R. A., Poskanzer, D. C., Bullock, J. D., and Boylen, G., Mercury poisoning and its treatment with N-acetyl-d,l-penicillamine, *New Eng. J. Med.,* 285, 10, 1971.

Piotrowski, J. and Bolanowska, W., Binding of phenyl-mercury acetate (^{203}Hg) in the body of rat, studied by molecular filtration technique, *Med. Pracy,* 21, 338, 1970, (Polish with English summary).

Piotrowski, J. K., Trojanowska, B., Wisniewska-Knypl, J. M., and Bolanowska, W., Further investigations on binding and release of mercury in the rat, in *Mercury, Mercurials and Mercaptans,* Miller, M. W. and Clarkson, T. W., Eds., from the 4th Rochester Conf. on Environmental Toxicity, June 17 to 19, 1971, Charles C Thomas, Springfield, Ill., in press.

Piper, R. C., Miller, V. L., and Dickinson, E. O., Toxicity and distribution of mercury in pigs with acute methylmercurialism, *Am. J. Vet. Res.,* 32, 263, 1971.

Piscator, M., Cadmium in the kidneys of normal human beings and the isolation of metallothionein from liver of rabbits exposed to cadmium, *Nord. Hyg. T.,* 45, 76, 1964, (Swedish).

Platonow, N., A study of the metabolic fate of methylmercuric acetate, *Occup. Health Rev.,* 20, 9, 1968a.

Platonow, N., Les effets compares des agents chélateurs sur la distribution du mercure organique chez les porcelets, *Can. Vet. J.,* 9, 142, 1968b.

Platonow, N., Etude toxicologique de la viande et des viscères de porcs empoisonnés à l'acetate d'ethylmercurique, *Can. Vet. J.,* 10, 202, 1969.

Poluektov, N. S., Vitkun, P. A., and Zelyukova, Y. V., Determination of milligram amounts of mercury by atomic absorption in the gaseous phase, *J. Anal. Khim.,* 19, 937, 1964, (Russian).

Prescott, L. F. and Ansari, S., The effects of repeated administration of mercuric chloride on exfoliation of renal tubular cells and urinary glutamic-oxaloacetic transaminase activity in the rat, *Toxicol. Appl. Pharmacol.,* 14, 97, 1969.

Powell, H. M. and Jamieson, W. A., Merthiolate® as a germicide, *Am. J. Hyg.,* 13, 269, 1931.

Prick, J. J. G., Sonnen, A. E. H., and Slooff, J. L., Organic mercury poisoning. I. *Proc. Koninklijke Nederlandse Akademie van Wetenschappen,* 70, 150, 1967a.

Prick, J. J. G., Sonnen, A. E. H., and Slooff, J. L., Organic mercury poisoning. II. *Proc. Koninklijke Nederlandse Akademie van Wetenschappen,* 70, 170, 1967b.

Prickett, C. S., Laug, E. P., and Kunze, F. M., Distribution of mercury in rats following oral and intravenous administration of mercuric acetate and phenylmercuric acetate, *Proc. Soc. Exp. Biol. Med.,* 73, 585, 1950.

Quino, E. A., Determination of dibutyl mercury vapors in air, *Am. Ind. Hyg. Ass. J.,* 23, 231, 1962.

Radaody-Ralarosy, M. P., Recherches histochimiques sur le passage de l'arsenic et du mercure dans le placenta, *Arch. Soc. Sci. Med. Biol.,* 19, 22, 1938.

Raeder, M. G. and Snekvik, E., Quecksilbergehalt mariner organismen, *Kgl. Norske Videnskab. Selskabs Forhandl.,* 13, 169, 1941.

Raeder, M. G. and Snekvik, E., Mercury determinations in fish and other aquatic organisms, *Kgl. Norske Videnskab. Selskabs Forhandl.,* 21, 102, 1949.

Rahola, T., Hattula, T., Korolainen, A., and Miettinen, J. K., The biological half-time of inorganic mercury (Hg^{2+}) in man, *Scand. J. Clin. Lab. Invest.,* Abstr., 27 (Suppl. 116), 77, 1971.

Ramel, C., Genetic effects of organic mercury compounds, *Hereditas,* 57, 445, 1967.

Ramel, C., Genetic effects of organic mercury compounds. I. Cytological investigations on allium roots, *Hereditas,* 61, 208, 1969a.

Ramel, C., Methylmercury as a mitosis disturbing agent, *J. Jap. Med. Ass.,* 61:9, 1072, 1969b.

Ramel, C., Tests of chromosome segregation in *Drosophila,* Abstr. 1st Ann. Meet. Env. Mutag. Soc., 22, 1970.

Ramel, C. and Magnusson, J., Genetic effects of organic mercury compounds. II. Chromosome segregation in *Drosophila melanogaster, Hereditas,* 61, 231, 1969.

Ramel, C. and Magnusson, J., The effects of methyl mercury on the segregation of chromosomes in *Drosophila melanogaster,* Abstr. Soc. Europ. Dros. Conf., 1971, in press.

Rapkine, L., Sur les processus chimiques au cours de la division cellulaire, *Ann. Physiol. Physicochim. Biol.,* 7, 383, 1931.

Rathje, A. O., A rapid ultraviolet absorption method for the determination of mercury in urine, *Am. Ind. Hyg. Ass. J.,* 30, 126, 1969.

Reba, R. C., Wagner, H. N., and McAfee, J. C., Measurement of Hg^{203} chlormerodrin accumulation by the kidneys for detection of unilateral renal disease, *Radiology,* 79, 134, 1962.

Reber, K., Blockierung der Speicherfunktion der Niere als Schutz bei Sublimatvergiftung, *Pathol. Bakt.* (Zurich), 16, 755, 1953.

Rentos, P. G. and Seligman, E. J., Relationship between environmental exposure to mercury and clinical observation, *Arch. Environ. Health*, 16, 794, 1968.

Ricker, G. and Hesse, W., Über den Einfluss des Quecksilbers, namentlich des Eingeatmeten, auf die Lungen von Versuchstieren, *Arch. Pathol. Anat.*, 217, 267, 1914.

Rissanen, K., Retention and distribution of mercury in cat, in Swedish-Finnish mercury symposium, Helsinki November 7, 1969, Department of Radiochemistry, University of Helsinki, Stencils, 1969, 21, (Swedish).

Rissanen, K., Erkama, J., and Miettinen, J. K., Experiments on microbiological methylation of mercury (2-) ion by mud and sludge in anaerobic conditions, FAO FIR MP/70/E-61, October 15, 1970, Rome.

Rissanen, K. and Miettinen, J. K., Use of mercury compounds in agriculture and its implications, in Mercury Contamination in Man and His Environment, FAO-IAEA-ILO-WHO, to be published in IAEA Technical Report Series.

Ritter, W. L. and Nussbaum, M. A., Occupational illnesses in cotton industries. III. The mercury hazard in seed treating, *Miss. Doctor*, 22, 262, 1945.

Riva, G., Zur Frage der chronischen Quecksilbernephrose, *Helv. Med. Acta*, 12, 539, 1945.

Rodger, W. J. and Smith, H., Mercury absorption by fingerprint officers using "Grey Powder," *J. Forensic Sci. Soc.*, 7, 86, 1967.

Rodin, A. E. and Crowson, C. N., Mercury nephrotoxicity in the rat. I. Factors influencing the localization of the tubular lesions, *Am. J. Pathol.*, 61, 297, 1962.

Rolfe, A. C., Russell, F. R., and Wilkinson, N. T., The absorptiometric determination of mercury in urine, *Analyst*, 80, 523, 1955.

Rosen, E., Mercurialentis, *Am. J. Opthalmol.*, 33, 1287, 1950.

Rosenthal, O., Quecksilber. Praxis der Behandlung, in Jadassohn, J., Ed., *Handbuch der Haut- und Geschlechtskrankheiten*, Springer Verlag, Berlin, 1928, Vol. 18, 105.

Rosenthall, L., Greyson, N. D., and Eidinger, S., Positive identification of lung neoplasms with $^{197}HgCl_2$, *J. Can. Ass. Radiol.*, 21, 181, 1970.

Ross, R. G. and Stewart, D. K. R., Movement and accumulation of mercury in apple trees and soil, *Can. J. Plant Sci.*, 42, 280, 1962.

Rost, G. S., Allergic reactions to catgut preservatives, *J. Iowa State Med. Soc.*, 43, 218, 1953.

Rothstein, A., Mercaptans – biological targets for mercurials, in *Mercury, Mercurials and Mercaptans*, Miller, M. W. and Clarkson, T. W., Eds., from the 4th Rochester Conf. on Environmental Toxicity, June 17 to 19, 1971, Charles C Thomas, Springfield, Ill., in press.

Rothstein, A. and Hayes, A. D., The metabolism of mercury in the rat studied by isotope technique, *J. Pharmacol. Exp. Ther.*, 130, 166, 1960.

Rothstein, A. and Hayes, A. D., The turnover of mercury in rats exposed repeatedly to inhalation of vapor, *Health Phys.*, 10, 1099, 1964.

Rottschafer, J. M., Jones J. D., and Mark, H. B., Jr., A simple, rapid method for determining trace mercury in fish via neutron activation analysis, *Environ. Sci. Technol.*, 5, 336, 1971.

Ruch, R. R., Gluskoter, H. J., and Kennedy, E. J., Illinois State Geological Survey Environmental Note, 43, 1971.

Rucker, R. R. and Amend, D. F., Absorption and retention of organic mercurials by rainbow trout and chinook and sockeye salmon, *Progr. Fish-Cult.*, 31, 197, 1969.

Ryazanov, V. A., Ed., *Limits of Allowable Concentrations of Atmospheric Pollutants*, Book 3, Moscow, 1957, (translated by B. S. Levine).

Saha, J. G., Lee, Y. W., Tinline, R. D., Chinn, S. H. F., and Austenson, H. M., Mercury residues in cereal grains from seeds or soil treated with organomercury compounds, *Can. J. Plant Sci.*, 50, 597, 1970.

Saito, M., Mizuno, M., Saito, S., and Maehara, Y., An autopsy case of acute encephalomyelopathia after organic mercury therapy, *Acta Pathol. Jap.*, 9, 544, 1959.

Saito, M., Osono, T., Watanabe, J., Yamamoto, T., Takeuchi, M., Ohyagi, Y., and Katsunuma, H., Studies on Minamata disease, establishment of the criterion for etiological research in mice, *Jap. J. Exp. Med.*, 31, 277, 1961.

Samsahl, K., Brune, D., and Wester, P. O., Simultaneous determination of 30 trace elements in cancerous and non-cancerous human tissue samples by neutron activation analysis, *Int. J. Appl. Radiat.*, 16, 273, 1965.

Sanchez-Sicilia, L., Seto, D. S., Nakamoto, S., and Kolff, W. J., Acute mercurial intoxication treated by hemodialysis, *Ann. Intern. Med.*, 59, 692, 1963.

Sass, J., Histological and cytological studies of ethyl mercury phosphate, *Phytopathology*, 27, 95, 1937.

Sato, H., A consideration on organic mercury poisoning, Ministry of Health and Welfare, Tokyo, Stencils, Dnr F 2346/68, 1968.

Schamberg, J. F., Kolmer, J. A., Raiziss, G. W., and Gavron, J. L., Experimental studies of the mode of absorption of mercury when applied by inunction, *J.A.M.A.*, 70, 142, 1918.

Schimmert, P. and Wanadsin, B., Die Wirkungsweise des Quecksilbers bei der Sublimatvergiftung, *Klin. Wochenschr.*, 28, 330, 1950.

Schmidt, H. and Harzmann, R., Humanpathologische und tierexperimentelle Beobachtungen nach Intoxikation mit einer organischen Quecksilberverbindung ("Fusariol"), *Int. Arch. Arbeitsmed.*, 26, 71, 1970.

Schulte, H. F., Mercury hazards in seed treating, *J. Ind. Hyg. Toxicol.*, 28, 159, 1946.

Schütz, A., Analytical method for small amounts of mercury in blood, urine and other biological material. Report 69 020 from Department of Occupational Medicine, University Hospital, S-221 85 Lund, Stencils, 1969, (Swedish).

Scott, A., The behavior of radioactive mercury and zinc after application to normal and abnormal skin, *Br. J. Dermatol.*, 71, 181, 1959.

Seifert, P. and Neudert, H., Zur Frage der gewerblichen Quecksilber-Vergiftung, *Zbl. Arbeitsmed.*, 4, 129, 1954.

Selter, O., Über Trophodermatoneurose, *Arch. Kinderheilkd.*, 37, 468, 1903.

Sera, K., Murakami, A., and Sera, Y., Studies on toxicity of organomercury compounds, *Kumamoto Med. J.*, 14, 65, 1961.

Sergeant, G. A., Dixon, B. E., and Lidzey, A. G., The determination of mercury in air, *Analyst*, 82, 27, 1957.

Seymour, A. H., The rate of loss of mercury by Pacific oysters, in Mercury in the Western Environment, abstract of papers presented at the Environmental Health Center Workshop, Oregon State University, Corvallis, Oregon, February 25 to 26, 1971.

Shacklette, H. T., Bryophytes associated with mineral deposits and solutions in Alaska, U.S. Geological Survey Bull. 1198-C, 1965.

Shacklette, H. T., Mercury content of plants, in *Mercury in the Environment*, U.S. Geological Survey Professional Paper, No. 713, Washington, D.C., U.S.G.P.O., 1970, 35.

Shacklette, H. T., Boerngen, J. G., and Turner, R. L., Mercury in the environment — surficial materials of the conterminous United States, U.S. Geological Survey Circular No. 644, Washington, D.C., 1971.

Shapiro, B., Kollmann, G., and Martin, D., Effects of mercurials on erythrocytes, *J. Nucl. Med.*, ,347, 1968.

Shelanski, M. L. and Taylor, E. W., Isolation of a protein subunit from microtubules, *J. Cell Biol.*, 34:2, 549, 1967.

Shepherd, M., Schumann, S., Flinn, R. H., Hough, J. W., and Neal, P. A., Hazard of mercury vapor in scientific laboratories, *J. Res. Natl. Bureau Standards (U.S.A.)*, 26, 357, 1941.

Silberberg, I., Prutkin, L., and Leider, M., Electron microscopic studies of transepidermal absorption of mercury, *Arch. Environ. Health*, 19, 7, 1969.

Sillén, L. G., The physical chemistry of seawater, *Am. Ass. Advan. Sci.*, 67, 549, 1961.

Sillén, L. G., Electrometric investigation of equilibrium between mercury and halogen ions. VIII. Survey and conclusions, *Acta Chem. Scand.*, 3, 539, 1949.

Silverberg, D. S., McCall, J. T., and Hunt, J. C., Nephrotic syndrome with use of ammoniated mercury, *Arch. Intern. Med.*, 120, 581, 1967.

Simonds, J. P. and Hepler, O. E., Experimental nephropathies. IX. A summary of experiments on dogs, *Quart. Bull. Northwestern Univ. Med. Sch.*, 19, 278, 1945.

Singerman, A. and Catalina, R. L., Exposure to metallic mercury, enzymatic studies, in Proc. 16th Int. Cong. Occupat. Health, Tokyo, 1970.

Sjöblom, V. and Hasänen, E., The amount of mercury in fish in Finland, *Nord. Hyg. T.*, 50, 37, 1969, (Swedish).

Sjöstrand, B., Simultaneous determination of mercury and arsenic in biological and organic materials by activation analysis, *Anal. Chem.*, 36, 814, 1964.

Skare, I., Microdetermination of mercury in biological samples. III. Automatic determination of mercury in urine, fish and blood samples, *Analyst*, in press, 1971.

Skerfving, S., Studies of consumers of methyl mercury contaminated fish, Report to the Research Board of the National Environmental Protection Board, National Institute of Public Health, Stockholm, 1971.

Skerfving, S., Hansson, A., and Lindsten, J., Chromosome breakage in human subjects exposed to methyl mercury through fish consumption, *Arch. Environ. Health*, 21, 133, 1970.

Skog, E. and Wahlberg, J. E., A comparative investigation of the percutaneous absorption of metal compounds in the guinea pig by means of the radioactive isotopes: ^{51}Cr, ^{58}Co, ^{65}Zn, ^{110m}Ag, ^{115m}Cd, ^{203}Hg, *J. Invest. Derm.*, 43, 187, 1964.

Slatov, I. V. and Zimnikova, Z. I., Certain clinical features of acute poisoning by ethylmercuric chloride, *Gig. Sanit.*, 33, 275, 1968, (Russian with English summary).

Smart, N. A., Use and residues of mercury compounds in agriculture, *Residue Rev.*, 23, 1, 1968.

Smart, N. A. and Hill, A. R. C., Determination of mercury residues in potatoes, grain and animal tissues using perchloric acid digestion, *Analyst*, 94, 143, 1969.

Smith, F. A., A preliminary survey of mercury tissue liver from gamebirds and fish in Utah, in Mercury in the Western Environment, abstract of papers presented at the Environmental Health Center Workshop, Oregon State University, Corvallis, Oregon, February 25 to 26, 1971.

Smith, H., *The Kidney: Structure and Function in Health and Disease*, Oxford Univ. Press, New York, 1951, 752.

Smith, J. C., Clarkson, T. W., Greenwood, M., Von Burg, R., and Laselle, E., The form of mercury in North American fish, *Fed. Proc.*, 30, 221 (Abst.), 1971.

Smith, J. C. and Wells, A. R., A biochemical study of the urinary protein of men exposed to metallic mercury, *Br. J. Ind. Med.*, 17, 205, 1960.

Smith, R. G., Vorwald, A. J., Patil, L. S., and Mooney, T. F., Effects of exposure to mercury in the manufacture of chlorine, *Am. Ind. Hyg. Ass. J.*, 31, 687, 1970.

Snyder, R. D., Congenital mercury poisoning, *New Eng. J. Med.*, 284, 1014, 1971.

Sobotka, T. J., Cook, M., and Brodie, R., Perinatal exposure to methyl mercury: brain development and neurotransmitter systems, *Pharmacologist*, 13, 469, 1971.

Sodee, D. B., A new scanning isotope, mercury^{197}, a preliminary report, *J. Nucl. Med.*, 4, 335, 1963.

Sollmann, T. and Schreiber, N. E., Chemical studies of acute poisoning from mercury bichloride, *Arch. Intern. Med.*, 57, 46, 1936.

Solomon, J. and Uthe, J. F., A rapid semimicromethod for methyl mercury residue analysis in fish by gas chromatography, in Mercury in the Western Environment, abstract of papers presented at the Environmental Health Center Workshop, Oregon State University, Corvallis, Oregon, February 25 to 26, 1971.

Somers, E., Mercury contamination of foods, in Mercury in Man's Environment, Proc. Roy. Soc. Can. Symp., February 15 to 16, 1971, 99.

Spyker, J. M. and Sparber, S. B., Behavioral teratology of methylmercury in the mouse, *Pharmacologist*, 13, 469, 1971.

Squire, J. R., Blainey, J. D., and Hardwicke, J., The nephrotic syndrome, *Br. Med. Bull.*, 13, 43, 1957.

Stahl, Q. R., Preliminary air pollution survey of mercury and its compounds, a literature review, U.S. Department of Health, Education and Welfare, Public Health Service, National Air Pollution Control Administration, Raleigh, N.C., 1969.

Steinwall, O. and Olsson, Y., Impairment of the blood-brain barrier in mercury poisoning, *Acta Neurol. Scand.*, 45, 351, 1969.

Stejskal, J., Acute renal insufficiency in intoxication with mercury compounds. III. Pathological findings, *Acta Med. Scand.*, 177, 75, 1965.

Stock, A., Die Gefährlichkeit des Quecksilberdampfes, *Z. Angew. Chem.*, 39, 461, 1926.

Stock, A., Über Verdampfung, Löslichkeit und Oxydation des metallischen Quecksilbers, *Z. Anorg. Chem.*, 217, 241, 1934.

Stock, A., Die chronische Quecksilber- und Amalgamvergiftung, *Arch. Gewerbepath.*, 7, 388, 1936.

Stock, A., Die mikroanalytische Bestimmung des Quecksilbers und ihre Anwendung auf hygienische und medizinische Fragen, *Svensk. Kemisk T.*, 50, 242, 1938.

Stock, A., Der Quecksilbergehalt des menschlichen Organismus. Mitteilung über Wirkung und Verbreitung des Quecksilbers, *Biochem. Z.*, 304, 73, 1940.

Stock, A. and Cucuel, F., Die Verbreitung des Quecksilbers, *Naturwiss.*, 22, 390, 1934a.

Stock, A. and Cucuel, F., Der Quecksilbergehalt der menschlichen Ausscheidungen und des menschlichen Blutes, *Z. Angew. Chem.*, 47, 641, 1934b.

Stock, A. and Lux, H., Die quantitative Bestimmung kleinster Quecksilbermengen, *Z. Angew. Chem.*, 44, 200, 1931.

Stock, A. and Zimmermann, W., Geht Quecksilber aus Saatgut-Beizmitteln in das geerntete Korn und in das Mehl über? *Z. Angew. Chem.*, 41, 1336, 1928a.

Stock, A. and Zimmermann, W., Geht Quecksilber aus Saatgut-Beizmitteln in das geerntete Korn und in das Mehl über? *Z. Angew. Chem.*, 41, 547, 1928b.

Storlazzi, E. D. and Elkins, H. B., The significance of urinary mercury. I. Occupational mercury exposure. II. Mercury absorption from mercury-bearing dental fillings and antiseptics, *J. Ind. Hyg. Toxicol.*, 23, 459, 1941.

Storrs, B., Thompson, J., Fair, G., Dickerson, M. S., Nickey, L., Barthel, W., and Spaulding, J. E., Organic mercury poisoning, *Morbidity and Mortality*, 19, 25, 1970a.

Storrs, B., Thompson, J., Nickey, L., Barthel, W., and Spaulding, J. E., Follow-up organic mercury poisoning, *Morbidity and Mortality*, 19, 169, 1970b.

Stråby, A., Analysis of samples of snow and water. Appendix to Westermark and Ljunggren, 1968, (Swedish).

Strunge, P., Nephrotic syndrome caused by a seed disinfectant, *J. Occup. Med.*, 12, 173, 1970.

Stuart, A. M., Trichomonas and vaginal discharge, *Br. Med. J.*, 1, 1078, 1936.

Sumari, P., Backman, A. -L., Karli, P., and Lahti, A., Health studies of Finnish consumers of fish, *Nord. Hyg. T.*, 50, 97, 1969, (Swedish).

Sumino, K., Analysis of organic mercury compounds by gas chromatography. Part I. Analytical and extraction method of organic mercury compounds, *Kôbe J. Med. Sci.*, 14, 115, 1968a.

Sumino, K., Analysis of organic mercury compounds by gas chromatography. Part II. Determination of organic mercury compounds in various samples, *Kôbe J. Med. Sci.*, 14, 131, 1968b.

Sunderman, F. W., Hawthorne, M. F., and Baker, G. L., Delayed sensitivity of the skin to phenylmercuric acetate, *Arch. Ind. Health*, 13, 574, 1956.

Surtshin, A., Protective effect of a sucrose diet in mercuric chloride poisoning, *Am. J. Physiol.*, 190, 271, 1957.

Suzuki, T., Neurological symptoms from concentration of mercury in the brain, in Chemical Fallout. Current Research on Persistent Pesticides, Miller, M. W. and Berg, G. G., Eds., Charles C Thomas, Springfield, Ill., 1969, 245.

Suzuki, T., Matsumoto, N., Miyama, T., and Katsunuma, H., Placental transfer of mercuric chloride, phenyl mercury acetate and methyl mercury acetate in mice, *Ind. Health*, 5, 149, 1967.

Suzuki, T., Miyama, T., and Katsunuma, H., Comparative study of bodily distribution of mercury in mice after subcutaneous administration of methyl, ethyl, and n-propyl mercury acetates, *Jap. J. Exp. Med.*, 33, 277, 1963.

Suzuki, T., Miyama, T., and Katsunuma, H., Comparative study of bodily distribution of mercury in mice after subcutaneous administration of propyl, butyl, and amyl mercury acetates, *Jap. J. Exp. Med.*, 34, 211, 1964.

Suzuki, T., Miyama, T., and Katsunuma, H., Affinity of mercury of the thyroid, *Ind. Health*, 4, 69, 1966.

Suzuki, T., Miyama, T., and Katsunuma, H., Mercury in the plasma after subcutaneous injection of sublimate or mercuric nitrate in rat, *Ind. Health*, 5, 290, 1967.

Suzuki, T., Miyama, T., and Katsunuma, H., Mercury contents in the red cells, plasma, urine and hair from workers exposed to mercury vapor, *Ind. Health*, 8, 39, 1970.

Suzuki, T., Miyama, T., and Katsunuma, H., Differences in mercury content in the brain after injection of methyl, phenyl or inorganic mercury, Proc. 16th Int. Cong. Occup. Health, Tokyo, September 22 to 27, 1969, 1971a, 563.

Suzuki, T., Miyama, T., and Katsunuma, H., Comparison of mercury contents in maternal blood, umbilical cord blood, and placental tissues, *Bull. Environ. Contam. Toxicol.* 5, 502, 1971b.

Suzuki, T., Takemoto, T., Kashiwazaki, H., and Miyama, T., Metabolic fate of ethylmercury salts in man and animal, in *Mercury, Mercurials and Mercaptans,* Miller, M. W. and Clarkson, T. W., Eds., from the 4th Rochester Conference on Environmental Toxicity, June 17 to 19, 1971, Springfield, Ill. Charles C Thomas, Springfield, Ill., in press.

Suzuki, T. and Tanaka, A., Absorption of metallic mercury from the intestine after rupture of Miller-Abbot balloon, *Ind. Med.*, 13, 52, 1971.

Suzuki, T. and Yoshino, Y., Effects of d-penicillamine on urinary excretion of mercury in two cases of methyl mercury poisoning, *Jap. J. Ind. Health*, 11, 487, 1969.

Swensson, A., Investigations on the toxicity of some organic mercury compounds which are used as seed disinfectants, *Acta Med. Scand.*, 143, 365, 1952.

Swensson, A., Lundgren, K. -D. and Lindström, O., Distribution and excretion of mercury compounds after single injection, *Arch. Ind. Health*, 20, 432, 1959a.

Swensson, A., Lundgren, K. -D. and Lindström, O., Retention of various mercury compounds after subacute administration, *Arch. Ind. Health*, 20, 467, 1959b.

Swensson, A. and Ulfvarson, U., Experiments with different antidotes in acute poisoning by different mercury compounds, effects on survival and on distribution and excretion of mercury, *Int. Arch. Gewerbepath.*, 24, 12, 1967.

Swensson, A. and Ulfvarson, U., Distribution and excretion of mercury compounds in rats over a long period after a single injection, *Acta Pharmacol.*, 26, 273, 1968.

Takahashi, H. and Hirayama, K., Accelerated elimination of methyl mercury from animals, *Nature*, 232, 201, 1971.

Takahashi, T., Kimura, T., Sato, Y., Shiraki, H., and Ukita, T., Time-dependent distribution of ^{203}Hg-mercury compounds in rat and monkey as studied by whole body autoradiography, *Jap. J. Hyg. Chem.*, 17, 93, 1971.

Takahata, N., Hayashi, H., Watanabe, B., and Anso, T., Accumulation of mercury in the brains of two autopsy cases with chronic inorganic mercury poisoning, *Folia Psychiatr. Neurol. Jap.*, 24, 59, 1970.

Takeda, Y., Kunugi, T., Hoshino, O., and Ukita, T., Distribution of inorganic, aryl and alkyl mercury compounds in rats, *Toxicol. Appl. Pharmacol.*, 13, 156, 1968a.

Takeda, Y., Kunugi, T., Terao, T., and Ukita, T., Mercury compounds in the blood of rats treated with ethylmercuric chloride, *Toxicol. Appl. Pharmacol.*, 13, 165, 1968b.

Takeda, Y. and Ukita, T., Metabolism of ethylmercuric chloride ^{203}Hg in rats, *Toxicol. Appl. Pharmacol.*, 17, 181, 1970.

Takeshita, A. and Uchida, M., The mercury compound in the brain and urine of the rat on oral administration of methyl methylmercuric sulfide, *Kumamoto Med. J.*, 16, 178, 1963.

Takeuchi, T., A pathological study of Minamata disease in Japan, Proc. VII. Int. Congr. Neurol., Rome, 1961, 1.

Takeuchi, T., Pathology of Minamata disease, in *Minamata Disease,* Kutsuna, M., Ed., Study group of Minamata disease, Kumamoto University, Japan, 1968a, 141.

Takeuchi, T., Experiments with organic mercury, particularly with methyl mercury compounds. Similarities between experimental poisoning and Minamata disease, in *Minamata Disease,* Kutsuna, M., Ed., Study group of Minamata disease, Kumamoto University, Japan, 1968b, 229.

Takeuchi, T., Biological reactions and pathological changes of human beings and animals under the condition of organic mercury contamination. Int. Conf. Environmental Mercury Contamination, Ann Arbor, 1970.

Takeuchi, T., Morikawa, N., Matsumoto, H., and Shiraishi, Y., A pathological study of Minamata disease in Japan, *Acta Neuropath.*, 2, 40, 1962.

Takizawa, Y., Studies on the Niigata episode of Minamata disease outbreak — investigation of causative agents of organic mercury poisoning in the district along the river Agano, *Acta Med. Biol.*, 17, 293, 1970.

Takizawa, Y. and Kosaka, T., Erforschung der Ursache von Vergiftungserscheinungen an organischem Quecksilber, welche in der Niigata Präfektur entlang des Agano Flusses aufgetreten sind, *Acta Med. Biol.*, 14, 153, 1966.

Tamaoki, T. and Miyazawa, F., Dissociation of *Escherichia coli* ribosomes by sulfhydryl reagents, *J. Mol. Biol.*, 23, 35, 1967.

Task group on lung dynamics: deposition and retention models for internal dosimetry of the human respiratory tract, *Health Phys.*, 12, 173, 1966.

Tati, M., The behaviour of mercury in the blood, *Proc. 14th Internat. Cong. Occup. Health*, 1963, 1964, 968.

Tatton, J. O. G. and Wagstaffe, P. J., Identification and determination of organomercurial fungicide residues by thin-layer and gas chromatography, *J. Chromatogr.*, 44, 284, 1969.

Taugner, R., Zur renalen Aufnahme und intrarenalen verteilung von Sublimat und Hg-Cystein, *Arzneim. Forsch.*, 16, 1120, 1966.

Taugner, R., zum Winkel, K., and Iravani, J., Zur Lokalisation der Sublimatanreicherung in der Rattenniere, *Virchows Arch. Pathol. Anat.,* 340, 369, 1966.

Taylor, E. L., Mercury poisoning in swine, *J. Am. Vet. Med. Ass.,* 111, 46, 1947.

Taylor, J. L., Chronic mercurial poisoning, *Guy's Hosp. Rep.,* 55, 171, 1901.

Taylor, N. S., Histochemical studies of nephrotoxicity with sublethal doses of mercury in rats, *Am. J. Pathol.,* 46, 1, 1965.

Taylor, W., Guirgis, H. A., and Stewart, W. K., Investigation of a population exposed to organomercurial seed dressings, *Arch. Environ. Health,* 19, 505, 1969.

Teisinger, J. and Fiserova-Bergerova, V., Pulmonary retention and excretion of mercury vapors in man, *Ind. Med. Surg.,* 34, 580, 1965.

Tejning, S., Mercury contents in blood corpuscles and blood plasma in persons who consumed an average of less than one meal of commercial saltwater-fish per week, Report 67 02 06 from Department of Occupational Medicine, University Hospital, Lund, Stencils, 1967a, (Swedish).

Tejning, S., The mercury content in blood corpuscles and plasma in workers exposed to methyl mercuric compounds by dressing seed. Report 67 02 10 from Department of Occupational Medicine, University Hospital, S-221 85 Lund, Stencils, 1967b, (Swedish).

Tejning, S., Mercury contents in blood corpuscles, blood plasma and hair in persons who had for long periods a high consumption of freshwater-fish from Lake Väner, Report 67 08 31 from Department of Occupational Medicine, University Hospital, S-221 85 Lund, Stencils, 1967c, (Swedish).

Tejning, S., Biological effects of methyl mercury dicyandiamide treated grain in the domestic fowl *Gallus gallus L., Oikos,* Suppl. 8, 1, 1967d.

Tejning, S., Mercury levels in blood corpuscles and in plasma in "normal" mothers and their new-born children, Report 68 02 20 from Department of Occupational Medicine, University Hospital, S-221 85 Lund, Stencils, 1968a, (Swedish).

Tejning, S., Mercury contents in blood corpuscles and in blood plasma in persons consuming large quantities of commercially sold saltwater-fish, Report 68 05 29 from Department of Occupational Medicine, University Hospital, S-221 85 Lund, Stencils, 1968b, (Swedish).

Tejning, S., International and Swedish standards concerning the mercury contents in food stuffs in the light of the amounts present in fish, in blood corpuscles, blood plasma and hair of man, *Nord. Hyg. T.,* 50, 103, 1969b, (Swedish).

Tejning, S., Mercury content in blood corpuscles, plasma and cerebrospinal fluid in man, Annual Report to the Swedish Medical Research Council, May 27, 1969c, Stencils, (Swedish).

Tejning, S., Mercury contents in blood corpuscles and in blood plasma in non-fisheaters, Report 70 04 06 from Department of Occupational Medicine, University Hospital, S-221 85 Lund, Stencils, 1970a, (Swedish).

Tejning, S., The mercury contents in blood corpuscles and in blood plasma in mothers and their new-born children, Report 70 05 20 from Department of Occupational Medicine, University Hospital, S-221 85 Lund, Stencils, 1970b, (Swedish).

Tejning, S. and Öhman, H., Uptake, excretion and retention of metallic mercury in chloralkali workers, Proc. 15th Int. Congr. Occup. Health, Wien, 1966, 239.

Teng, C. T. and Brennan, J. C., Acute mercury vapor poisoning. A report of four cases with radiographic and pathologic correlation, *Radiology,* 73, 354, 1959.

Tennant, R., Johnston, H. J., and Wells, J. B., Acute bilateral pneumonitis associated with the inhalation of mercury vapor, *Conn. Med.,* 25, 106, 1961.

Thorpe, V. A., Determination of mercury in food products and biological fluids by aeration and flameless atomic absorption spectrophotometry, *J. Ass. Offic. Anal. Chem.,* 54, 206, 1970.

Tillander, M., Miettinen, J. K., and Koivisto, I., Excretion rate of methyl mercury in the seal *(Pusa Hispida),* FAO, FIR: MP/70/ E-67 October 21, 1970.

Timm, F. and Arnold, M., Der Celluläre Verblieb Kleiner Quecksilbermengen in der Rattenniere, *Arch. Pharmakol. Exp. Pathol. Naunyn-Schmiedebergs,* 239, 393, 1960.

Timm, F., Naundorf, Ch., and Kraft, M., Zur Histochemie und Genese der chronischen Quecksilbervergiftung, *Int. Arch. Gewerbepath.,* 22, 236, 1966.

Tokunaga, A., Medical rehabilitation for motor disturbance in Minamata disease, *Kumamoto Med. J.,* 19, 220, 1966.

Tokuomi, H., Clinical investigations on Minamata disease. Minamata disease in human adult, in *Minamata Disease,* Kutsuna, M., Ed., Study group of Minamata disease, Kumamoto University, Japan, 1968, 37.

Tokuomi, H., Organic mercury poisoning, *Jap. J. Med.,* 8, 260, 1969.

Tokuomi, H., Okajima, T., Kanai, J., Tsunoda, M., Ichiyasu, Y., Misumi, H., Shimomura, K., and Takaba, M., Minamata disease — an unusual neurological disorder occurring in Minamata, Japan, *Kumamoto Med. J.,* 14, 47, 1961.

Tomizawa, C., Studies on residue analysis of pesticides in plant materials. I. Behavior of mercury in rice plants applied with 103-Hg-organomercury disinfectants, *Shokuhin Eiseigaku Zasshi,* 7, 26, 1966, (Japanese).

Tomizawa, C., Kobayashi, A., Shibuya, M., Koshimizu, Y., and Oota, Y., Studies on residue analysis of pesticides in plant materials. II. Neutron activation analysis of mercury in rice grains, *Shokuhin Eiseigaku Zasshi,* 7, 33, 1966, (Japanese).

Tonomura, K. and Kanzaki, F., The reductive decomposition of organic mercurials by cell-free extract of a mercury-resistant pseudomonas, *Biochem. Biophys. Acta,* 184, 227, 1969.

Tonomura, K., Maeda, K., and Futai, F., Studies on the action of mercury-resistant microorganisms on mercurials II. The vaporization of mercurials stimulated by mercury-resistant bacterium, *J. Ferment. Technol.,* 46, 685, 1968.

Tonomura, K., Maeda, K., Futai, F., Nakagami, T., and Yamada, M., Stimulative vaporization of phenylmercuric acetate by mercury-resistant bacteria, *Nature*, 217, 644, 1968a.

Tonomura, K., Nakagami, T., Futai, F., and Maeda, K., Study on the action of mercury-resistant microorganisms on mercurials. I. The isolation of mercury-resistant bacterium and the binding of mercurials to the cells, *J. Ferment. Technol.*, 46, 506, 1968b.

Tornow, E., Die heute gebräuchlichen Saatbeizmittel und ihre Gefahren, *Zbl. Arbeitsmed.*, 3, 137, 1953.

Trachtenberg, I. M., The chronic action of mercury on the organism, current aspects of the problem of micromercurialism and its prophylaxis, ZDOROV'JA, Kiev, 1969, (Russian with German translation).

Trachtenberg, I. M., Goncharuk, G. A., and Balashov, V. E., Nature of the cardiotoxic effect of mercury and organic mercury compounds, *Chem. Abstr.*, 65, abstr. no. 15969, 1966.

Trachtenberg, I. M., Savitskij, I. V., and Sternhartz, R. Y., The effect of low mercury concentrations on the organism (the problem of combined action of toxic and thermal factors), *Gig. Tr. Prof. Zabol.*, 9, 7, 1965, (Russian with German translation).

Trenholm, H. L., Paul, C. L., Baer, H., and Iverson, F., Methyl mercury ^{203}Hg excretion by lactation in guinea pigs (abstr.), *Toxicol. Appl. Pharmacol.*, 18, 97, 1971.

Troen, P., Kaufman, S. A., and Katz, K. H., Mercuric bichloride poisoning, *New Eng. J. Med.*, 224, 459, 1951.

Truhaut, R. and Boudène, C., Microdosage du mercure dans les milieux biologiques d'origine animale, *Bull. Soc. Chim. Fr.*, 11-12, 1850, 1959.

Tryphonas, L. and Nielsen, N. O., The pathology of arylmercurial poisoning in swine, *Can. J. Comp. Med.*, 34, 181, 1970.

Tsubaki, T., Clinical features of organic mercury intoxication, *Jap. J. Med.*, 8, 262, 1969.

Tsubaki, T., Clinical and epidemiological aspects of organic mercury intoxication, in Mercury in Man's Environment, Proc. Roy. Soc. Can., Symp., February 15 to 16, 1971, 131.

Tsubaki, T., Sato, T., Kondo, K., Shirakawa, K., Kambayashi, K., Hiroda, K., Yamada, K., Morune, I., Ueki, S., Kawakami, K., Okada, K., Chujo, S., and Kobayashi, H., Diagnosis of mercury poisoning. I. Circumstances connected with the outbreak of the illness, in Report on the cases of mercury poisoning in Niigata, Ministry of Health and Welfare, Tokyo, Stencils, 1967a, 5, (Japanese with Swedish translation).

Tsubaki, T., Shirakawa, K., Sato, T., Kambayashi, K., Kondo, K., Hiroda, K., Yamada, K., Murone, I., Ueki, S., Nemoto, H., Sato, S., Izumitani, H., and Honda, T., Diagnosis of mercury poisoning. II. Clinical observations. 1. Clinical observations and diagnosis, in Report on the cases of mercury poisoning in Niigata, Ministry of Health and Welfare, Tokyo, Stencils 1967b, 23, (Japanese with Swedish translation).

Tsuchiya, K., Epidemic of mercury poisoning in the Agano river area – an introductory review – *Keio J. Med.*, 18, 213, 1969.

Tsuda, M., Anzai, S., and Sakai, M., Organic mercury poisoning – a case report – *Yokohama Med. Bull.*, 14, 287, 1963.

Tugsavul, A., Merten, D., and Suschny, O., The reliability of low-level radiochemical analysis: results of intercomparisons organized by the agency during the period 1966-1969, IAEA, Vienna, March 30, 1970, Stencils.

Turk, J. L. and Baker, H., Nephrotic syndrome due to ammoniated mercury, *Br. J. Derm.*, 80, 623, 1968.

Turrian, H., Grandjean, E., and Turrian, V., Industriehygienische und Medizinische Untersuchungen in Quecksilberbetrieben, *Schweiz. Med. Wochenschr.* 86, 1091, 1956.

Ueda, K., Methylmercury in food in relation to its content in hair in general population, *J. Jap. Med. Ass.*, 61, 1034, 1969.

Ueda, K., Aoki, H., and Nishimura, M., Alkylmercury levels in fish and hair of the general population in Japan, Proc. 16th Int. Cong. Occup. Health, Tokyo, September 22 to 27, 1969, 1971, 557.

Ui, J. and Kitamura, S., Mercury in the Adriatic, *Mar. Pollut. Bull.*, 26, 56, 1971.

Ukita, T., Hoshino, O., and Tanzawa, K., Determination of mercury in urine, blood and hair of man in organic mercurial poisonings, *J. Hyg. Chem.*, 9, 138, 1963, (Japanese with English summary, and English translation).

Ukita, T., Takeda, Y., Sato, Y., and Takahashi, T., Distribution of ^{203}Hg labeled mercury compounds in adult and pregnant mice determined by whole-body autoradiography, *Radioisotopes*, 16, 439, 1967, (Japanese with English summary).

Ukita, T., Takeda, Y., Takahashi, T., Yoshikawa, M., Sato, Y., and Shiraki, H., Distribution of ^{203}Hg-mercury compounds in monkey studied by whole-body autoradiography, Proc. 1st symp. Drug Metabolism and Action, November 14 to 15, 1969, Chiba, Japan, Pharmaceut. Soc. Jap., Hongo, Tokyo, 1970.

Ulfvarson, U., Distribution and excretion of some mercury compounds after long term exposure, *Int. Arch. Gewerbepath.*, 19, 412, 1962.

Ulfvarson, U., Mercury, aldrine and dieldrin in pheasants, *Sv. Kem. T.*, 77, 235, 1965, (Swedish).

Ulfvarson, U., The effect of the size of the dose on the distribution and excretion of mercury in rats after single intravenous injection of various mercury compounds. *Toxicol. Appl. Pharmacol.*, 15, 1, 1969a.

Ulfvarson, U., The absorption and distribution of mercury in rats, fed with organs from other rats injected with various mercury compounds, *Toxicol. Appl. Pharmacol.*, 15, 525, 1969b.

Ulfvarson, U., Transportation of mercury in animals, *Studia Laboris et Salutis*, 6, 1, 1970.

Umeda, M., Saito, K., Hirose, K., and Saito, M., Cytotoxic effect of inorganic phenyl and alkyl mercuric compounds on HeLa cells, *J. Exp. Med.*, 39, 47, 1969.

Underdal, B., Mercury in foods determined by activation analysis: I. Egg, *Nord. Vet. Med.*, 20, 9, 1968a, (English with German and Norwegian summary).

Underdal, B., Mercury in foods determined by activation analysis: II. Pork and pig's liver, *Nord. Vet. Med.*, 20, 14, 1968b, (English with German and Norwegian summary).

Underdal, B., Studies of mercury in some food stuffs, *Nord. Hyg. T.*, 50, 60, 1969, (Norwegian).

Underwood, G. B., Gaul, L. E., Collins, E., and Mosby, M., Overtreatment dermititis of the feet, *J.A.M.A.*, 130, 249, 1946.

Ünlü, Y., Heyraud, M., and Keckes, S., Mercury as hydrospheric pollutant 1. Accumulation and excretion of 203-$HgCl_2$ in *Tapes decusatus L*. Paper presented to FAO Technical Conference on Marine Pollution and its effects on living resources and fishing, December 8 to 18, 1970, Rome, *Stencils*.

U.S. Geological Survey Professional Paper No. 713: *Mercury in the Environment,* a compilation of papers on the abundance, distribution and testing of mercury in rocks, soils, waters, plants and the atmosphere, U.S. G. P. O., Washington, D.C., 1970.

Uthe, J. F., Armstrong, F. A. J., and Tam, K. C., Determination of trace amounts of mercury in fish tissues: results of a North American check sample study, *J.A.O.A.C.*, 54, 866, 1971.

Valek, A., Acute renal insufficiency in intoxication with mercury compounds, I. Aetiology, clinical picture, renal function, *Acta Med. Scand.*, 177, 63, 1965.

Veilchenblau, L., Neuartige Berufskrankheiten in der Landwirtschaft, *Münch. Med. Wochenschr.*, 79, 432, 1932.

Verich, G. E., Cardiotoxic effect of thiol poisons, *Chem Abstr.*, (18021r), 75, 269, 1971.

Vintinner, F. J., *Dermatitis venenata* resulting from contact with an aqueous solution of ethyl mercury phosphate, *J. Ind. Hyg. Toxicol.*, 22, 297, 1940.

Viola, P. L. and Cassano, G. B., The effect of chlorine on mercury vapor intoxication. Autoradiographic study, *Med. Lavoro*, 59, 437, 1968.

Voege, F. A., Levels of mercury contamination in water and its boundaries, in Mercury in Man's Environment, Proc. Roy. Soc. Can., Symp. February 15 to 16, 1971, 107.

Voigt, G. E., Histochemische Untersuchung über die Verteilung des Quecksilbers bei experimenteller Sublimatvergiftung, *Acta Pathol. Microbiol. Scand.*, 43, 321, 1958.

Vostal, J., Renal excretory mechanisms of mercury compounds, Working Paper, MAC-symposium on mercury, Stockholm, Stencils, 1968.

Vostal, J., Mercuric ion, organomercurials and renal diuresis, in *Mercury, Mercurials and Mercaptans,* Miller, M. W. and Clarkson, T. W., Eds., from the 4th Rochester Conf. on Environmental Toxicity, June 17 to 19, 1971. Charles C Thomas, Springfield, Ill., in press.

Vostal, J. and Clarkson, T. W., The release of inorganic mercury from diuretic and non-diuretic organomercurials in dogs, *Fed. Proc.*, 29, 481, 1970.

Vostal, J. and Heller, J., Renal excretory mechanisms of heavy metals. I. Transtubular transport of heavy metal ions in the avian kidney, *Environ. Res.*, 2, 1, 1968.

Vouk, V. B., Fugas, M., and Topolnik, Z., Environmental conditions in the mercury mine of Idria, *Br. J. Ind. Med.*, 7, 168, 1950.

Wada, O., Toyokawa, K., Suzuki, T., Suzuki, S., Yano, Y., and Nakao, K., Response to a low concentration of mercury vapor, *Arch. Environ. Health*, 19, 485, 1969.

Wagner, H. N., Weiner, I. M., McAfee, J. G., and Martinez, J., 1-mercuri-2-hydroxypropane (MHP), *Arch. Intern. Med.*, 113, 696, 1964.

Wahlberg, J. E., Percutaneous toxicity of metal compounds. A comparative investigation in guinea pigs, *Arch. Environ. Health*, 11, 201, 1965a.

Wahlberg, J. E., "Disappearance measurements", a method for studying percutaneous absorption of isotope-labeled compounds emitting gamma-rays, *Acta Dermatovener.*, 45, 397, 1965b.

Wahlberg, P., Henriksson, K., and Karppanen, E., Migrating birds as poison vectors, *Nord. Med.*, 84, 889, 1970, (Swedish).

Wallace, R. A., Fulkerson, W., Shults, W. D., and Lyon, W. S., Mercury in the environment. The human element, Environmental Program, Oak Ridge National Laboratory, ORNL-NSF-EP-1, 1971.

Wanntorp, H., Borg, K., Hanko, E., and Erne, K., Mercury residues in wood-pigeons (*Columba p. palumbus L.*) in 1964 and 1966, *Nord. Vet. Med.*, 19, 474, 1967, (English with German and Swedish summaries).

Warkany, J., Acrodynia — postmortem of a disease, *Am. J. Dis. Child.*, 112, 147, 1966.

Warkany, J. and Hubbard, D. M., Mercury in the urine of children with acrodynia, *Lancet*, 29, 829, 1948.

Warkany, J. and Hubbard, D. M., Adverse mercurial reactions in the form of acrodynia and related conditions, *Am. J. Dis. Child.*, 81, 335, 1951.

Warkany, J. and Hubbard, D. M., Acrodynia and mercury, *J. Pediatr.*, 42, 365, 1953.

Warren, H. V. and Delavault, R. E., Mercury content of some British soils, *Oikos*, 20, 537, 1969.

Watanabe, S., Mercury in the body 10 years after long term exposure to mercury, Proc. 16th Int. Cong. Occup. Health, Tokyo, September 22 to 27, 1969, 1971, 553.

Webb, J. L., *Enzyme and Metabolic Inhibitors*, Academic Press, New York, 1966, Vol. 2, 729.

Weed, L. A. and Ecker, E. E., The utility of phenyl-mercury-nitrate as a disinfectant, *J. Infect. Dis.*, 49, 440, 1931.

Weed, L. A. and Ecker, E. E., Phenyl-mercuric compounds. Their action on animals and their preservative values, *J. Infect. Dis.*, 52, 354, 1933.

Weiner, I. M. and Müller, O. H., A polarographic study of mersalyl (Salyrgan®) thiol complexes and of the excreted products of mersalyl, *J. Pharmacol. Exp. Ther.*, 113, 241, 1955.

Welander, E., Investigations on the uptake and elimination of mercury in the human body, *Nord. Med. Ark.*, 18, 1, (no. 12); 1, (no. 15), 1886, (Swedish).

Wershaw, R. L., Sources and behavior of mercury in surface waters, in *Mercury in the Environment*, U.S. Geological Survey Professional Paper, No. 713, Washington, D. C., U.S. G. P. O., 1970, 29.

Wessel, W., Georgsson, G., and Segschneider, I., Elektronenmikroskopische Untersuchungen über Weg und Wirkung hochdosierten Sublimats nach Injektion in die Arteria Renalis, *Virchows Arch. (Zellpathol.)*, 3, 88, 1969.

West, I. and Lim, J., Mercury poisoning among workers in California's mercury mills, *J. Occup. Med.*, 10, 697, 1968.

Westermark, T., Mercury in aquatic organisms, in *The Mercury Problem in Sweden*, Royal Swedish Ministry of Agriculture, Stockholm, 1965, 25, (Swedish).

Westermark, T. and Ljunggren, K., Development of analytical methods for mercury and studies on its dissemination from industrial sources, in Report to the Swedish Technical Research Council, Isotope Technics Laboratories, S-114 28 Stockholm, Stencils, 1968, (Swedish).

Westermark, T. and Sjöstrand, B., Activation analysis of mercury, *Int. J. Appl. Radiat.*, 9, 1, 1960, (English with French, Russian, and German summaries).

Westöö, G., Mercury in our foodstuffs, in *The Mercury Problem in Sweden*, Royal Swedish Ministry of Agriculture, Stockholm, 1965a, 77, (Swedish).

Westöö, G., Mercury in eggs, *Vår Föda.*, 5, 1, 1965b, (Swedish).

Westöö, G., Mercury in foodstuffs — is there a great risk of poisoning? *Vår Föda.*, 4, 1, 1965c, (Swedish).

Westöö, G., Determination of methylmercury compounds in foodstuffs: I. Methylmercury compounds in fish, identification and determination, *Acta Chem. Scand.*, 20, 2131, 1966a.

Westöö, G., Mercury in eggs of Swedish hens, in meat of Swedish hens, broilers and chickens, and in Swedish chicken liver, *Vår Föda.*, 7, 85, 1966b, (Swedish).

Westöö, G., Mercury in meat and liver of pigs, calves, oxen and reindeer, *Vår Föda.*, 7, 88, 1966c, (Swedish).

Westöö, G., Determination of methylmercury compounds in foodstuffs: II. Determination of methylmercury in fish, egg, meat, and liver, *Acta Chem. Scand.*, 21, 1790, 1967a.

Westöö, G., Total mercury in fish, *Vår Föda.*, 1, 1, 1967b, (Swedish).

Westöö, G., Total mercury and methylmercury levels in eggs bought in Sweden, June 1966 to September 1967, *Vår Föda.*, 9, 121, 1967c, (Swedish).

Westöö, G., Methylmercury compounds in fish, *Oikos*, Suppl. 9, 11, 1967d.

Westöö, G., Determination of methylmercury salts in various kinds of biological material, *Acta Chem. Scand.*, 22, 2277, 1968a.

Westöö, G., Comparison between Japanese and Swedish methods for determining methylmercury compounds in fish flesh, National Institute of Public Health, S-104 01 Stockholm 60, Stencils, 1968b, (Swedish).

Westöö, G., Comparison between Japanese and Swedish methods for determining methylmercury compounds in fish, National Institute of Public Health, S-104 01 Stockholm 60, Stencils, 1968c, (Swedish).

Westöö, G., Methylmercury, ethylmercury and total mercury found in Japanese fish, National Institute of Public Health, S-104 01 Stockholm 60, Stencils, 1968d, (Swedish).

Westöö, G., Methylmercury compounds in animal foods, in *Chemical Fallout. Current Research on Persistent Pesticides*, Miller, M. W. and Berg, G. G., Eds., Charles C Thomas, Springfield, Ill., 1969a, 75.

Westöö, G., Mercury and methylmercury levels in some animal foods, August 1967 to October 1969, *Vår Föda.*, 7, 137, 1969b, (Swedish with English summary).

Westöö, G., Mercury compounds in animal foods, *Nord. Hyg. T.*, 50, 67, 1969c, (Swedish).

Westöö, G., Mercury and methylmercury levels in pig's brain, *Vår Föda.*, 9-10, 147, 1970, (Swedish with English summary).

Westöö, G. and Rydälv, M., Mercury and methylmercury in fish and crayfish, *Vår Föda.*, 3, 19, 1969.

Westöö, G. and Rydälv, M., Methylmercury levels in fish caught March 1968-April 1971, Report on mercury in foods by the Joint FAO/WHO Expert Committee on Food Additives 1970, *Vår Föda.*, 23, 179, 1971, (Swedish).

White, D. E., Hinkle, M. E., and Barnes, I., Mercury contents of natural thermal and mineral fluids, in *Mercury in the Environment*, U.S. Geological Survey Professional Paper, No. 713, Washington, D. C., U.S. G. P. O., 1970, 25.

Whitmore, F. C., *Organic Compounds of Mercury*, The Chemical Catalog Co., New York, 1921.

WHO (World Health Organization), Meeting of investigators for the international study of normal values for toxic substances in the human body, WHO, Occ. Health, 66.39, Geneva, 1966, 6.

Wien, R., The toxicity of parachlorometacresol and of phenyl-mercuric nitrate, *Quart. J. Pharm.*, 12, 212, 1939.

Wiklander, L., Mercury in subsoil water and river water, *Grundförbättring*, 4, 151, 1968, (Swedish with English summary).

Wilkening, H. and Litzner, S., Über Erkrankungen insbesondere der Niere durch Alkyl-Quecksilberverbindungen, *Dtsch. Med. Wochenschr.*, 77, 432, 1952.

Williston, S. H., Mercury in the atmosphere, *J. Geophys. Res.*, 73, 7051, 1968.

Wilson, A. J., Eruption due to hydophen ointment, *Nebr. State Med. J.*, 24, 70, 1939.

Winter, D., Sauvard, S., Stanescu, C., Nitelea, I., Nestorescu, B., and Vrejoiu, G., The protective action of glutamic acid in experimental mercury poisoning, *Arch. Environ. Health*, 16, 626, 1968.

Wishart, W., A mercury problem in Alberta's game birds, *Alberta: Lands-Forests-Parks—Wildlife,* 13(2), 4, 1970.

Wiśniewska, J. M., Trojanowska, B., Piotrowski, J., and Jakubowski, M., Binding of mercury in the rat kidney by metallothionein, *Toxicol. Appl. Pharmacol.,* 16, 754, 1970.

Witschi, H. P., Untersuchungen über die intestinale Ausscheidung von Quecksilber bei Ratten, *Beitr. Gerichtl. Med.,* 23, 288, 1965.

Wood, J. M., Kennedy, F. S., and Rosen, C. G., Synthesis of methyl-mercury compounds by extracts of a methanogenic bacterium, *Nature,* 220, 173, 1968.

Wood, R. W. and Weiss, A. B., An analysis of hand tremor induced by industrial exposure to inorganic mercury, *Fed. Proc.,* 30, 221 Abst., 1971.

Woodson, T. T., A new mercury vapor detector, *Rev. Sci. Instrum.,* 10, 308, 1939.

Working team for the coordination of mercury analyses in fish, final report for grant 3-13-1 concerning comparison between different analytical methods for mercury. Report December 6, 1968 to the Royal Commission on Natural Resources of 1964, National Environment Protection Board, S-171 20 Solna 1, Stencils, 1968, (Swedish).

Yamada, T., Uptake of phenyl mercuric acetate through the root of rice and distribution of mercury in rice plants, *Nippon Nogei Kaguku Kaishi,* 435, 1968.

Yamaguchi, S. and Matsumoto, H., Diagnostic significance of the amount of mercury in hair, Proc. 15th Int. Congr. Occup. Health, Vienna, 1966, 255.

Yamashita, M., Distribution in organs and excretion of mercury on some experimental organic mercury compounds poisoning, *J. Jap. Soc. Int. Med.,* 53, 529, 1964, (Japanese with English summary, and a partial Swedish translation).

Yoshino, Y., Mozai, T., and Nakao, K., Distribution of mercury in the brain and in its subcellular units in experimental organic mercury poisonings, *J. Neurochem.,* 13, 397, 1966a.

Yoshino, Y., Mozai, T., and Nakao, K., Biochemical changes in the brain in rats poisoned with an alkylmercury compound, with special reference to the inhibition of protein synthesis in brain cortex slices, *J. Neurochem.,* 13, 1223, 1966b.

Young, E., Ammoniated mercury poisoning. I. The absorption of mercury from ointments, *Br. J. Derm.,* 72, 449, 1960.

Zahorsky, J., Three cases of erythroderma (acrodynia) in infants, *Med. Clin. N. Am.,* 6, 97, 1922.

Zautashvili, B. Z., Problem of mercury hydrogeochemistry as illustrated by the mercury deposits of Abkhazia, *Geokhimiya,* 3, 357, 1966, (Russian).

Zeyer, H. G., Methoxäthylquecksilberoxalatvergiftung, *Zbl. Arbeitsmed.,* 2, 68, 1952.

Zollinger, H. V., Autoptische und experimentelle untersuchungen über Lipoidnephrose, hervorgerufen durch chronische Quecksilbervergiftung, *Schweiz. Z. Allg. Pathol.,* 18, 155, 1955.

SECONDARY REFERENCES

If the reader wishes to study more thoroughly those references in the Russian Language which are mentioned in this report and which are known to us through our translation of the 1969 monograph of Trachtenberg, the Medved and Kosmider articles, and through Dr. Nordberg's discussions in the U.S.S.R., the following list may be of assistance. Otherwise, secondary references are not listed.

Alekseeva, M. B., New methods for the study of atmospheric air — determination of mercury, in Ryazanov, V. A., Ed., *Limits of Allowable Concentrations of Atmospheric Pollutants*, Book 3, translated by B. S. Levine, U. S. Dep. Comm., Washington, D. C., 1957, 129.

Drogitjina, E. A., Toksitjeskie polinevrity i encefalomielopolinevrity, M., 1959.

Drogitjina, E. A., in *Promyelennaja toksikologija i klinika professional nych zabolevanij chimiceskoj etiologii*, M. 28, 1962.

Gabelova, N. A., in *Trudy po primeneniju radioaktivnych isotpov v medicine*, M., 139, 1953.

Galojan, S. A., in *Tiolovyjo soedinenja v medicine*, 79, 1959.

Gimadejev, M. M., K gigieniceskoj i toksikologiceskoj charakteristike vlijanija malych koncentracij rtuti na organizm. Avtoref. Cand. diss. Kazan, 1958.

Gimadejev, M. M., The effect of fumes of mercury on producing conditioned reflexes of rabbits, *Pharmacol. Toxicol.*, 2, 136, 1962.

Ginzberg, S. L., *Gig. Sanit.*, 8, 24, 1948.

Ivanov-Smolenskij, A. G., in *Trudy Ukrainskoge instituta gigieny truda i profzabolevanij*. XX. Charkov, 1939.

Ivanov-Smolenskij, A. G., in *Referaty naucno-issledovatel skich rabet*, 7. Mediko-biologiceskie nauki Izd-vo., AMN, SSSR, M., 1949.

Ochnjanskaja, L. G., in Klinika i patologija professional nych nejrointoksikatsij, *Tr. AMN SSSR*, XXXI, M., 28, 1954.

Poleshajev, N. G., K metodike opredelenija parov rtuti, *Gig. Tr. Tech. Bezopasnosti*, No. 6, 1936.

Poleshajev, N. G., Izvlecenije iona rtuti iz raztvorov azotnoj i sernoj kyslot i kolicestvennoje opredelenije jego, *Gig. Sanit.*, No. 5, 1946.

Poleshajev, N. G., K metodike opredelenija rtuti v atmosfernom vozduche, *Gig. Sanit.*, 6, 74-76, 1956.

Ryazanov, V. A., Alekseeva, M. V., and Senderikhina, D. Ya., Methods for the collection and study of air samples in the control of atmospheric cleanliness of inhabited localities, in *Limits of Allowable Concentrations of Atmospheric Pollutants*, Book 1, translated by B. S. Levine, U. S. Dept. Comm., Washington, D. C., 1952, 89.

Sadcikova, M. N., Klinika, rannjaja diagnostika i terapija chroniceskoj intoksikatsii rtutju (kliniko-fiziologiceskie issledovanija). Avtoref. kand diss. M., 1955.

Salimov, V. A., Izmenenie tkanevych belkov pri eksperimental noj rtutnoj intoksikatsii. Avtoref. kand. diss. M., 1956.

Sanotskij, I. V., Avchimenko, M. M., Ivanov, N. G., and Timodzevskaja, L. A., in *Obscie voprosy promyslennoj toksikologii*, M., 65, 1967.

RA 1231
.M5
F74

RA
1231
.M5
F74